COMPUTER SIMULATION IN PHYSICAL METALLURGY

ISPRA — COURSES

ON MATERIALS, ENGINEERING AND MECHANICAL SCIENCE

A series devoted to the publication of courses and educational seminars given at the Joint Research Centre, Ispra Establishment, as part of its education and training program. Published for the Commission of the European Communities, Directorate-General Information Market and Innovation.

Already published:

ELASTIC-PLASTIC FRACTURE MECHANICS

Edited by L. H. LARSSON

1985, viii + 528 pp. ISBN 90-277-2017-7

COMPUTER SIMULATION IN PHYSICAL METALLURGY

Lectures delivered at the ISPRA-Course held at Ispra, Italy, 21-25 May 1984, organized by the Joint Research Centre of the Commission of the European Communities in cooperation with the University of Trento

Edited by

GIANNI JACUCCI
University of Trento, Trento, Italy

D. REIDEL PUBLISHING COMPANY

A MEMBER OF THE KLUWER ACADEMIC PUBLISHERS GROUP

DORDRECHT / BOSTON / LANCASTER / TOKYO

Library of Congress Cataloging in Publication Data
Main entry under title:

Computer simulation in physical metallurgy.

(Ispra courses on materials, engineering, and mechanical science)
"Ispra–Course on Computer Simulation in Physical Metallurgy"–Pref.
1. Physical metallurgy–Mathematical models–Addresses, essays, lectures. 2. Physical metallurgy–Data processing–Addresses, essays, lectures. I. Jacucci, Gianni. II. Ispra–Course on Computer Simulation in Physical Metallurgy (1984) III. Commission of European Communities. Joint Research Centre. IV. Università degli studi di Trento. V. Series.
TN690.2.C65 1986 669'.9'0285 86-31246
ISBN 90-277-2192-0

Commission of the European Communities Joint Research Centre Ispra (Varese), Italy

Publication arrangements by
Commission of the European Communities
Directorate-General Information Market and Innovation, Luxembourg

EUR 9833

Published by D. Reidel Publishing Company
P.O. Box 17, 3300 AA Dordrecht, Holland

Sold and distributed in the U.S.A. and Canada
by Kluwer Academic Publishers,
190 Old Derby Street, Hingham, MA 02043, U.S.A.

In all other countries, sold and distributed
by Kluwer Academic Publishers Group,
P.O. Box 322, 3300 AH Dordrecht, Holland

Printed in The Netherlands

TABLE OF CONTENTS

PREFACE

In the past 25 years the understanding of many-body effects in condensed matter has received an enormous impulse from computer simulation. Well established in the field of statistical mechanics of simple liquids and solids, numerical experiments on model systems using the computer have more recently brought important advances in the field of materials science. A variety of aspects ranging from defects in crystals to surface behaviour, to radiation induced effects, have been the object of computer simulation in research groups mostly working in close contact with experiments in materials research laboratories.

This recent progress provided the impetus for the ISPRA-Course on Computer Simulation in Physical Metallurgy proposed and coordinated by Antonio Scotti of the Physics Division of the Joint Research Centre of the European Community Commission at Ispra.
This book contains articles corresponding to the lectures delivered at the ISPRA-Course.

The methods described consist mainly of Monte Carlo sampling of configuration space of many particle systems, molecular dynamics simulation of model substances, and the numerical solution of the equations describing continuous models. Applications include computer studies in statistical thermodynamics of crystalline and amorphous solids, computer simulation of defects and diffusion in crystals, of radiation effects, of the dynamics of extended defects, and numerical studies of macroscopic models.

The presence among the speakers of experts in the theory and phenomenology of solid state physics and physical metallurgy offered an excellent occasion for a critical review of computer simulation research in this field.
In this respect, of a special value has been the participation of C.P. Flynn, Director of the Materials Research Laboratory of the University of Illinois at Urbana Champaign.

Because of the interconnection between subjects, the order of the articles is more or less arbitrary. They appear in the same order as the lectures were delivered. Hence the Course Programme may offer a guide-line to the reader:

Computer studies in the statistical thermodynamics of crystalline and amorphous solids, including

- Polymorphic transformation in crystals
- Nucleation and precipitation in alloys
- Segregation at boundaries
- "Defects" in amorphous structures
- Surface properties

Computer simulation of defects and diffusion in crystals

- Bulk diffusion
- Surface diffusion
- Grain boundary diffusion

Computer simulation of radiation effects

- Swelling
- Creep
- Amorphization

Computer simulation of the dynamics of extended defects

- Dislocating climbing
- Fracture
- Shock wave compaction

Numerical studies on macroscopic models

- Heat and particle fluxes during irradiation
- Radiation creep and thermal creep using rate theory
- Phase stability under irradiation

The articles are the fruits of rather intensive labour by all of the contributors (except for three which appear only as an extended abstract). The authors are to be complimented and thanked for such contributions.
The scientific, organizational and financial support of the Course by the ISPRA Joint Research Centre is gratefully acknowledged.

Special thanks are due to the Course attendees for very valuable contributions and for their patience.

Gianni Jacucci

It is with great pleasure that I take this opportunity to express my gratitude to my friend Gianni Jacucci who was the real soul of this Course, and made possible through his competence and international reputation the organization of the Course in the form which proved to be so successful.

Antonio Scotti

MOLECULAR DYNAMICS AT CONSTANT EXTERNAL STRESS: STUDIES OF CRYSTAL STRUCTURE TRANSFORMATIONS

M. PARRINELLO
Dipartimento di Fisica Teorica
Universita' di Trieste
and
International School for Advanced Studies
Trieste, Italy

A major issue in solid state theory is the study of the mechanical behavior of solids under various conditions of temperature and stress. At low temperatures traditional solid state theories based on the harmonic approximation give a rather good account of the observed properties. However a large number of systems are so much affected by anharmonicity that the harmonic approach is bound to fail. Under such circumstances molecular dynamics (MD) is expected to be of invaluable help (1). Crystal structure transformation are a clear example of a highly anharmonic phenomena and a such a natural candidate for MD studies. However until very recently such studies were believed to be impossible. This was a consequence of the periodic boundary conditions (pbc) that in MD one is forced to assume in order to minimize boundary effects. The reason for this difficulty lies in the fact that structural transformations are often associated to changed in the shape of the crystal. Fixed pbc prevent shape fluctuations and inhibit the occurrence of structural changes. If the number of particles is very large the effect of the boudary becomes irrelevant since by forming appropriate defects the system will eventually crystallize in the structure of lowest free energy. However even with modern computers one is restricted to small systems especially for complex molecular crystals.
Recently however new and powerful tecniques have been develloped that allow variations in the shape of the MD cell (2) under conditions of constant stress or hydrostatic pressure. The usefulness of this approach has been demonstrated in a large variety of cases (3) leading to a deeper understanding of the mechanism underlying the transition and of the details of the associated kinetics.
In these lectures the theoretical foundations of these new MD methods will be reviewed and several applications will be discussed.

G. Jacucci (ed.), Computer Simulation in Physical Metallurgy, 1–2.

REFERENCES

1. E.R. Cowley, G. Jacucci, M.L. Klein and I.R. McDonald, Phys. Rev. B14 1758 (1976)

2. M. Parrinello and A. Rahman, Phys. Rev. Lett. 45 1196 (1980)

3. M. Parrinello and A. Rahman, J. Appl. Phys. 52, 7158 (1981)

4. M. Parrinello and A. Rahman, J. de Phys. C6, 511 (1981)

5. M. Parrinello, A. Rahman and P. Vashishta, Phys. Rev. Lett. 50, 1073 (1983)

MONTE CARLO STUDIES OF INTERFACE FREE ENERGIES IN NUCLEATION THEORY.

Anna Perini
Istituto per la Ricerca Scientifica e Tecnologica
38050 Povo
Italy

ABSTRACT. An efficient Monte Carlo technique of free energy computation is used to measure the free energy of solute clusters in a binary alloy. The Ising model for the binary alloy is considered. The accuracy of these free energy measurements allows to perform a detailed check of the validity of the phenomenological models commonly used when nucleation phenomena are studied.

1. INTRODUCTION

Lattice models seem to give a satisfactory rapresentation of real systems particularly when the triggering of first order phase transition is studied. Monte Carlo (M.C.) simulation of the kinetics of processes like gas condensation, decomposition of a supersaturated alloy, crystal growth, were developed on Ising type lattice models [2-4].

In this lesson (sec.3.) we will focus on the applications of the M.C. technique to measurements of the interesting thermodynamical quantities from statistical averages. Some examples of this kind of computation on lattice models will be briefly recalled (sec.2.). The dynamical interpretation of M.C. simulation which is at the base of the simulations of the kinetics of the processes mentioned above will be briefly reported.

Free energy estimates cannot be directly obtained by M.C. methods. In sec.4. we will present a particularly efficient technique of measuring free energy differences developed by Bennett and Torrie , Valleau [10] and applied to several problems [8,9,11].

Section 5. is devoted to the description of the application of this technique to cluster free energy computation . It is well known that the cluster free energy is a crucial quantity in the nucleation theory. From this application very accurate estimates of the cluster free energy are obtained and a detailed control of the validity of phenomenological models ("capillarity models") results.

G. Jacucci (ed.), Computer Simulation in Physical Metallurgy, 3–20.

2. LATTICE MODELS

In this section we will describe some lattice models, widly used in computer simulation: the lattice gas model and the binary alloy model which are equivalent to the ferromagnetic Ising model, and the solid on solid model which maps into the X-Y model.
Let us describe in more details the ferromagnetic Ising model and give the principal relations of the equivalent models to it. It consists of a rigid lattice the sites of which are occupied either by a "spin up" or a "spin down". The site variable σ_i assumes the values ± 1 depending on the orientation of the spin laying on the i-th site: $\sigma_i=+1$ "spin up", $\sigma_i=-1$ "spin down". Nearest neighbour interactions are considered. The energy of the system in the configuration specified by $\{\sigma_i\}$ is defined as:

$$E^I \{\sigma_i\} = -J \sum_{\langle ij \rangle} \sigma_i \ \sigma_j \ - B \sum_i \sigma_i \qquad (1)$$

J represents the interaction between spins: J>0 for a ferromagnetic system since parallel nearest neighbour spins are energetically favored, J<0 for the antiferromagnetic system since antiparallel nearest neighbour spins are energetically favored; B is the external magnetic field (it is supposed to be constant); the symbol $\langle ij \rangle$ denotes a nearest neighbour pair of spins.
As is well known this model gives a crude description of a physical ferromagnetic substance. Its importance is due to the fact that it is the simplest model, treatable with mathematical rigor, which shows the transition from the non ferromagnetic state to the ferromagnetic state when the system is quenched under the Curie temperature. The literature about this model is vast. Let us remember that the 2-D model (for B=0) was solved exactly by Onsager and that series expansions for the thermodynamical quantities on several lattices are available[1]. As an example of M.C. simulation on this model let us cite the work of Fosdick [6] where order parameters at different concentrations and temperatures were computed for the square lattice and the simple cubic lattice.

2.1 The lattice gas

The lattice gas model can be obtained from the ferromagnetic model considering "spins up " as holes and "spin down" as molecules. Supposing that nearest neighbour molecules have interaction energy $-\varepsilon$ this system represents a gas of molecules which can take only discrete positions and interact through a two body potential $v(|\bar{r}_i - \bar{r}_j|)$

$$\begin{aligned} v(r) &= \infty && r=0 \\ v(r) &= -\varepsilon && r=\text{nearest neighbour distance} \qquad (2) \\ v(r) &= 0 && \text{otherwise} \end{aligned}$$

The equivalence with the ferromagnetic system is obtained in terms of the following relations (Z is the lattice coordination number and M the number of sites):

Ising model	Lattice gas
N_+	N_a
N_-	$M-N_a$
$4J$	ε
B	$-\mu/2 - ZJ$
$F_I\ (T,B) + ZJM/2+BM$	$\Omega_{LG}\ (T,M,\mu)$

This model is usually adopted when vapor condensation is studied. Binder [2] has simulated the dynamics of this system by M.C. method giving a satisfactory description of the nucleation processes of the condensing vapor.

2.2 Binary alloy.

In the binary alloy model atoms af two species, A and B occupy the lattice sites. At each site there shall be only one atom. The bond energy for the different nearest neighbour pairs are given: ε_{AA}, ε_{BB}, ε_{AB}. The equivalence to the ferromagnetic model requires the following relations :

Ising model	Binary alloy
N_+	N_A
N_-	$N_B=M - N_A$
$4J$	$2\varepsilon_{AB} - \varepsilon_{AA} - \varepsilon_{BB}$
B	$(\mu_A - \mu_B)/2 + \varepsilon_{BB} - \varepsilon_{AA}$
$F_I(T,B)$	$M(\mu_A+\mu_B-(2\varepsilon_{AB}+\varepsilon_{AA}+\varepsilon_{BB})/2) + \Omega_A(T,M,\mu_A,\mu_B)$

The phase decomposition of a binary alloy following quenching was studied by Kalos et al. [3] with M.C. simulation on this model. As reported in sec. 5. we used M.C. methods to compute the free energy of B atoms on this model.

2.3 Solid on solid model.

It is a typical fluid-solid interface model, widely used in the study of crystal growth processes. In this model the fluid-solid interface is divided in cells of equal size and shape which can be either in the solid or in the fluid state with the following restrictions: solid cells on the top of fluid cells are excluded. Different configurations for this systems can be characterized by the different heights (h_i) of the "towers" of solids cells and given the interaction energies of two adjacent cells ε_{ss}, ε_{ff}, ε_{sf}, the configuration energy writes:

$$E_{sos} = - \sum_{\langle ij \rangle} V (h_i - h_j) , \quad V=2\ \varepsilon_{sf} - \varepsilon_{ss} - \varepsilon_{ff} \qquad (3)$$

It has been shown that this model is mappable in the X-Y model [5]. M.C. simulation of crystal growth shown both the linear growth kinetics and the non linear growth kinetics in correspondence of suitable values of the parameter V [4].

3. THE MONTE CARLO THECNIQUE

The thermodynamical quantities of a system of interacting particles can be expressed in the framework of the statistical mechanics as suitable ensamble averages. In this section we will describe the use of M.C. techniques in computing canonical averages, (system with number of particles, temperature and volume fixed).
Let us consider one of the model described in sec.2 , the canonical average of a parameter A which depends on the state of the system is expressed as:

$$\langle A \rangle = \Sigma_\alpha A_\alpha \exp(-\beta E_\alpha) \,/\, \Sigma_\alpha \exp(-\beta E_\alpha) \qquad (4)$$

where α indicates a generic configuration $\{\sigma_i\}$ and the sum goes up to the 2^M possible configurations of the system of M spins. For example A could be the potential energy or the magnetisation. The exact computation of $\langle A \rangle$ imply to consider all the 2^M configurations , but still for a square lattice having ten sites on the edge it becomes an impossible task ($2^{100} \sim 1.3\ 10^{30}$).
An estimate of the average value can be obtained computing the sum on a randomly selected subset of $\mathcal{N}$ configurations, (this is the standard M.C. method for the evaluation of multidimensional integrals):

$$\langle A \rangle_{\mathcal{N}} = \sum_{\alpha=1}^{\mathcal{N}} A_\alpha \exp(-\beta E_\alpha) \,/\, \sum_\alpha \exp(-\beta E_\alpha) \qquad (5)$$

This simple way to use the M.C. requires an enormous value for $\mathcal{N}$ because the principal contributions to the average come from a very small fraction of the 2^M configurations. A more clever sampling of the configurational space is usually employed. The configurations are choosen with a probability P_α which favors those configurations which give the important contributions to the sum (eq.4). This second technique is called "importance sampling". An estimate of $\langle A \rangle$ now is given by:

$$\langle A \rangle_{\mathcal{N}}^{II} = \sum_\alpha^{\mathcal{N}} A_\alpha \exp(-\beta E_\alpha)\, P_\alpha^{-1} \,/\, \sum_\alpha^{\mathcal{N}} \exp(-\beta E_\alpha)\, P_\alpha^{-1} \qquad (6)$$

<u>How to choose P_α ?</u>

- the simplest choise is $P_\alpha \propto \exp(-\beta E_\alpha)$.This is the classical importance sampling of Metropolis et al.[13]. In this case eq.6 reduces to a simple arithmetic average:

$$\langle A \rangle_{\mathcal{N}}^{M} = 1/\mathcal{N} \cdot \sum_{\alpha=1}^{\mathcal{N}} A_\alpha \qquad (7)$$

- Another reasonable criterion is to choose the probability P_α which allows one to obtain the minimum value of the variance of the quantity

$\langle A \rangle_{\mathcal{N}} / \langle A \rangle$. Fosdick [6] found that the probability satisfying this requirement is:

$$P_\alpha = | A_\alpha / \langle A \rangle - 1 | \cdot \exp(-\beta E_\alpha) \,/\, \Sigma_\alpha | A_\alpha / \langle A \rangle - 1 | \cdot \exp(-\beta E_\alpha) \qquad (8)$$

This expression depends explicitly on ⟨A⟩, which is the quantity being estimated. In practise one computes a first estimate of ⟨A⟩ and uses it in eq.8 in order to obtain a better measure of ⟨A⟩. This procedure can be repeted several times. A disadvantage of this second choise respect to the one adopted in eq.7 is that we obtain the optimum estimate of just one parameter at a time, on the other side it can be usefull in those cases where fluctuations on the estimate cannot easily reduced increasing the number of samples.

Practical realisation of the importance sampling.

The problem is now the following : how can we extract $\mathcal{N}$ configurations, from the ensamble of system configurations, with probability P_α?

Once an initial configuration α is buid, (that is once assigned the sites to the N_+ "spins up" and to the $M-N_+$ "spins down"), we define what is called M.C. step, that is we define the rule which allows us to obtain a new configuration from the old one. For example we choose at random two nearest neighbour spins, exchange them and accept the new configuration with probability $W_{\alpha\alpha'}$ which depends on the old and on the new configuration .This move is performed on choosing a random number r $(0<r<1)$ from the uniform distribution, comparing it with $W_{\alpha\alpha'}$: if $W_{\alpha\alpha'}>r$ we accept the new configuration otherwise we restore the old one. $W_{\alpha\alpha'}$ can be interpreted as the conditional probability of finding the system in the configuration α' at time t+1 if at time t it is in the configuration α . The described procedure is a Markov process characterized by the transition probability $W_{\alpha\alpha'}$ which satisfies the normalisation condition:

$$\sum_{\alpha'} W_{\alpha\alpha'} = 1 \qquad (9)$$

Now the probability of finding the system in the state α at time t+1 is:

$$\psi_\alpha(t+1) = \sum_{\alpha'} W_{\alpha\alpha'} \; \psi_{\alpha'}(t) \qquad (10)$$

From eq.10 it is clear that the requirement on $W_{\alpha\alpha'}$ in order to obtain a sample from the distribution P_α , must be:

$$\lim_{t\to\infty} \psi_\alpha(t) = P_\alpha \qquad (11)$$

independently on the initial configuration.

In order to obtain a solution of the limit in eq.11 it is sufficient, (but not necessary),to require the validity of the detailed balance condition:

$$P_\alpha \, W_{\alpha\alpha'} = P_{\alpha'} \, W_{\alpha'\alpha} \qquad (12)$$

which gives an expression for the ratio of the conditional probability:

$$W_{\alpha\alpha'} / W_{\alpha'\alpha} = \exp(-\beta(E_{\alpha'} - E_\alpha)) \qquad (13)$$

$W_{\alpha\alpha'}$ is not uniquely specified from eq.12. Typical choises are reported below:

$$W_{\alpha\alpha'} = \exp(-\beta(E_{\alpha'} - E_\alpha)) \quad \text{if } E_{\alpha'} - E_\alpha > 0 \qquad (14)$$
$$W_{\alpha\alpha'} = 1 \quad \text{otherwise}$$

$$W_{\alpha\alpha'} = \exp(-\beta(E_{\alpha'} - E_\alpha)) / (1 + \exp(-\beta(E_{\alpha'} - E_\alpha))) \qquad (15)$$

Let us reproduce some intuitive arguments [13] in order to prove that eq.14 yields the solution of the eq.11.
Let us consider a large number of systems on which the same Markov process (for example that of eq.14), is performed. Suppose that n_α systems are in the configuration α and $n_{\alpha'}$ in the configuration α' at the step t of the Markov processes. Now let us try to change two spins in the systems in order to realize the transition from the actual configuration α into a new configuration α' with the "a priori" probability $w_{\alpha\alpha'} = w_{\alpha'\alpha}$,(the "a priory" probability is the probability of moving from α to α', before discriminating on $\exp(-\beta(E_{\alpha'} - E_\alpha))$).
If $E_{\alpha'} < E_\alpha$ the number of systems moving from α to α' is:

$$n_{\alpha\to\alpha'} = n_\alpha \, w_{\alpha\alpha'} \qquad (16)$$

while the number of systems moving from α' to α is:

$$n_{\alpha'\to\alpha} = n_{\alpha'} \, w_{\alpha\alpha'} \exp(-\beta(E_\alpha - E_{\alpha'})) \qquad (17)$$

the net number of systems moving from the state α to the state α' is:

$$\Delta n_{\alpha\alpha'} = n_{\alpha\to\alpha'} - n_{\alpha'\to\alpha} = n_\alpha \, w_{\alpha\alpha'} (n_\alpha / n_{\alpha'} - \exp(-\beta(E_\alpha - E_{\alpha'}))) \qquad (18)$$

As long as $n_\alpha / n_{\alpha'}$ is smaller than the canonical value we have $\Delta n_{\alpha\alpha'} < 0$ which means that systems tend to go to the state α hence $n_\alpha / n_{\alpha'}$ increases; if $n_\alpha / n_{\alpha'}$ is too big $\Delta n_{\alpha\alpha'} > 0$, hence $n_\alpha / n_{\alpha'}$ tends to decrease. Asyntotically for $t \to \infty$ our ensamble of systems will approach the steady state corresponding to the canonical distribution. Let us observe that the above considerations do not specify how rapidly the canonical distribution is approached. This problem is related to the definition of the M.C. step.
In conclusion eq.11 allows us to obtain estimate of the average $\langle A \rangle$ using configurations of a suitable Markov process (i.e. with asymptotic limit P_α) and we can write:

$$\lim_{\mathcal{N}\to\infty} 1/\mathcal{N} \sum_{\alpha}^{\mathcal{N}} A_\alpha = \sum_\alpha A_\alpha \, P_\alpha \qquad (19)$$

Let us stress the difference between the expression of eq.7 and eq.19. While in eq.7 it was assumed that each α-configuration was chosen with the Boltzmann probability, here (eq.19) successive configurations are correlated and the required probability distribution appears only in the limit $t \to \infty$.

Two last considerations concerning the correlation of successive configurations will be pointed out, that is:
- how the persistence of this correlation affects the variance of the average ⟨A⟩ ,and
- how the dynamical interpretation of the M.C. processes comes from the presence of this correlation.

3.1 Variance estimation

Let us consider the usual expression for the variance of the average value of $\mathcal{N}$ independent measures A_i in terms of the single measure variance $\sigma^2 = \langle A^2 \rangle - \langle A \rangle^2$

$$\sigma_{\mathcal{N}}^2 = \sigma^2/\mathcal{N} \qquad (20)$$

the persistence of correlations in τ subsequent configurations of a $\mathcal{N}$ steps M.C. chaine reflects on the variance of the average ⟨A⟩ as follows:

$$\sigma_{\mathcal{N}}^2 = \sigma^2\,\tau/\mathcal{N} \qquad (21)$$

$\tau/\mathcal{N}$ represents the effective number of independent configurations. Hence the variance on the average is increased by the factor τ. Eq. 21 gives an operative expression for τ :

$$\tau = \lim_{\mathcal{N} \gg \tau} \mathcal{N}\sigma_{\mathcal{N}}^2/\sigma^2 \qquad (22)$$

On the assumption that τ is small respect to $\mathcal{N}$ the limit is realized dividing the configurations of the chaine in $\mathcal{N}/n$ groups of n configurations, n is let to increase. One computes:

$$\sigma_n^2 = n/\mathcal{N}\cdot\sum_{i=1}^{\mathcal{N}/n} [\,A_i - \langle A\rangle_{n,i}]^2 \quad \text{and} \quad \sigma^2 = 1/\mathcal{N}\cdot\sum_{j=1}^{\mathcal{N}} (A_j - \langle A\rangle)^2 \qquad (23)$$

where $\langle A\rangle_{n,i}$ denotes the averages on the n values of the i-th group.
τ is obtained as :

$$\tau = \lim_{n\to\mathcal{N}} n\,\sigma_n^2/\sigma^2 \qquad (24)$$

Substituting the obtained value of τ in eq.21 the desidered variance results [10], [11].

3.2 Dynamical interpretation of the M.C. simulation.

Usually M.C. simulation is considered a typical technique which gives ensamble averages computations in contrast to the molecular dynamics method which yields time averages. Due to the use of Markov processes in which subsequent configurations are highly correlated even M.C. averages can be interpreted as time averages.
A time scale can be associated to a given number of succesive configurations of the Markov process (often the transition probability

of eq.14 and eq.15 are written with a multiplying factor $1/\tau_s$). A reasonable choise is τ_s equal to the time necessary for performing M single spin exchanges.
The probability that the system is in the state α is governed by the master equation :

$$d\, P_\alpha(t)/d\, t = - \Sigma_\alpha W_{\alpha\alpha'} P_\alpha(t) + \Sigma_{\alpha'} W_{\alpha'\alpha} P_{\alpha'}(t) \qquad (25)$$

The convergence of $P_\alpha(t)$ toward the equilibrium distribution (see limit eq.11) ensures that the model is ergodic.
Time dependent average ⟨A(t)⟩ are hence defined as :

$$\langle A(t) \rangle = \Sigma_\alpha P_\alpha(t)\, A_\alpha \qquad (26)$$

where A depends on time only implicitly through the probability distribution. Equivalently we can average on the initial state $P_\alpha(t_0)$ while the time dependent behavior of the observable A comes from the described MarKov process :

$$\langle A(t) \rangle = \Sigma_\alpha P_\alpha(t_0)\, A_\alpha(t) \qquad (27)$$

Let us now point out a relevant difference between M.C. and molecular dynamics simulation: the Kinetics of the M.C. simulation is often a rather artificial one ,(for example in the case of vapor condensation on lattice gas molecules appear and disappear randomly on lattice sites),on the other side molecular dynamics yields the true Kinetics which follows from the Hamiltonian. Nevertheless the two dynamics lead to the same thermal distribution. This Kind of M.C. simulation are generally applied when computational efficiency requires the use of simple models which contains only a part of the degrees of freedom of the system while the others act as a heat bath inducing random transitions in the considered variables. This is the case of the vapor condensation. As Binder observes [12] for a M.C. run one can observe 10^8 individual cluster reactions ,(like single particle absorption or emission from a cluster of molecules), while in molecular dynamics runs the number of cluster reactions is by several order of magnitude smaller, hence molecular dynamics study of realistic model over a time scale equivalent to the M.C. studies seems not feasible.

4. THE "OVERLAPPING DISTRIBUTIONS " (O.D.M.) TECHNIQUE FOR FREE ENERGY COMPUTATIONS.

On writing eq.7 , in the previous section, we obtained a usefull technique for canonical average estimations,but we lost the informatios about the denominator of eq.4, that is about the configurational integral Q :

$$Q = \Sigma_\alpha \exp(-\beta U_\alpha) \qquad (28)$$

Hence we cannot obtain reasonable estimates of the free energy F :

$$F = -1/\beta \; \lg(Q) \qquad (29)$$

from a direct application of the described method.
The typical approch to the problem of free energy measurements by M.C. (or by molecular dynamics) methods rests on the idea that free energy differences between two thermodynamical states can be computed integrating the appropriate differential along a reversible path connecting the two states. For example using the relation between internal energy and free energy ($U = \partial/\partial\beta(\beta F)$) one can obtain the free energy at a given temperature integrating the internal energy between a reference state (where F is known) and the interesting state :

$$\beta F - \lim_{\beta \to 0} \beta F = \int_0^{\beta} U(\beta')\, d\beta' \qquad (30)$$

This method becomes impracticable when $U(\beta)$ must be computed for a broad range of temperature between the two extremes of integration.
This procedure is generalized to the case where the two states refer to systems with different Hamiltonians [10]. A description of this method as well as several example of applications can be found in the paper of Jacucci and Quirke [11]. Here we recall the two versions of this approch : the first, called "one sided evaluation",which yields free energy difference estimations through averages of suitable quantities on one of the two canonical ensamble, the second which implies averages computed on both canonical ensambles, (it is called "two sided evaluation"). The limits of applicability of these two versions will be discussed. The details of the application of this method to the cluster free energy computation will be presented in sec.5.

4.1 One sided evaluation

Let us consider two different systems having Hamiltonians $H=K+U_I$ and $H=K+U_{II}$ respectively and suppose that a one to one correspondence between their configurations does exist (the two configurational spaces must be congruent). The free energy difference between the two systems can be expressed as follows:

$$\exp{-\beta(F_{II}-F_I)} = \sum_i \exp(-\beta(U_{II}{}^i - U_I{}^i))\exp(-\beta(U_I{}^i)) / \sum_i \exp(-\beta(U_I{}^i))$$
$$= \langle \exp(-\beta(U_{II} - U_I)) \rangle_I \qquad (31)$$

equivalently we obtain:

$$\exp{-\beta(F_{II}-F_I)} = \sum_i \exp(-\beta(U_{II}{}^i)) / \sum_i \exp(-\beta(U_I{}^i - U_{II}{}^i))\exp(-\beta(U_{II}{}^i))$$
$$= 1/\langle \exp(\beta(U_{II} - U_I)) \rangle_{II} \qquad (32)$$

These averages can also be expressed as sums over the domain of values of the energy difference $\Delta_i = U_{II} - U_I$ through the function $f(\Delta)$ which represents the fraction of configurations having a given value of Δ in one of the canonical ensamble:

$$f_\alpha(\Delta) = \langle \delta(\Delta - U_{II} + U_I)\rangle_\alpha = \Sigma_i \delta(\Delta - U_{II}{}^i + U_I{}^i) \exp(-\beta U_\alpha^i) / Q_\alpha \qquad (33)$$
$$\alpha = I \text{ or } II$$

Eq.31 and eq.32 rewrite:

$$\exp(-\beta(F_{II}-F_I)) = \langle \exp(-\beta\Delta)\rangle_I = \sum_{\{\Delta_i\}} f_I(\Delta_i)) \exp(-\beta\Delta_i) \qquad (34)$$

$$\exp(-\beta(F_{II}-F_I)) = 1/\langle \exp(-\beta\Delta)\rangle_{II} = 1/\sum_{\{\Delta_i\}} f_{II}(\Delta_i) \exp(-\beta\Delta_i) \qquad (35)$$

Good estimates of $F_{II}-F_I$ are expected when a reasonable measure of $f_\alpha(\Delta)$, for a wide range of values of Δ, is available. Otherwise the estimates can be affected by a large error as will result from the example depicted in fig.1.

4.2 Two sided evaluation

Let us handle eq.33 in order to obtain the expression for the two sided evaluation:

$$f_I(\Delta) = \Sigma_i \delta(\Delta_i - U_{II}{}^i + U_I{}^i) \exp(-\beta(U_I{}^i - U_{II}{}^i)) \exp(-\beta U_{II}{}^i) / Q_I \cdot Q_{II}/Q_{II}$$
$$= \exp(-\beta\Delta) Q_{II}/Q_I \, \Sigma_i \delta(\Delta_i - U_{II}{}^i + U_I{}^i) \exp(-\beta U^i)/Q_{II}$$
$$= \exp(-\beta\Delta) Q_{II}/Q_I \, f_{II}(\Delta) \qquad (36)$$

here use is made of the one to one correspondence between the two configurational spaces. We obtain :

$$Q_I/Q_{II} = \exp(-\beta\Delta) f_I(\Delta) / f_{II}(\Delta) \qquad (37)$$

The advantage of the two sided evaluation respect to the one sided evaluation consists in the possibility of obtaining accurate estimates of $F_{II}-F_I$ applying eq.37 only to those values of Δ for which the two distributions are numerically appreciable. Fig.1 reports an example of histograms of $f_I(\Delta)$ and $f_{II}(\Delta)$ which can be obtained by M.C. computation (note that they will be histograms also when Δ is a continuum variable). It is immediate to see that the one sided evaluation will be strongly in error in fact the area under the histograms representing $f_I(\Delta)\exp(-\beta\Delta)$ in fig.1b is not proportional to the area under the histograms $f_{II}(\Delta)$:

$$\sum_{\{\Delta_i\}} f_I(\Delta_i) \exp(-\beta\Delta_i) \neq \sum_{\{\Delta_i\}} \text{cost } f_{II}(\Delta_i) \qquad (38)$$

(use is made of eq.37).

Only when the two distributions of fig.1b overlap for a wide range of values of Δ the two sided evaluation and the one sided evaluation will agree.

5. APPLICATION OF THE O.D.M. TO CLUSTER FREE ENERGY COMPUTATION

In the classical nucleation theory which describes the beginning of a first order phase transition in a supersaturated system the free energy of "drops" of the nucleating phase (clusters), and expecially the interfacial energy between the original and the nucleating phase , are crucial parameters. Phenomenological models (called "capillarity models") consider these clusters as macroscopic drops and give their free energy as sum of a bulk and a surface contribution which are defined using the macroscopic value for bulk and surface energy per particle. Due to the thoretical difficulties for giving precise theoretical predictions for this quantity several attempts have been made to compute the cluster free energy using computer simulation. Indirect free energy estimations for clusters of Lennard Jones atoms were obtained by Abraham [14] with molecular dynamics method. Binder et al. [7] performed a M.C. simulation of a cluster surrounded by monomers at at a concentration higher than the equilibrium value on the lattice gas model. They obtained estimates for the cluster free energy by measuring the average cluster size and the average monomer concentration.

Direct measurements of the cluster free energy were obtained applying the overlapping distributions technique to clusters in the Ising model in 2D and 3D [8]. The high precision of these data allowed us to check the validity of the capillarity expression term by term and to state the range of validity of these phenomenological models.

We will describe first the model and how the O.D.M. apply to give the free energy difference between clusters containing respectivly N and N+1 solute atoms. The principals outcomings of the comparison between these measured values and the predictions of the capillarity expressions will be briefly recalled for the 2D and the 3D cases.

5.1 The model

Let us consider the Ising model for the binary alloy (sec.2) : a cluster of N solute atoms (B-atoms) on a lattice of M sites forms when each B-atom is connected by at least one nearest neighbour bond to any one of the remaining N-1 B-atoms. The energy for each configuration can be written as:

$$E_N = n_{AB}\, w_{AB} / 2 + 2 (M - N) \varepsilon_{AA} + 2 N \varepsilon_{BB} \qquad (39)$$

where n_{AB} is the number of A-B bonds; $w_{AB}=2\varepsilon_{AB}-\varepsilon_{AA}-\varepsilon_{BB}$; M-N is the number of A-atoms. In this model the only contribution to the cluster free energy comes from the configurational partition function Q

$$Q_N = \sum_i \exp(-\beta E_N{}^i) \qquad (40)$$

The sum extends to all configurations of a cluster of indistinguishable particles counting only one of the traslationally equivalent configurations. For small values of N Q_N can be computed exactly , but because of the rapid increase of the configurations with N it becomes very soon an unsolvable task and one should turn to machine calculations.

5.2 Overlapping distributions method applied to cluster free energy.

This method has been used to compute the free energy difference between a cluster of N+1 and a cluster of N solute atoms. The N-cluster and the N+1-cluster correspond respectively to the system I and to the system II of sec.4. In order to state the required one to one correspondence between the two systems we must consider distinguishable particles . Each configuration for the M distinguishable particles where the first N are of B-type will be considered as configurations for the N-cluster (I-system). The correspondent configurations of the N+1-cluster (II-system) is obtained changing the N+1-th particle ,which is of A-type, with a B-particle. Note that the configurational integral for indistinguishable particles are simply related to that for distinguishable particles:

$$Q_N^{dist} = (M - N)!\, N!\, Q_N^{ind} \qquad (41)$$

Once stated the one to one correspondence we can define the energy difference for the i-th configuration:

$$\Delta_i = E_{N+1}^{\,i} - E_N^{\,i} \qquad (42)$$

and the energy distribution :

$$f_\alpha(\Delta) = \Sigma_i\, (\Delta, \Delta_i) \exp(-\beta E_\alpha^{\,i}) \,/\, Q_\alpha \qquad \alpha = N \text{ or } N+1 \qquad (43)$$

Let us observe that for this system Δ assumes only few discrete values. In fact from eq.39 writes:

$$\Delta = E_{N+1} - E_N = \Delta n_{AB}\, w_{AB} / 2 + 2 (\varepsilon_{BB} - \varepsilon_{AA}) \qquad (44)$$

where Δn_{AB} is the net number of A-B bonds created by going from the N-cluster configuration to the corresponding N+1- cluster configuration. As a consequence the energy distributions will be histograms on five values in 2D and in principle four estimates of $F_{N+1}-F_N$, one for each value of Δ for which the two distributions overlap, could be obtained:

$$\exp(- \beta(F_{N+1}-F_N)) = Q_{N+1}/ Q_N = \exp(-\beta \Delta) f_N(\Delta)/f_{N+1}(\Delta) \qquad (45)$$

Let us give some more details about the computation of $f_N(\Delta)$ (suppose $\Delta/\varepsilon_{AB}=2$). Once given a chaine of configurations of the canonical ensamble of the N-cluster, $f_N(\Delta)$ will receive a contribution only from those configurations where the N+1 particle is connected to the N-cluster by one A-B bond . If we consider as the N+1-th particle each A particle which is adjacent to the cluster we can obtain several contributions to $f_N(2)$ from each configuration. That is we increase the statistics considering indistinguishable particles and expressing

$f_N(\Delta)$ as a canonical average over configurations of indistinguishable particles:

$$f_N^{dist}(\Delta) = \langle \sum_k^{M-N} \delta\Delta_k, \Delta \quad / \; M - N \rangle_{ind} \qquad (46)$$

The validity of an analogous expression for $f_{N+1}(\Delta)$ can be prooved:

$$f_{N+1}^{dist}(\Delta) = \langle \sum_k^{N+1} \delta\Delta_k, \Delta \quad / \; N + 1 \rangle_{ind} \qquad (47)$$

Computer time was saved by measuring $f_N(\Delta)$ for $\Delta/\varepsilon_{AB}=2$, indeed changing into A a B atom which is linked to the cluster by one bond only, surely conserves the cluster and no other check is needed. The uncertainty on the estimated values of $F_{N+1}-F_N$ was computed measuring the variance of $f_{N+1}(\Delta)$ and $f_N(\Delta)$ as explained in sec.3. Typical runs for the 2D case performed 250000 exchanges per B-atom requiring a CPU time of the order of 40 mn on a CDC 7600 computer for a cluster of 45 B-atoms. For the 3D case 10^5 exchanges per B-atom were considered requiring a CPU time of the order of 20mn for a cluster of 20 B-atoms (on the same computer).

5.3 Numerical results of $F_{N+1}-F_N$: test of the capillarity expression.

5.3.1 Square lattice.

On the two dimensional square Ising model the free energy difference was computed with an accuracy better than 2.5% for clusters containing up to 50 solute atoms, at three reduced temperature: $T/T_c=.40$, $T/T_c=.59$, $T/Tc=.80$ [8]. The agreement with phenomenological expression like eq.(48) is surprisingly good down to small sizes (N~10) provided the capillarity terms are computed with the prescriptions specified below:

$$F_N = \mu\tilde{n} + \Sigma\tilde{n}^{1/2} + \beta^{-1} \tau \lg(\tilde{n}) + c \qquad (48)$$

$\tilde{n}$ is the number of atoms (A and B) in the cluster ($\tilde{n} = N/C_{coex}$, C_{coex} is the concentration along the coexistence line between dense and dilute phases).

Prescriptions on the terms of eq. (48):

<u>Bulk contribution</u> to F : μ is the chemical potential at the coexistence line. It is known from low temperature expansion [1].

<u>Surface contribution</u>: Σ is the specific surface free energy of a macroscopic drop having the equilibrium shape. This quantity is known from exact computations [15].

<u>Logarithmic term</u> : τ takes into account the fluctuations of the cluster shape respect to the equilibrium shape : is given the value 1.25 [16].

<u>Size independent term</u>: it is obtained fitting $F = \sum_{m=1}^{N-1} (F_{m+1}-F_m) - F_1$. This analysis gives for c a positive value, temperature dependent. This term can be interpreted as a curvature correction to Σ since on writing

$$\Sigma = \Sigma_{\infty}(1 + \varepsilon N^{1/2}) \qquad (49)$$

yields:

$$c = \varepsilon \, \Sigma_\infty \, c_{coex}^{-1/2} \qquad (50)$$

(Σ_∞ is the surface specific free energy for an infinite surface).
The numerical values of the capillarity coefficients are reported in table I, the contribution of each term is depicted in fig.2.

5.3.2 Simple cubic lattice.

On the 3 D simple cubic Ising model the free energy difference was computed for clusters up to 20 solute atoms, at T/Tc =.40, T/Tc=.59. The accuracy of the data is better than .5%.
The analysis previously described for the 2-D case gives the following capillarity expression as best fit :

$$F_N = \mu \tilde{n} + \Sigma \, \tilde{n}^{2/3} + c \, \tilde{n}^{1/3} + \beta^{-1} \, \tau \, \lg(\tilde{n}) + d \qquad (51)$$

The values of the coefficients are reported in table II, and have the explained meaning. Σ is not known analitically but the two limits corresponding to the spherical and the cuboidal shapes can be estimated using low temperature expansion for the surface tension [1].
τ is given the value -1/9 [16]. The term $c \, \tilde{n}^{1/3}$ ("edge" free energy) corresponds to the curvature correction to Σ introduced in the 2-D case since substituting $N^{1/2}$ with $N^{2/3}$ in eq.(42) gives:

$$c = \varepsilon \Sigma_\infty c_{coex}^{-2/3} \qquad (52)$$

c is found to be negative in agreement with phenomenological models [17].
The constant is found to be positive, temperature dependent. This term can be interpreted as a "corner" contribution.
An intuitive way to understand the physical meaning of the surface, edge, corner contribution is to consider the cluster shape at 0 Kelvin. For most N the shape would be a parallelepiped and sites located on surface,edges and corners have different numbers of nearest neighbors outside the cluster therefore give different contributions to the free energy. The single contributions are depicted in fig. 3.

CONCLUSIONS

The application of M.C. methods in computing thermal averages on lattice models are described in details (sec.3). Particular attention is given to the problem of free energy measurements. Section 4 concerns with an efficient method (the "overlapping distributions method") which yields accurate estimates of free energy differences in terms of suitable energy distributions computed as canonical averages with M.C. or molecular dynamics simulations.
A particular application of this method to the computation of free energy of solute clusters on the Ising binary alloy model is presented in sec.5.

The accuracy of these results allows a detailed check of the validity of the capillarity expression for the cluster free energy. A good agreement ,down to small sizes (N~10) is found provided a curvature correction to the surface term and a "corner" contribution in 3D are taken into account.

REFERENCES

[1] C.Domb in '<u>Phase transitions and critical phenomena</u>', vol.3 ed.C.Domb and M.S.Green (London: Academic 1974) 357

[2] K.Binder and H.Muller-Krumbhaar Phys. Rev. B9,2328, (1974)
K.Binder,D.Stauffer and H.Muller-Krumbhaar Phys.Rev.**B10**, 5261,(1974)
K.Binder,D.Stauffer and H.Muller-Krumbhaar Phys.Rev. **B12**,5261,(1975)

[3] A.B. Bortz ,M.H. Kalos ,J.L.Lebovitz and M.A. Zendejas Phys. Rev. **B10**, 535 (1974)
J.Marro, A.B.Bortz , M.H.Kalos and J.L.Lebovitz, Phys. Rev. **B12**, 2000 (1975)
M.Rao, M.H.Kalos, J.L.Lebowitz and J.Marro, Phys.Rev.**B13**,4328 (1976)
Amit Sur and J.L.Lebowitz, J.Marro, M.H.Kalos Phys. Rev. **B15**, 3014, (1977)

[4] G.H. Gilmer, Science, **208** 355 (1980)

[5] R.H.Swendsen Phys. Rev. **B17** 3710, (1978)

[6] L.D. Fosdick in '<u>Method incomputational physics</u>' eds. B.Alder, S.Fernback, M. Rotenberg (N.Y. Academic 1963) p.245

[7] K. Binder and M.H. Kalos, J. Stat.Phys. **22**, 363 (1980)

[8] G.Jacucci, A.Perini, and G.Martin, J.Phys. A:Math Gen 16, 369 (1983)
A.Perini, G.Jacucci, and G.Martin, Phys.Rev. B, 1, March (1984)

[9] G. Jacucci, N. Quirke Surf.sci. **144**, 92 (1984)

[10] C.H. Bennet J. Comp.Phys. **22**, 245 (1976)
J.P. Valleau and G.M. Torrie in '<u>Modern Theoretical Chemistry</u>' Ed. Berne B. J. Plenum, New York , p.169 and ref. therein

[11] A rewiew of these methods and several exposed application examples can be found in: G.Jacucci and N.Quirke, '<u>Computer simulation of</u> solids' Lecture Notes in Physics, Springer Berlin (1982)

[12] K.Binder, Adv. in Coll. and Interf. Sci., **7**, 279 (1977)

[13] N.Metropolis, A.W.Rosenbluth, N.N.Rosenbluth, A.H.Teller, E.Teller, J.Chem.Phys. **21** ,1087 (1953)

[14] F.F. Abraham J. Chem. Phys. **58**, 3166 (1973)

[15] R.K.P. Zia and J.E.Avron, Phys. Rev. **B25**, 2042 (1982)
C.Rottman and M.Wortis Phys. Rev. **B24**, 6274 (1981)

[16] M.J.Lowe and D.J.Wallace, J.Phys. A :Math.Gen.**13**, L381 (1980)
D.J.Wallace , in Proceedings of the Cargese Summer Institute on Phase Transitions 1980 ed. J.C.Guillon and J.Zinn - Justin, New York Plenum (1982)

[17] R.C.Tolman, J. Chem. Phys. **17**, 333 (1949)
P.P. Buff, J. Chem. Phys. **23**, 419, (1955)

Table I : Parameter value for the 2D model (eq.48); in this computation $\varepsilon_{AA} = \varepsilon_{BB} = 0$

T/T_c	.801	.588	.401
W_{AB}	2.2	3.0	4.4
μ_{coex}/W_{AB}	$-7.20\ 10^{-3}$	$9.\ 10^{-4}$	$3.5\ 10^{-5}$
C_{coex}	.978	.997	.9998
a (eq.48)	$-7.4\ 10^{-3}$	$-9.0\ 10^{-4}$	$-3.5\ 10^{-5}$
b (eq.48)	.665	1.251	1.645
c (eq.48)	1.426	.715	.301
$\tau/\beta W_{AB}$	.568	.417	.284
ε (eq.48)	2.14	.57	.18

Table II : Parameter values for the 3D model (eq.51); in this computation $\varepsilon_{AA} = \varepsilon_{BB} = 0$

T/T_c	.590	.400
W_{AB}	1.5030	2.2170
μ_{coex}/W_{AB}	$8.33\ 10^{-3}$	$6.02\ 10^{-4}$
C_{coex}	.985	.999
a (eq.51)	$-8.45\ 10^{-3}$	$-6.03\ 10^{-4}$
b (eq.51)	2.00	2.95
c (eq.51)	-1.41	-1.35
d (eq.51)	3.442	1.713
$\tau/\beta W_{AB}$	$-7.39\ 10^{-2}$	$-5.01\ 10^{-2}$
ε (eq.51)	-.831	-.458

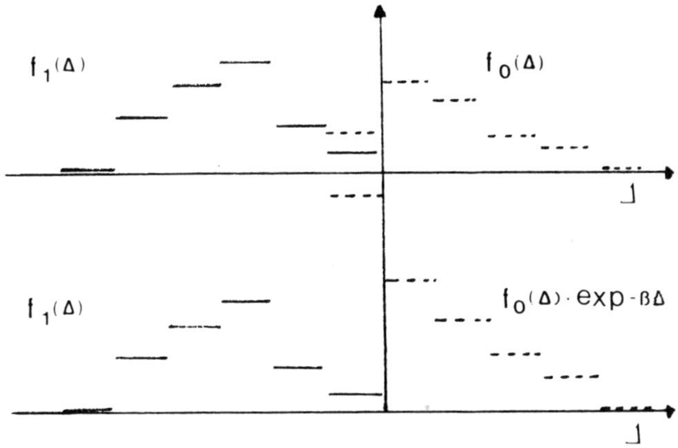

Fig.1a) example of energy distributions $f_\alpha(\Delta)$ obtained by M.C.sampling
Fig.1b) due to the bad estimate of $f_0(\Delta)$ for negative value of Δ, eq.31 cannot be applied (here o corresponds to I and 1 to II).

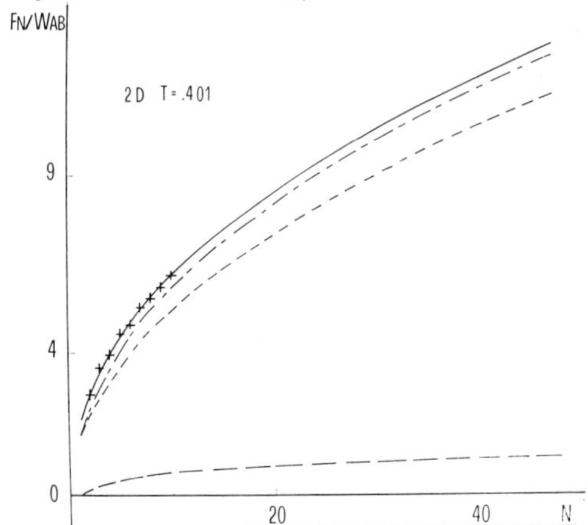

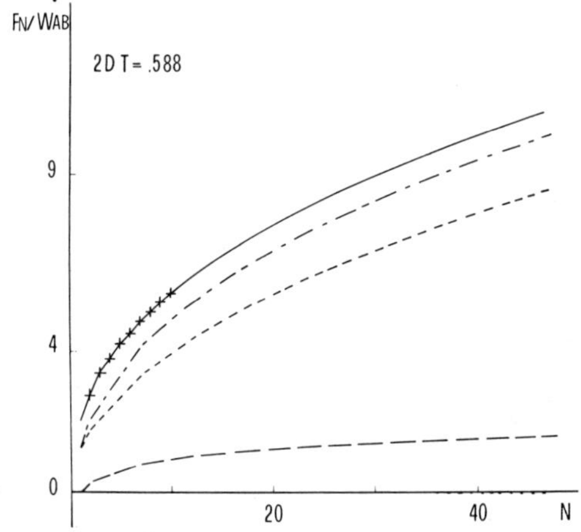

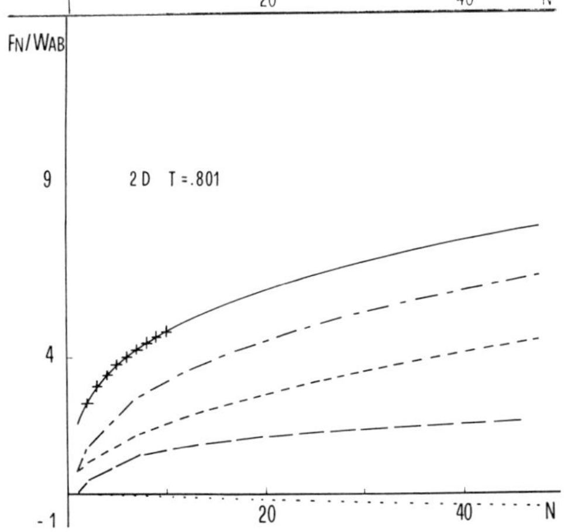

Fig 2 : 2D M.C. values of the free energy F_N (in units of $w_{AB} = 2\varepsilon_{AB} - \varepsilon_{AA} - \varepsilon_{BB}$) as a function of the number of B atoms contained in the cluster:
--- capillarity express. (eq.48); + M.C.data;...bulk;- - - surface; —— logarithmic term;—-— sum of the previous term. Temperature is expressed in T_c units.

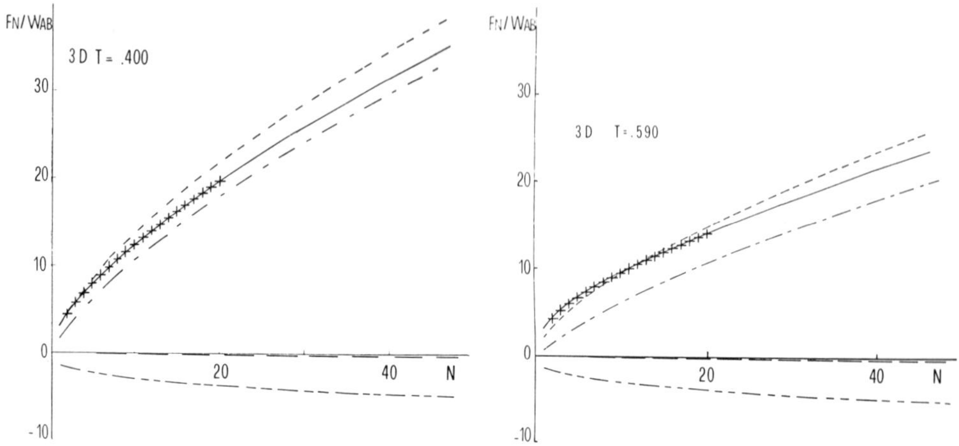

Fig 3 : 3D M.C. values of the free energy F_N (in units of $w_{AB}=2\varepsilon_{AB}-\varepsilon_{AA}-\varepsilon_{BB}$) as a function of the number of B atoms contained in the cluster:--- capillarity expression (eq. 51); + M.C. data; ...bulk; - - -surface;—— logarithmic term;——— —curvature correction; ———sum of the previous term. Temperature is expressed in T_c units.

SURFACES AND INTERFACES IN IONIC MATERIALS

P. W. Tasker
Theoretical Physics Division
A.E.R.E. Harwell
Oxon. OX11 ORA
United Kingdom

Abstract. Ionic materials have a wide and growing range of applications but their behaviour can be complex due to the competing effects of intrinsic defects and impurities which are often present in relatively large amounts. Computer calculations are a valuable way of interpreting and predicting the properties of oxide ceramics. In this paper, the application of static lattice simulations is reviewed with particular emphasis on the characteristics of surfaces and grain boundaries. The calculations can predict reliably the structure and energy of interfaces, the segregation of impurities and the consequent changes in material properties.

Introduction

Ionic materials are those in which the Coulomb interaction between charged ions plays a predominant role in determining the cohesion of the crystal. Many important ceramic materials come into this class including most oxides. There is growing enthusiasm for using ceramics in many applications including gas sensors, catalyst supports, devices,[1−2] as well as utilising their high temperature capabilities as engineering materials in, for example, engines and turbines. The processing and fabrication of ceramics is very different from the corresponding treatments of metals. Because of their high melting points ceramic products are fabricated by high temperature treatment of powders in a process known as sintering. The mechanism and role of impurities in this process is poorly understood at a fundamental level[3]. Knowledge of the behaviour of oxides at an atomistic level would also be valuable to the study of processes such as the corrosion of metals. For example, the high temperature corrosion rate of many metals is determined by enhanced ionic diffusion along grain boundaries in the growing oxide scale[4]. Thus, the processing of ceramics, their application in wide variety of technologies, and the materials behaviour of oxides in determining other phenomena such as corrosion all require understanding and interpretation at a basic level. Computer simulation can play a particularly vital function in aiding the interpretation of experiment. It enables the underlying mechanisms to be determined and the often complex, competing effects of, for example, impurities to be analysed into their components. As understanding is gained, the simulations can be used to extrapolate experimental results into regimes where measurement is difficult or impossible. Ultimately, the simulations are of value in limiting the amount of empirical searching that is made when developing new materials. For example, a theoretical study of a wide range of dopants can identify those most likely to give a desired property and are most promising for experimental investigation. Of course, the

G. Jacucci (ed.), Computer Simulation in Physical Metallurgy, 21–40.

increased fundamental understanding may also reveal new important effects not previously suspected in empirical studies.

In this paper, we describe the application of static lattice computer simulations to the study of oxide materials. Although much work has been done on the study of bulk defects, we shall concentrate, for examples, on the properties of surfaces and interfaces since these play a particularly vital role in many of the important phenomena.

The Theoretical Model

A static calculation starts with a model of the crystalline material in which each atom interacts with the others via an interatomic potential. A perfect crystal should be in equilibrium with no forces acting on any atom. When a defect, such as a point defect (e.g. a vacancy) or an extended defect (e.g. a surface), is introduced this equilibrium is disturbed. The calculation proceeds by searching for a new atomic configuration in which all the atoms are again in equilibrium. At the end of the relaxation, the structure and energy of the defect is determined. In a real crystal, the atoms are vibrating about their mean positions so a static calculation is approximate in that it neglects dynamical effects. In particular, the energy calculated is an internal energy and knowledge of the entropy is required if comparison is to be made with the free energy at high temperatures.

Simulations on ionic materials differ in technique from those on metals because of the predominant Coulomb interaction[5–6]. The long-range nature of this interaction requires that the summations are not just carried out in real space. However, the true pair-like nature of this leading term in the interatomic potential means that the interactions are often more reliably known for ionic materials than for either metals or semiconductors. Thus, the simulations can tackle a wider range of useful problems. Since the component atoms of an ionic material are charged, point defects such as vacancies or impurities are, in general, also charged. This produces a long-range polarization of the surrounding crystal that cannot be simulated explicitly. A generalization of the Mott–Littleton approximation[7] is used and this is illustrated in figure 1. This shows schematically how a calculation of a point defect near an ionic grain boundary would be carried out. The grain boundary structure is calculated first using the two-dimensional periodicity of the system. Examples of actual calculations will be discussed in a later section. The point defect (e.g. an impurity) is introduced at a site near the boundary. A spherical region of crystal around the defect is modelled explicitly with every ion relaxed to equilibrium. Beyond that, the crystal is treated as a dielectric continuum which is polarized by the defect charge. The size of the inner, explicit region can be increased until the defect energy does not vary; 150 ions in the inner region is generally sufficient. The accuracy of the simulation depends finally on the reliability of the interatomic potentials. These consist of three parts.

1. The Coulomb interaction between ions of charges Z_1 and Z_2 respectively i.e.

$$V(R) = \frac{Z_1 Z_2}{R}$$

2. The short-range interaction. This is required for crystal stability and to reproduce the elastic properties of the material. It may take a simple analytic form, for example a Buckingham potential;

$$V(R) = A\exp(-R/\rho) - C/R^6$$

or be given in numerical form. In either case, its range usually extends no further than the next-nearest neighbour distance and so its summation presents no technical problems.

3. The polarizability of each atom must be included if dielectric properties of the material are to be reproduced. This is usually incorporated via the shell model[8] in which each ion is represented as as a core and shell of opposite charge coupled by an harmonic potential, i.e.

$$V(R) = \tfrac{1}{2}kR^2.$$

Displacements of the shell relative to the core lead to the development of a dipole moment on the ion sites. The short-range potential acts between shells thus coupling the intra-atomic and inter-atomic interactions. A good description of lattice polarization is essential to the reliable calculation of the relaxation around a charged defect.

The potentials for a particular material are obtained in one of two ways. Firstly, they can be obtained empirically by fitting the parameters (e.g. the A, ρ, C, and k defined above) to the known crystal bulk properties such as structure, lattice energy, elastic and dielectric constants[9–10]. An important point is that only perfect crystal, bulk properties are used in the fitting so the calculation of defect properties is predictive. A compilation of empirical potentials for many ionic materials is available[11]. An alternative approach is to calculate the short-range potential by some quantum mechanical method. This is particularly valuable when the bulk properties have not been measured or when impurity potentials are needed. The electron-gas method has been applied to a wide range of problems[12]. The shell parameters that determine the ionic polarizabilities cannot be reliably calculated so they are still determined empirically. The bulk structure and elastic properties are used to test these calculated potentials and generally very good agreement is obtained. Thus, in either case, we should start the defect calculations with a model that describes well the bulk properties of the material.

The minimisation of the energy and relaxation of the structure is carried out by an efficient numerical procedure. Usually a quasi–Newton–Raphson method is employed which uses both the first and second derivatives of the potentials[13]. The calculations are not restricted to simple materials or structures but most of the examples that are discussed below are of simple oxides since more experimental data are available to validate the model for these cases.

The Ideal Surface

The first stage in the calculation of the surface properties of a material is the determination of the structure of the ideal, planar surface. This is achieved by terminating a crystal block so as to produce the face of interest. The ions in the surface region are then relaxed to give the equilibrium structure. A distinction is usually made between surface reconstructions in which the symmetry of the surface is altered from that of a simple termination of the bulk structure, and surface relaxations in which the ions are distorted from their usual sites but the symmetry is maintained. Highly ionic materials differ from the more covalently bonded materials such as semiconductors in that the low-energy faces are rarely reconstructed and often show only very small relaxations. Typically, faces that contain both anions and cations in the surface plane show a small outward displacement of the anion and a small inward displacement of the cation, known as rumpling[14]. For example, the most recent Low Energy Electron Diffraction (LEED) analysis of the structure magnesium oxide showed that the (001) face was rumpled by $2 \pm 2\%$ of an interatomic spacing with a net outward relaxation of $0 \pm 0.75\%$[15]. The most recent theoretical simulation gave 2.5% and −0.7% for the rumple and relaxation respectively[16]. Both experiment and

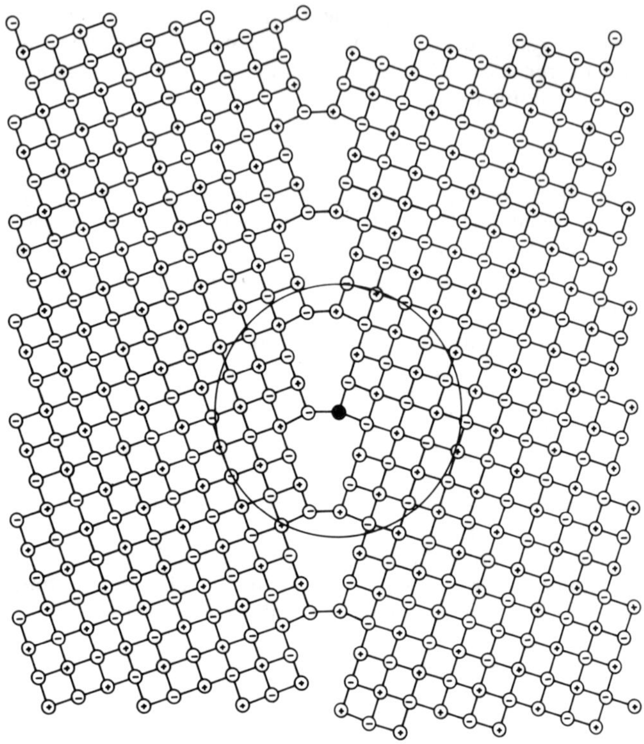

Figure 1. The explicitly relaxed region around a point defect near an ionic grain boundary.

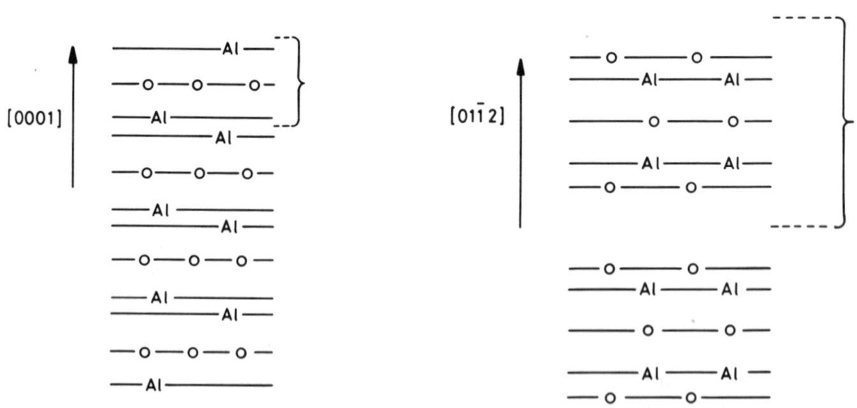

Figure 2. The planar stacking near (a) the (0001) and (b) the (01$\bar{1}$2) surfaces of α–alumina[16].

theory agree that this lowest energy surface of MgO closely resembles a simple termination of the crystal bulk.

The Coulomb interaction in ionic materials does, however, prevent certain surfaces from existing without substantial reconstruction[17]. In certain crystallographic directions, the stacking sequence of planes is dipolar, i.e. the planes alternate in charge. An example is the (111) direction in rocksalt structured crystals along which a plane of all anions alternates with a plane of all cations. If a neutral crystal terminated in this direction, the electric field created would polarize the whole crystal giving rise to an infinite surface energy. Where such surfaces exist, they must be restructured to remove the dipole moment. One example is the (100) surface of urania whose observed structure has only half the normal density of oxygen ions in the surface plane[18]; exactly what is needed to remove the long-range field. Ensuring that there are no dipole moments often enables the likely configuration of the surface to be identified. Figure 2 shows the stacking sequence of planes in two low energy directions in α–alumina. In the basal plane direction, (0001) shown figure 2(a) the sequence consists of close-packed oxygen planes interleaved with puckered planes of aluminium ions filling two-thirds of the octahedral sites. The non-dipolar surface must terminate with a plane of aluminium ions in just one-third of exposed, octahedral sites and, indeed, that structure has been observed[19]. Similarly, in the $(01\bar{1}2)$ face shown in figure 2(b) the terminating plane must consist of oxygen ions.

The simulations also, of course, give a value for the surface energy of each face. As mentioned above, these are internal energies and should be compared with experimental measurements extrapolated to absolute zero temperature. This is shown for a range of ionic materials in Table 1.

Crystal surface	Calculated [16][20–21]	Experimental[22–28] extrapolated to 0K
NaCl (001)	0.212	0.190, 0.276, 0.300, 0.227
NaBr	0.187	0.178
KCl	0.170	0.173, 0.252, 0.110
KI	0.139	0.139
CaF_2 (111)	0.476	0.450, 0.510
SrF_2	0.407	0.260–0.465
BaF_2	0.349	0.280, 0.350
UO_2 (111)	1.06	1.05, 1.4
ThO_2	1.02	
CeO_2	1.05	
MgO (001)	1.16	1.04, 1.2, 1.15
NiO	1.15	
α-Al_2O_3 $(01\bar{1}2)$	2.57	
α-Al_2O_3 (0001)	2.97	

Table 1. Calculated and measured surface energies in Jm^{-2}

The thermodynamic properties of the surface are valuable in understanding the stability of surfaces, the growth of bubbles and voids and the energetics of sintering, wetting etc. Although these are not finite temperature calculations, the static result is often sufficient since it enables the relative surface and interface energies to be assessed. The agreement between calculation and experiment in the table is very good particularly considering the uncertainties in the experimental values obtained from either cleavage experiments or liquid drop contact angles. The calculation of surface energy is very staightforward and when the interatomic potentials give a good description of the bulk properties it is probably a more reliable determination than the experimental measurement.

Surface energies can be extrapolated to higher temperatures by calculating separately the surface entropy. This can be done using the same model for the material (i.e. interatomic potentials etc.) via a calculation of the surface phonons. A finite slab of crystal is used with two surfaces which are sufficiently far apart not to interact. The dynamical matrix is calculated and diagonalised for this large system[29]. For example, a fifteen layer slab of MgO gives rise to a dynamical matrix with 90 eigenvalues. These are shown as a function of surface wave-vector in figure 3[30]. Most of the vibrational modes are the bulk phonons projected into the plane of the surface Brillouin zone because of the lack periodicity in the third dimension. However, some surface modes, which are localised in the surface region, can be identified. For example, the mode below the bottom of the bulk bands is the Rayleigh, surface accoustic mode. The slab entropy is obtained by sampling the vibrational frequencies throughout the Brillouin zone and evaluating,

$$S = k\sum_{i=1}^{N_f}\{-\ln(1-e^{-x}) + (x/e^{x}-1)\}$$

where $x = h\omega_i/kT$ with ω_i a sample vibrational frequency and N_f is the number sampling points in the zone. The bulk vibrational entropy is obtained similarly and the surface entropy can be obtained by subtraction of the two values appropriately normalised[31]. Figure 4 shows the vibrational surface entropy and specific heat capacity as a function of temperature for the same (001) surface of MgO. The calculated values are rather lower than those determined experimentally[32] but this may well be due to difficulties in determining the surface area of the finely divided samples[31]. Since a calculation based on vibrational modes makes, at best, a quasi-harmonic approximation these calculations will become more approximate at very high temperatures.

Surface extended defects such as steps can also be calculated by considering the structure of the vicinal surfaces[33]. The faces with indices such as (n01) are equivalent to (001) surfaces with a monatomic step every n interatomic spacings. They can also be viewed as surfaces deviating by an angle $\theta = \tan^{-1}(1/n)$ from the (001) face. Figure 5 shows the surface energy of such faces as a function of angle for MgO. In ionic materials, in contrast to metals, the surface energy varies strongly with angle and this has consequences, for example, for the interpretation of thermal grooving experiments. The interaction between surface steps can be extracted from plots such as figure 5. In MgO the step energy is 3.72×10^{-10} Jm^{-1} and like steps repel each other with a short-range exponential repulsion. Unlike the perfect surface, the step is substantially distorted from its ideal configuration and this is shown in figure 6. Sites near such steps may be particularly important in many surface processes, such as adsorption sites in catalysis.

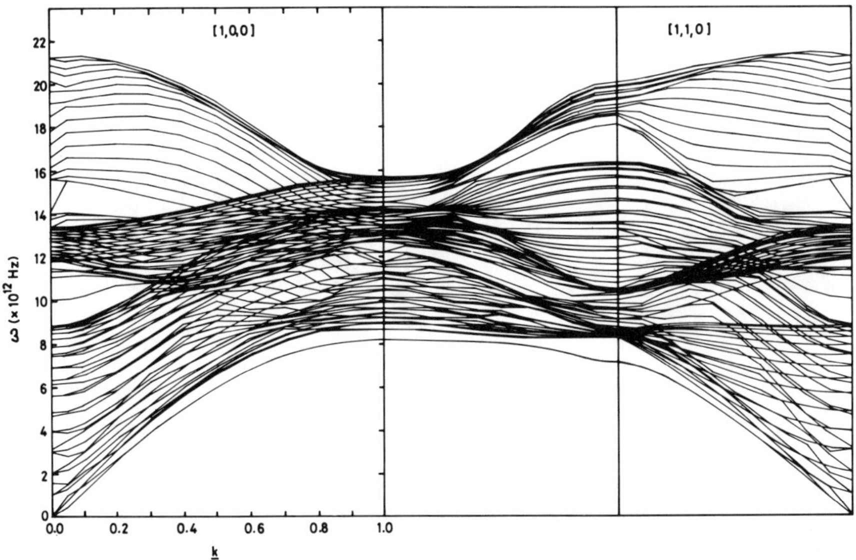

Figure 3. Phonon dispersion curves calculated for a 15 layer slab of MgO terminating with (001) surfaces[30].

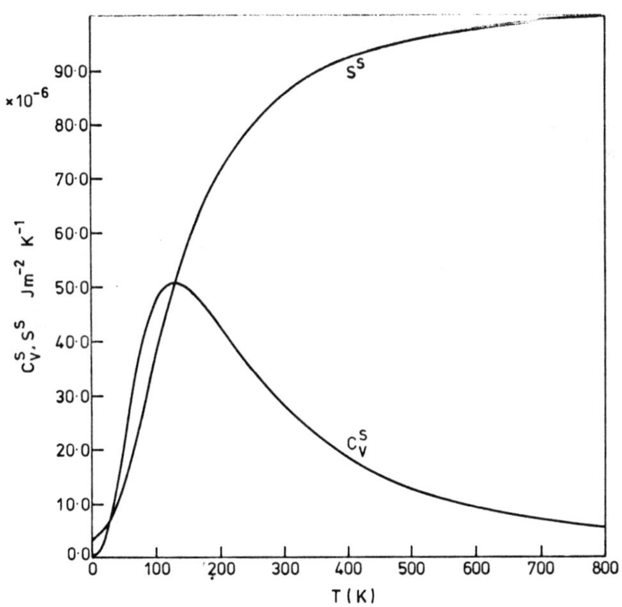

Figure 4. Vibrational contribution to the surface entropy, S^s, and specific heat capacity at constant volume, C_V^s, for MgO (001)[30].

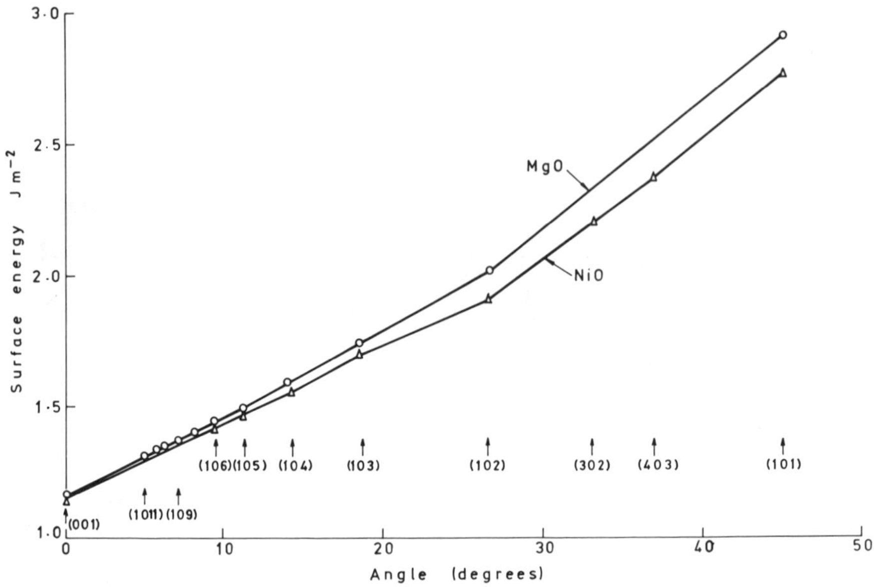

Figure 5. Variation of surface energy with angular deviation from (001) for MgO (10n) surfaces[33].

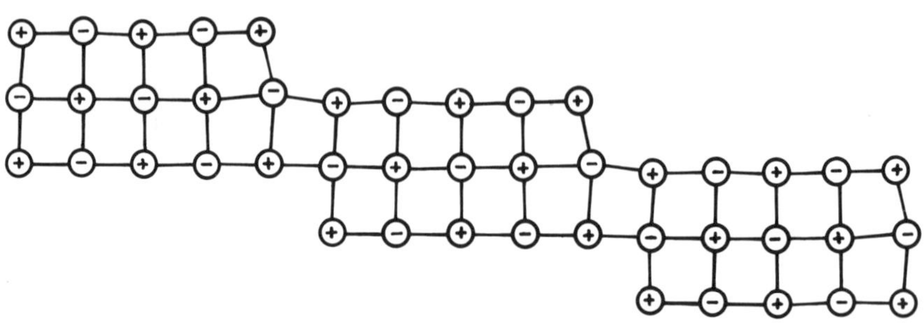

Figure 6. Atomic structure of a step on an MgO (105) surface[33].

Defects and Impurities Near Surfaces.

Near the crystal surface the usual constraint of charge neutrality is relaxed and the surface can aquire a charge compensated by an extended space-charge region below the interface[34]. This is caused by the difference in formation energies of charged point defects and by segregation energies of charged impurities. In the case of MgO, calculations indicate that the energy of both anion and cation vacancies at the surface is very close to that in the bulk[35]. Figure 7 shows the variation of energy for a cation vacancy with depth below the surface. The maximum which is a general feature in ionic materials provides an energetic barrier to diffusion between the bulk and surface. These calculations indicated that while both vacancy energies are higher at the surface than at the bulk, the cation vacancy energy is closer to the bulk value. This implies that at finite temperatures the surface charge would be negative. However, for stoichiometric oxides such as MgO the space-charge screening lengths are enormous due to the low concentrations of intrinsic defects. In these cases the surface charge and space-charge region is likely to be determined by impurity levels.

The segregation of impurity ions to the free surfaces not only generates possible surface charges but also can alter the structure and chemistry of the interface. If we assume that there is only one type of surface site (monolayer model), neglect defect interactions for the moment (ideal solution), and assume that the impurities are distributed randomly over their available sites (regular solution), then the surface numbers of any two impurities in a multi-component mixture is given by;

$$\frac{n_1{}^s}{n_2{}^s} = \frac{n_1}{n_2}\exp(-\Delta G/kT)$$

where n_i is the number of atoms of species i in the bulk and the superscript s indicates numbers in the surface. When there are only two components (i.e. one impurity) this becomes the usual Langmuir–McLean isotherm with ΔG as the free energy of segregation[36]. Neglecting again the vibrational entropy component, the energy of segregation can be calculated from the difference in the energy to substitute an impurity ion near the surface and in the bulk. High concentrations of segregating species can be incorporated by considering substitutional arrays of defects in the surface. When the impurity ions do interact with each other (the normal occurence) the energy of segregation, ΔG, becomes coverage dependent. This is illustrated in Table 2 for some isovalent impurities in MgO[37–38].

Coverage	Ni	Ca	Sr	Ba
isolated	+0.01	−1.08	−2.23	−3.90
$\frac{1}{4}$	+0.02	−0.84	−1.68	−2.99
$\frac{1}{2}$	+0.03	−0.48	−0.65	−1.11
full	+0.03	−0.48	−1.12	−1.41

Table 2. Segregation energies (eV) for impuities in MgO[37–38].

The positive value for the segregation energy of nickel indicates that this impurity does not segregate and this has been confirmed experimentally. The larger impurities all segregate with an energy that increases with increasing ion size. The experimental heat of segregation for calcium in magnesia of −0.52eV[39] agrees very well with the

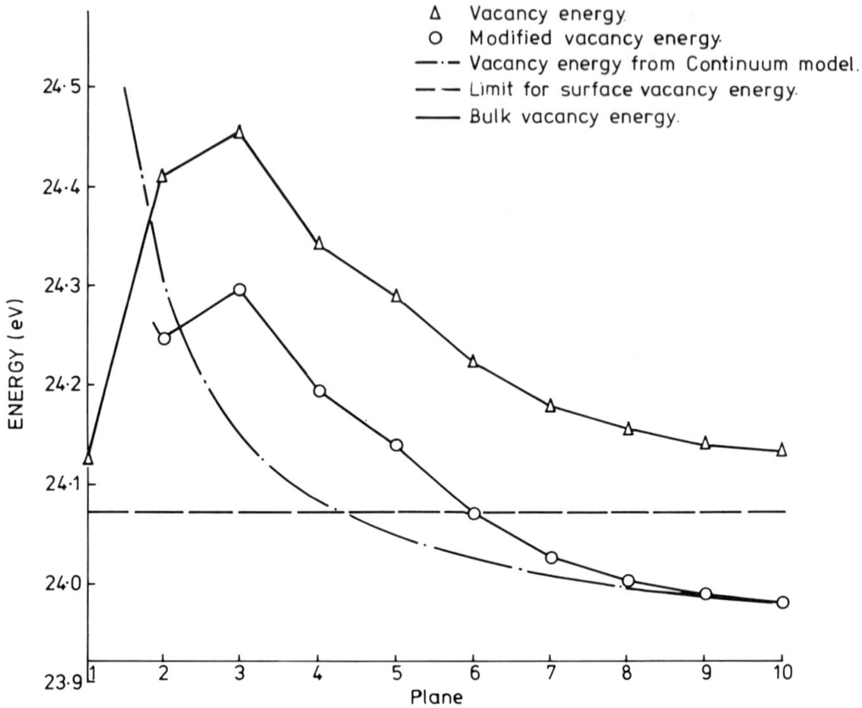

Figure 7. Cation vacancy formation energy as a function of depth below an (001) MgO surface with the bulk limit shown as a dashed line[35].

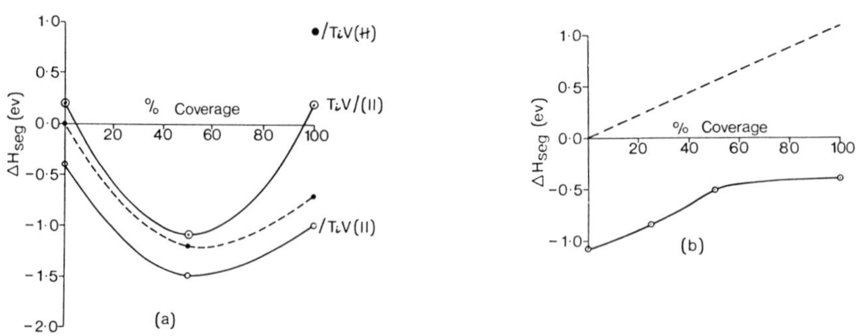

Figure 8. Calculated heats of segregation for (a) titania in MgO and (b) calcia in MgO. The broken line in each case shows the energy of aggregation in the crystal bulk. In (a) data are shown for Ti–Vacancy complexes in the surface plane (TiV/) and in the plane below the surface[41].

calculated values. The energy of segregation can be seen from table 2 not to be constant but to vary strongly with coverage thus giving rise to non–Langmuir type behaviour. In the case of the larger ions such as strontium or barium the energy even goes through a maximum at about half coverage. This is due to a restructuring that occurs at high surface concentrations induced by the mutual repulsions of the large ions. These larger scale restructurings are difficult to find by static calculations which assume some periodicity parallel to the surface. In this case, the new structure was first inferred from an instability in the surface phonons[40]. The energy function will lead to a bimodal surface concentration since at temperatures at which the surface concentration of strontium will be about a half, the distribution of impurities should separate into regions of less than and regions of greater than 50% coverage.

An example of an aliovalent impurity is that of titanium in MgO. Titania dissolves in magnesia to introduce substitutional Ti^{4+} ions and compensating cation vacancies. Thus, using the conventional Kroger–Vink notation in which the subscript indicates the lattice site on which the defect sits and the superscripts $\times$, $\cdot$, and $'$ indicate neutral, positive and negative charges respectively relative to normal lattice charges,

$$TiO_2 + 2Mg_{Mg}^{\times} = Ti_{Mg}^{\cdot\cdot} + V_{Mg}^{''} + 2MgO.$$

The situation is simplified since the vacancies and substitutional ions are strongly bound to form neutral pairs,

$$Ti_{Mg}^{\cdot\cdot} + V_{Mg}^{''} = (Ti_{Mg}\text{--}V_{Mg})^{\times}$$

and therefore behave rather like isovalent impurities in that they do not contribute to the surface charge. However, the coverage dependence of the heat of segregation is very different from that calculated for the alkaline earth impurities[41]. Figure 8(b) shows the dependence for calcium in MgO, with the dashed line showing the increase in energy with impurity aggregation in the bulk. Figure 8(a) shows the corresponding case for titania in MgO. Here in both the surface and bulk the curves show a pronounced minimum indicating second phase formation of magnesium titanate ($MgTiO_3$). At low surface concentrations (high temperature) the energy of segregation is $\approx$0.3eV and the behaviour is Langmuir-like. At higher surface concentrations precipitation of the titanate is expected and the concentration of free impurity pairs is determined by the solubilty of $MgTiO_3$.

Finally, another way of representing the heat of segregation of an impurity is as a change in the surface energy. The surface energy is lowered by a segregating species and this is illustrated for isovalent impurities in MgO in figure 9[38]. We see that for surface concentrations of strontium of greater than about 75% or of barium of greater than about 10% the surface energy of the crystal becomes negative. This plot considers only the internal energy and to determine the effect on the surface energy at high temperatures both the vibrational entropy and the difference between the bulk and surface configurational entropy must be considered. This last term, for example, will prevent the surface segregated with strontium having a negative surface energy at temperatures above 700K. In addition, precipitation of the impurity may have occurred before this, but the surface energy can certainly be very substantially lowered by impurity segregation. A negative surface energy would imply stabilisation of the crystal in finely divided form and a thermodynamic barrier to sintering. These examples illustrate not only the diversity of segregation behaviour that can be observed in even a simple material like MgO, but also how impurities can substantially alter surface properties and behaviour.

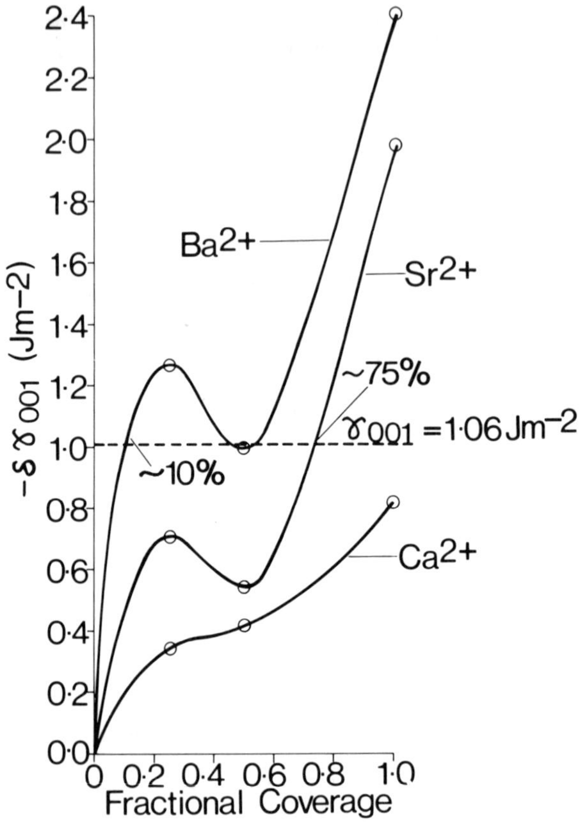

Figure 9. Calculated decrease in the (001) surface energy, $\delta\gamma_{001}$, of MgO as a function of impurity coverage[38] at low temperature.

Grain Boundary Structure.

The same computational techniques can be used to study interfaces either as extended defects within a material or between different materials. The most common interface is the grain boundary since almost all real materials are polycrystalline. It also plays a vital role in many important oxide processes. An example is the high temperature gas-phase corrosion of many metals. In this regime, an oxide scale grows on the metal surface and the growth kinetics are expected to follow a parabolic rate law in the classical theory of Wagner[42]. Once the scale has formed further oxidation can only occur if metal or oxygen is transported through the oxide layer. Thus the rate of scale growth is inversly proportional to its thickness giving the expected time$^{\frac{1}{2}}$ dependence. Nickel oxide is a non-stoichiometric, semiconducting oxide which is generally deficient in cations (i.e. $Ni_{1-x}O$). The enhanced concentration of cation vacancies means that cation diffusion is likely to be rate determining in the oxidation. Measurements of the corrosion rate of nickel indeed show parabolic kinetics but the independently measured rate of cation diffusion in the bulk lattice cannot account for the observed rate except at very high temperature[43]. The experiments have shown, however, that the nickel diffuses very much more quickly down grain boundaries in the growing oxide scale[44] and that this rate accords with the overall corrosion rate[45]. Furthermore, impurities such as ceria that are incorporated in the oxide as it grows can very much reduce the rate of oxidation[46]. Therefore, understanding of the mechanism of enhanced grain boundary diffusion and the influence of impurities can give insight into the corrosion process and its control.

Grain boundaries can be generally described as the rotation of one crystal grain with respect to another. The rotation can be resolved about axes parallel and perpendicular to the interface. Rotations about a perpendicular axis give rise to pure twist boundaries while those about a parallel axis produce pure tilt boundaries. A general grain boundary about an arbitrary axis then contains components of both twist and tilt. This resolution is equivalent to the description of dislocations as screw or edge. Indeed, low angle twist boundaries can be described in terms of arrays of screw dislocations while the corresponding tilt boundaries can be described as arrays of edge dislocations. For certain angles of rotation, the two grains have a periodic registry on a finite scale and such boundaries are known as coincidence interfaces. If the crystals on either side of the interface are imagined as filling all space, the overlapping lattices coincide at certain regular sites which define the coincidence lattice. Coincidence boundaries are often designated from this construction by a parameter Σn where n is the reciprocal density of coincidence sites[47].

Because of the periodicity of coincidence boundaries, it is possible to calculate their properties in a relatively straightforward manner. Furthermore it is expected that these boundaries will be particularly stable, being represented as cusps in the energy versus angle of misorientation plots. There is some evidence from studies on metal boundaries that the structure of more general boundaries may be made up from low-energy units derived from the coincidence structures[48]. Figure 10 shows four, typical, coincidence, tilt boundaries calculated for nickel oxide (NiO)[49–50]. In figure 10(a) a schematic diagram of the $\Sigma 5$ (310) 36.9°/[001] tilt boundary is shown. It resembles an array of [001] dislocations with open 'pipes' running down the boundary. Similarly, the 73.7°/[001] boundary (figure 10(b)) resembles an array of [011] dislocations also with a substantial open space in the boundary plane. These open 'pipes' are unlike the structures found in metal boundaries which are generally more close-packed. They are caused by the requirement when optimising the structure of not bringing like-charged ions into nearest neighbour positions. They do suggest possible pathways for enhanced diffusion which would be expected to be anisotropic. Figures 10(c) and

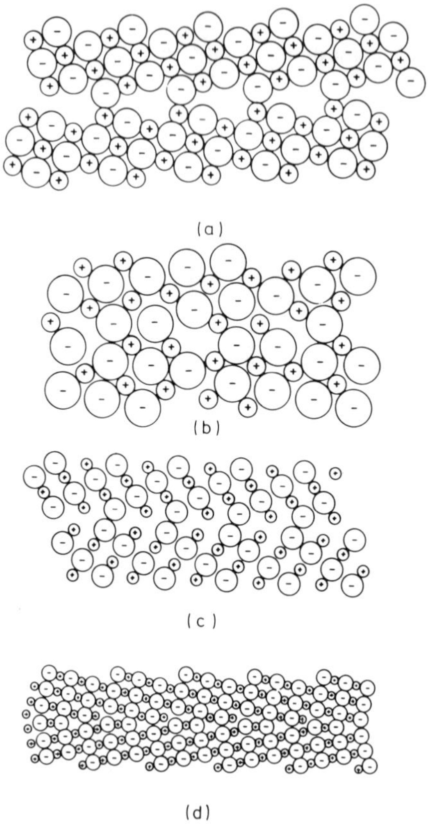

Figure 10. Calculated structures of coincidence tilt grain boundaries in NiO[49–50]; (a) $\Sigma5$, 36.9°/[001]; (b) $\Sigma25$, 73.7°/[001] (c) $\Sigma11$, 129.5°/[011]; (d) $\Sigma19$, 26.5°/[011]

10(d) show two different structural types found symmetric tilt boundaries about a [011] axis. In the first case, (the 129.5° Σ11 boundary), open channels are also present while the 26.5° Σ19 boundary is fairly densely packed. This illustrates the range of structures, which may be expected to show different properties, that are found even for coincidence boundaries in a simple structure such as rocksalt.

Figure 11 shows the corresponding structure for a Σ5 36.9°/[001] *twist* boundary in the same material(51). The boundary is shown above the lower cystal grain (drawn with dashed lines). A simple twist of one grain above another does not produce a stable boundary since like-charged ions are brought into close proximity(52–53). The stable structure shown here has a reduced number of ions in the boundary plane. This structure can be generalised to derive a complete series of twist boundaries all with a reduced ion density, which maximises the amount of lattice in the boundary plane that lies at a rotation intermediate between that of the upper and lower grain(54). This lower ion density is also likely to make diffusion easier in the boundary plane but in this case the behaviour is expected to be isotropic.

Defects and Impurities Near Boundaries.

As for surfaces, grain boundaries can become charged with an excess concentration of one type of defect or impurity in the boundary plane. A non-stoichiometric oxide such as NiO has a high concentration of intrinsic point defects which therefore play a more important role in determining boundary charge than, for example, the vacancies in MgO. The nickel vacancies in NiO are found experimentally to be singly charged at moderate temperatures. The simple vacancy would be expected to be doubly charged so a hole must be closely bound to reduce the apparent charge. The holes are assumed in the calculations discussed below to be localised on cation sites (i.e. as Ni^{3+} ions). This implies electronic conduction in NiO by a polaron mechanism. The singly charged vacancy is then modelled as a bound pair of a vacancy with a Ni^{3+} ion.

Since there are substantial differences in the structures of different grain boundaries and we wish to draw general conclusions about grain boundary properties, the calculations have been carried out for each of the tilt boundary types illustrated in figure 10. These represent all the structural types that were calculated for the [001] and [011] tilts. An important difference between calculations on boundaries and surfaces is the range of different possible defect sites in the boundary plane. This leads to a range of defect energies for each interface. In Table 3, we record the lowest defect energy for vacancies and holes at each of the four boundaries(55).

Boundary	Cation vacancy	Anion vacancy	Hole (Ni^{3+})
36.9°/[001]	–0.36	–0.30	–0.25
67.4°/[001]	–1.61	–1.25	–0.45
109.5°/[011]	–0.95	–0.98	–0.23
38.9°/[011]	–1.54	–1.56	–0.54

Table 3. Vacancy and hole interaction energies (eV) near grain boundaries in NiO(55).

The interaction energies given in the table are the difference between bulk and interface defect energies and negative values indicate that boundary sites are favoured over bulk sites. For each defect there is at least one site in the boundary of lower

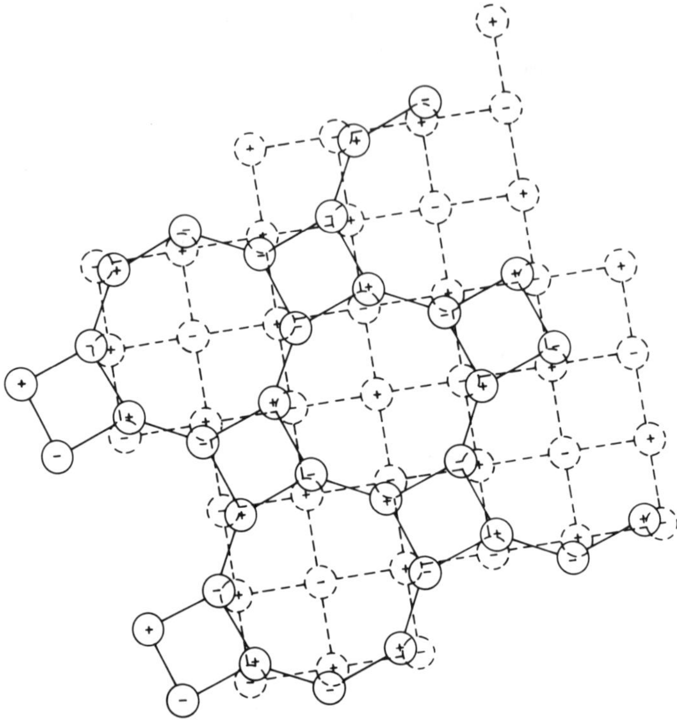

Figure 11. The relaxed structure of the boundary plane of a Σ5, 36.9°/[001] twist grain boundary in NiO[51].

energy than the bulk. There is also always at least one unfavoured site with positive interaction energy. Thus the concentration of all intrinsic defects will be enhanced on the boundary. In order to calculate the defect concentrations, the free energy must be minimised allowing also for the electrostatic interaction between the charged defects as was done by Kliewer and Koehler for surface space-charges[56–57]. In the case of nickel oxide boundaries, although the defect energies vary strongly with boundary structure, the qualitative behaviour is similar in all cases. All defect concentrations are enhanced but cation vacancy attraction is greater than that for the hole. Therefore the boundary will have a negative charge with a compensating positive space-charge region with excess holes. For example, in the case of the 109.5°/[011] tilt boundary, the vacancy concentration at 1000K is increased by a factor of about forty in the boundary region over the normal bulk concentration. This will contribute to an enhanced diffusion rate. The space-charge region has a screening length only of about 2nm at this temperature[55].

Impurity behaviour near boundaries makes a contrast with the behaviour near surfaces. As with the defects, there is a range of possible sites and thus a range of segregation energies. Table 4 shows the lowest segregation energies for three differently charged impurities near the four tilt boundaries.

Boundary	Co^{2+}	Al^{3+}	Ce^{4+}
36.9°/[001]	−0.09	−0.18	−1.15
67.4°/[001]	−0.22	−0.39	−1.73
109.5°/[011]	−0.12	−0.24	−0.41
38.9°/[011]	−0.06	−0.53	−0.74

Table 3. Impurity segregation energies (eV) near grain boundaries in NiO[58].

For all impurities studied whether they are larger or smaller than the host, or positively or negatively charged, their concentrations are enhanced near any tilt grain boundary. They do nòt all segregate to the same sites in the boundary and it is the diversity in available sites that produces this universal behaviour. Thus in contrast to the behaviour at surfaces, *all impurities segregate to tilt grain boundaries.* Charged impurities will influence the boundary charge, but allowance must also be made for the compensating defect. Thus, Al^{3+} ions are positively charged but generate an equal quantity of compensating negative vacancies which may segregate more strongly to the interface. So although the aluminium concentration is enhanced at the boundary, there is no isoelectric temperature at which the boundary charge becomes neutral[58].

Conclusions.

We have reviewed in this paper some of the recent applications of static lattice computer simulations of ionic oxides to the study of surfaces and grain boundaries. A particular theme has been the role of impurities in concentrating at interfaces and modifying the observed properties. Several important areas have been neglected; for example, the combination of structure determination by these methods coupled with quantum mechanical calculations has led to many interesting results in the study of chemisorption and surface reactivity[59–60]. Electronic properties of grain boundaries are being exploited in devices such as varistors. These properties may also be related, in part, to special structural effects of ionic or partly ionic materials[61–62]. In all these areas, computer simulation plays a particularly vital role in aiding the interpretation of experiments at an atomistic level. Impurities affect the properties of oxides so radically, and even the 'purest' material may have a substantial impurity concentration, that calculation may provide the best way of sorting out the various competing effects.

References

(1) Williams D.E. and McGeehin P. 1983 A.E.R.E.–R 11022 and *Royal Society of Chemistry, Specialist Periodical Report on Electrochemistry* to be published.

(2) Levinson C.M. (ed.) 1981 *Grain Boundary Phenomena in Electronic Ceramics, Adv. Ceramics* **1**

(3) Kingery W.D., Bowen H.K. and Uhlmann D.R. 1976 *Introduction to Ceramics* 2nd edition (Wiley, New York).

(4) Hughes A.E., Atkinson A. and Chadwick A.T. 1984 *Proc. of Symposium on 'Defect Properties and Processing of High Technology Non-metallic Materials', Materials Research Society Meeting, Boston, 1983* to be published.

(5) Catlow C.R.A. and Mackrodt W.C. (eds) 1982 *Computer Simulation of Solids, Lecture Notes in Physics* **166** (Springer–Verlag, Berlin).

(6) Mackrodt W.C. 1983 *Mass Transport in Solids (eds: Beniere F. and Catlow C.R.A.)* page 107, (Plenum, NewYork).

(7) Mott N.F. and Littleton M.J. 1938 *Trans. Faraday Soc.* **34** 485.

(8) Dick B.G. and Overhauser A.W. 1958 *Phys. Rev.* **112** 90.

(9) Catlow C.R.A., Diller K.M. and Norgett M.J. 1977 *J. Phys. C: Solid State Physics* **10** 1395.

(10) Sangster M.J. and Stoneham A.M. 1981 *Phil. Mag.* **B43** 597.

(11) Stoneham A.M. 1981 *A.E.R.E.–R 9598.*

(12) Mackrodt W.C. and Stewart R.F. 1979 *J. Phys. C: Solid State Physics* **12** 431.

(13) Norgett M.J. and Fletcher R. 1970 *J. Phys. C: Solid State Physics* **3** L190.

(14) Prutton M., Walker J.A., Welton–Cook M.R., Felton R.C. and Ramsey J.A. 1979 *J. Phys. C: Solid State Physics* **12** 5271.

(15) Welton–Cook M.R. and Berndt W. 1982 *J. Phys. C: Solid State Physics* **15** 5691.

(16) Tasker P.W. 1984 *Advances in Ceramics* in press.

(17) Tasker P.W. 1979 *J. Phys. C: Solid State Physics* **12** 4977.

(18) Ellis W.P. and Taylor T.N. 1980 *Surface Science* **91** 409.

(19) Chang C.C. 1968 *J. Appl. Phys.* **39** 5570.

(20) Tasker P.W. 1979 *Phil. Mag.* **A39** 119.

(21) Tasker P.W. 1980 *J. Phys. (Paris)* **41** C6–488.

(22) Jaeger F.M. 1917 *Z. Anorg. Chem.* **101** 1.

(23) Tosi M.P. 1964 *Solid State Physics* **16** 1.

(24) Becher P.F. and Freeman S.W. 1978 *J. Appl. Phys.* **49** 3779.

(25) Kraatz P. and Zoltai T. 1974 *J. Appl. Phys.* **45** 4741.

(26) Gilman J.J. 1960 *J. Appl. Phys.* **31** 2208.

(27) Nikolopoulos P., Nazare S. and Thummler F. 1977 *J. Nucl. Mater.* **71** 89.

(28) Hodkin E.M. and Nicholas M.G. 1977 *J. Nucl. Mater.* **67** 171.

(29) Chen T.S. and DeWette F.W. 1978 *Surface Science* **74** 373.

(30) Masri P. and Tasker P.W. 1983 *A.E.R.E.–R 10752.*

(31) Chen T.S. Alldredge G.P. and DeWette F.W. 1977 *Surface Science* **62** 675.

(32) Rieder K.H. 1971 *Surface Science* **26** 637.

(33) Tasker P.W. and Duffy D.M. 1984 *Surface Science* **137** 91.

(34) Frenkel J. 1946 *Kinetic Theory of Liquids* p.36 (O.U.P., Oxford).

(35) Duffy D.M., Hoare J.P. and Tasker P.W. 1984 *J. Phys. C: Solid State Physics* **17** L195.

(36) McLean D. 1957 *Grain Boundaries in Metals* p.131 (O.U.P., Oxford).

(37) Colbourn E.A., Mackrodt W.C. and Tasker P.W. 1983 *J. Mater. Sci.* **18** 1917.

(38) Tasker P.W., Colbourn E.A. and Mackrodt W.C. 1984 *J. Am. Ceram. Soc.* in press.

(39) McCune R.C. and Wynblatt P. 1983 *J. Am. Ceram. Soc.* **66** 111.

(40) Masri P. and Tasker P.W. 1984 *Surface Science* in press.

(41) Mackrodt W.C., Colbourn E.A. and Tasker P.W. 1984 *Surface Science* in press.

(42) Wagner C. 1933 *Z. Physik. Chem.* **B21** 25.

(43) Atkinson A., Hughes A.E. and Hammou A. 1981 *Phil. Mag.* **A43** 1071.

(44) Atkinson A. and Taylor R.I. 1981 *Phil. Mag.* **A43** 979 .

(45) Atkinson A., Taylor R.I. and Hughes A.E. 1982 *Phil. Mag.* **A45** 823.

(46) Chadwick A.T. 1982 *A.E.R.E.–R 10352.*

(47) Bollmann W. 1970 *Crystal Defects and Crystalline Interfaces* (Springer–Verlag, Berlin).

(48) Sutton A.P. 1981 *The Structure and Properties of Grain Boundaries,* Doctoral thesis, University of Pennsylvania.

(49) Duffy D.M. and Tasker P.W. 1983 *Phil. Mag.* **A47** 817.

(50) Duffy D.M. and Tasker P.W. 1983 *Phil. Mag.* **A48** 155.

(51) Tasker P.W. and Duffy D.M. 1983 *Phil. Mag.* **A47** L45.

(52) Wolf D. 1980 *J. Phys. (Paris)* **41** C6–142.

(53) Wolf D. and Benedek R. 1981 *Adv. Ceramics* **1** 107.

(54) Duffy D.M. and Tasker P.W. 1984 *Comm. Am. Ceram. Soc.* in press.

(55) Duffy D.M. and Tasker P.W. 1984 *Phil. Mag.* in press.

(56) Kliewer K.L. and Koehler J.S. 1965 *Phys. Rev.* **140** 1226.

(57) Kliewer K.L. 1965 *Phys. Rev.* **140** 1241.

(58) Duffy D.M. and Tasker P.W. 1984 *Phil. Mag.* in press.

(59) Colbourn E.A. and Mackrodt W.C. 1982 *Surface Science* **117** 571.

(60) Pope S.A., Guest M.F., Hillier I.H., Colbourn E.A., Mackrodt W.C. and Kendrick J. 1983 *Phys. Rev.* **B28** 2191.

(61) Duffy D.M. and Stoneham A.M. 1983 *J. Phys. C: Solid State Physics* **16** 4087.

(62) Duffy D.M. and Tasker P.W. 1984 *J. Appl. Phys.* **56** 971.

MACHINE CALCULATIONS IN SOLID STATE DIFFUSION

G. JACUCCI
Centro del Consiglio Nazionale delle Ricerche
e
Dipartimento di Fisica dell'Universita' di Trento
38050 Povo, Trento, Italy

Computer simulation and sampling methods have been very successful in the study of thermal and dynamical properties of disordered many body systems, like liquids and high temperature solids. The application of these methods to point defect diffusion (1) has extendend traditional defect calculations beyond the harmonic model (2). Concentration and migration parameters of point defects were the goal uniquely of Lattice Statics and Lattice Dynamics. Today's machine calculations correctly take into account all consequences of a given force law. Anharmonic effects can be precisely isolated comparing Monte Carlo results to accurate Lattice Dynamics calculations (3). Dynamical effects related to the memory of the system (4) can be measured by Molecular Dynamics and compared to analytical theories of jump rates which consider curved saddle surfaces (5,6,7).

The lecture reviews recent developments in this area, with particular reference to the case of a vacancy in a fcc lattice. Methods will be outlined and account will be given of what can be learned from such studies. In addition, Molecular Dynamics generations of jump histories will be described. Relevant jump mechanism and specific dynamical events can be assessed in this way. Particular attention will be devoted to double jumps of the vacancy observed close to melting in bulk monoatomic crystals (8). Computer simulation of time resolved experiments, like inelastic neutron scattering, NMR and Mössbauer, directed to the investigation of point defect diffusion will be illustrated. Deviations from random walk and the isolated point defect picture, observed in simulation studies of superionic conductors (9) and of diffusion on crystal surfaces (10,11) will be also described.

G. Jacucci (ed.), Computer Simulation in Physical Metallurgy, 41–42.

REFERENCES

1. C.H. Bennett, in "Diffusion in Solids: Recent Developments, edited by J.J. Burton and A.S.Novick (Academic, New York, 1975), Ch. 2

2. G. Jacucci, in "Diffusion in Crystalline Solids" edited by G.E. Murch and A.S. Nowick (Academic, New York, 1984), Ch.8

3. G. De Lorenzi, G. Jacucci and M. Ronchetti, to be published

4. C.P. Flynn and G. Jacucci, Phys. Rev. B10, 6225 (1982)

5. C.P. Flynn, Phys. Rev. Letters 35, 1721 (1975)

6. G. Jacucci, M. Toller, G. De Lorenzi and C.P. Flynn Phys. Rev. Letters 52, 295 (1984); Phys. Rev. B32, 2082 (1985)

7. G. De Lorenzi, C.P. Flynn and G. Jacucci, Phys. Rev. B30 5430 (1984)

8. A. Da Fano and G. Jacucci, Phys. Rev. Letters 39, 950 (1977)

9. G. Jacucci and A. Rahman, J. Chem. Phys. 69, 4117 (1978)

10. G. De Lorenzi, G. Jacucci and V. Pontikis, Surf. Sci. 116, 391 (1982)

11. G. de Lorenzi and G. Jacucci, Surf. Sci. to appear (1985)

MOLECULAR DYNAMICS STUDIES OF SURFACE DIFFUSION

Giulia De Lorenzi
Materials Research Laboratory and Physics Department
University of Illinois at Urbana-Champaign
Urbana, Illinois 61801

ABSTRACT. Molecular Dynamics simulations of the diffusion of vacancies and adatoms on various low index surfaces of face centered cubic (FCC) and body centered cubic (BCC) crystals are discussed. It is shown how the structure of the surface and the softness of the potential repulsion play the central role in determining the diffusion properties. Complicated migration mechanisms are active for the migration of adatoms: they can be clarified by supporting Molecular Dynamics with detailed investigations of the topography of the potential energy surface in the many dimensional configuration space of the crystal. As an example, the exchange mechanism for adatom migration on channelled surfaces is presented.

I. INTRODUCTION

Although in many physical phenomena the presence of surfaces delimiting blocks of condensed matter can be ignored or treated as a negligible boundary problem, nevertheless a variety of processes occuring at crystal surfaces are of great physical and technological interest: not only are they expressions of basic properties of condensed matter, but also they can determine properties of the bulk (as for example do crystal growth, evaporation, phase transitions and catalysis).

Most processes occuring at a crystal surface are influenced, if not determined, by the way atoms move on the surface. The mobility of atoms on a crystal surface can indeed be orders of magnitude larger than in the bulk at the same temperature and become the determining step in many processes such as the approach to equilibrium shape, the rate of chemical reaction and surface reconstruction.

Still, little is known about the microscopic mechanisms with which atoms move on a crystal surface. It is not known, for example, how the surface diffusivity is influenced by the presence of steps or kinks, nor by chemical specificity. As a matter of fact, the simple and intuitive fact that the diffusivity can vary significantly on

G. Jacucci (ed.), Computer Simulation in Physical Metallurgy, 43–65.

different faces of the same crystal, was established experimentally for the first time just a few years ago. It was with the development of the Field Ion Microscope (FIM)[1] that the direct observation of atomic displacements on almost perfectly clean and well characterized surfaces became possible, at least for metals at low temperatures (about 0.10 of the melting temperature). With FIM quantitative measurements of surface mobility are feasible, thus making it possible to learn a great deal about atomic and molecular behavior on solid surfaces. In a sense an atom moving over a crystal can be viewed as a convenient probe of local properties.[2] From the work of Bassett,[3] Ehrlich[4] and many others (for a review see the paper by Ehrlich in Physics Today[5]) some information is now available about specific migration mechanisms of atoms adsorbed on the surfaces of transition metals like W, Pt, Rh, Ir and Au.

Two of the basic questions addressed by this work are:
i) the relation between surface diffusivity and surface structure, and
ii) the nature of the interaction forces between atoms on a surface. Among many interesting findings we mention here a few facts relevant to our discussion:

a) the diffusivity has been found to be strongly dependent on the surface structure; on FCC metals, like Pt, Rh, Ir, and Ag, the diffusivity decreases with surface roughness, while on the BCC metal W roughness and diffusivity are not so simply related;
b) on channelled surfaces like FCC(110), FCC(311) and BCC(211), only one dimensional diffusion along the channels is generally observed, but on Pt(110) and Rh(110) diffusion occurs also across the channels; for the case of W and Rh adsorbed on Ir an exchange mechanism has been proved to be responsible for this cross-channel diffusion[6];
c) for single adatoms moving on an otherwise flat surface, a random walk model seems to describe the jumping process fairly well, at least for the low temperatures of these experiments.

Contributions to questions i) and ii) have been attempted in some general terms with computer simulation on model substances. These calculations are however still at a very primitive stage, particularly for the case of metals. The problem is that the complexity of the dynamics of a many particle system makes it almost necessary to use a simple pair potential, and pair potentials are not adequate to describe the rapid change in electronic density at a surface.

The Lennard-Jones potential has been used in molecular dynamics simulations of the migration of point defects, vacancies and adatoms on FCC crystal surfaces. Some information about the dependence of the diffusivity from surface structure has been obtained and will be discussed in Part II of this lecture.

The BCC structure, which is the one exhibited by many simple metals like the alkalis, is not stable with the Lennard-Jones potential. We are soon faced with the fact that the structure and potential are two closely related properties of condensed matter. This aspect, in relation to the Molecular Dynamics simulation of point defects diffusion on stable BCC surfaces, will be discussed in Part III.

Finally, Part IV will deal with the study of complicated migration mechanisms active in surface diffusion. The crystal may be represented by a point in configuration space travelling on the potential energy surface. This surface resembles a mountain landscape, with the notable difference of being 3N dimensional (with N the number of particles) instead of 3 dimensional. There are deep valleys (equilibrium configurations) and canyons connected by mountain passes (saddle points). The representative point spends long intervals wandering in the bottom of a valley, and eventually it will jump over a mountain pass and reach a neighboring valley. These passages are generally quite rapid. The rate of transition is determined by the properties of the saddle point region, not only by its height with respect to the equilibrium configuration (migration energy) but also by the topography of the potential energy surface in the neighborhood of the saddle point. This topography may be very complicated and may involve more than one saddle point, in parallel or in sequence, leading to the same equilibrium configuration. For the understanding of such mechanisms Molecular Dynamics can be supported with detailed investigations of the topography of the potential energy surface around the key configurations along the diffusion path. An example, the exchange mechanism for adatom migration on channelled surfaces, will be illustrated in some detail.

II. SURFACE DIFFUSION AND SURFACE STRUCTURE

The simulation of a crystal surface can be obtained by removing the usual periodic boundary conditions in one direction, while keeping them in the other two directions. The simulated model consists then of an infinite periodic slab, typically 10 to 20 crystal planes thick. The crystallite can then be oriented so that the free surfaces are in the desired orientation. The usual Molecular Dynamics (MD) is generally employed. It differs however from the constant volume MD by the fact that the system can expand in the direction along which the periodic boundary conditions have been removed. An attractive tail in the potential is then clearly necessary to hold the slab together. The Lennard-Jones potential was used.

A vacancy is created by removing one atom from the surface of the slab, while an adatom is created by placing an additional atom on top of the surface. At the beginning of the simulation the system is allowed to relax and then the dynamics of all the particles is followed, typically for 2 nsec (100000 steps). The trajectory of the migrating particle, when plotted, nicely reproduces the images produced by FIM. (see fig. 1)

Simulations of three different low index surfaces have been reported by various authors[7] with the results all in qualitative agreement. These are the FCC (111), (100) and (110) faces, which differ substantially in their structures. They represent, respectively, the most densely packed surface, a square, loosely packed surface, and a channelled surface. Figs. 2, 3, and 4 show trajectories of adatoms on these three surfaces at about the same temperature. The striking feature of these pictures is the diversity of diffusion paths on the three surfaces. Also, for the vacancy (not represented) very different behaviours are observed for the three cases.

On the densely packed (111) face the adatom moves freely over the surface. The random walk model, for the particle hopping from one site to the other, is clearly not applicable in this case. On the other hand, in this compact substrate, the vacancy is almost immobile. The space between atoms, even in the presence of a vacant site, is too tight, and in order to allow a nearby atom to jump into the vacancy another atom has to jump out of the surface. This process turns out to be energetically quite unfavorable.

On the looser (100) surface, the adatom is observed to spend quite long intervals in equilibrium sites and eventually to jump. The characteristic of this case with respect to a jump model is that jumps to second and third nearest neighbours are observed in addition to simple jumps to next nearest neighbours. The same kind of behaviour is found for the vacancy, where the second nearest neighbour jumps correspond to the simultaneous displacement of two atoms in the same direction. These events are not rare; they account for almost 50% of all jumps at .50 T_M. They are probably of the same kind as the double jumps in Al found by DaFano and Jacucci.[8]

On the channelled (110) face again a new pattern of migration paths is observed. The adatom runs along the channels, as could be expected, but it also diffuses in the direction normal to the channels. This occurs not by direct jumps over the ridges but by performing exchanges with the substrate atoms.

This variety of diffusion behaviours is consistent with the reported FIM migration energies. The difference in temperature between simulation and experiment can account for the absence of multiple jumps at the low temperatures of the experiment. What is reassuring is the presence of the exchange mechanism both in the simulation and in some of the FIM samples. It implies that indeed some features of the diffusion do depend essentially on the structure of the surface only.

III. SURFACE DIFFUSION, SURFACE STABILITY AND INTERATOMIC POTENTIAL

III.1 Instability of a bcc slab with hard potentials.

The attempt to extend use of the Lennard-Jones potential to the BCC slab meets with catastrophe. The system initially put into a geometrically perfect structure collapses under shearing stress. The slab continues straining until a symmetric compact structure bearing

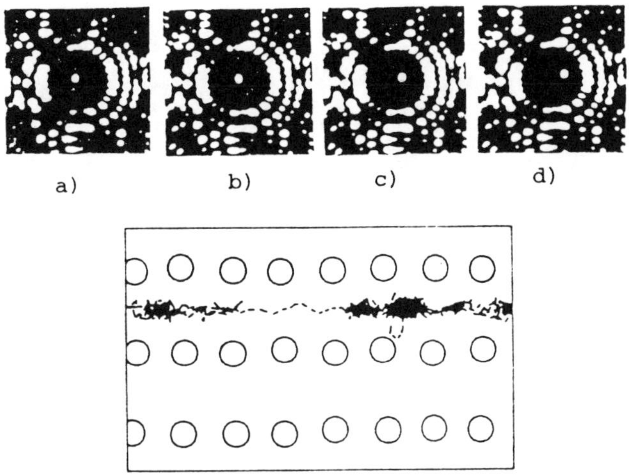

Fig. 1: Adatom diffusion along a channel on a BCC (211) surface, from FIM (a,b,c,d) and from MD simulation (e). The real case corresponds to Re atom on W surface at T = 327 K (about .10 T_M); images are taken after 60 sec diffusion intervals. (from G. Ehrlich (1982))[4] The white spot in the middle of each picture indicates the position of the adsorbed atom, which is moving from a) to d) from the left to the right along a channel; the channels direction is deduced from the positions of the spots surrounding the central circular surface. The simulated case e) corresponds to a Na atom on Na surface at T = 150K (about .40 T_M); the adatom trajectory is shown for an interval of 20 psec; the positions of the surface atoms at the beginning of the run are represented with open circles. (from G. DeLorenzi, G. Jacucci (1984[14])

no resemblance to the original one is attained. It is well known that BCC crystals with Lennard-Jones forces are unstable to shear. Computer simulations of bulk BCC Lennard-Jones crystals at normal densities can only be carried out on small crystallites subject to periodic boundary conditions along all three cartesian directions.[9] The boundary conditions provide artificial constraints stabilizing the structure against shear. If these constraints are removed, e.g. by applying flexible boundary conditions thereby allowing shear deformations, the Lennard-Jones crystal collapses.[10] To prevent this catastrophe, artificial constraints have also been employed in surface studies[11]. The model system is reduced to two layers of unconstrained atoms placed on top of rigid lattice layers. However, investigations of the properties of the BCC structure can be more satisfactorily carried out using interaction potentials that make the structure stable.

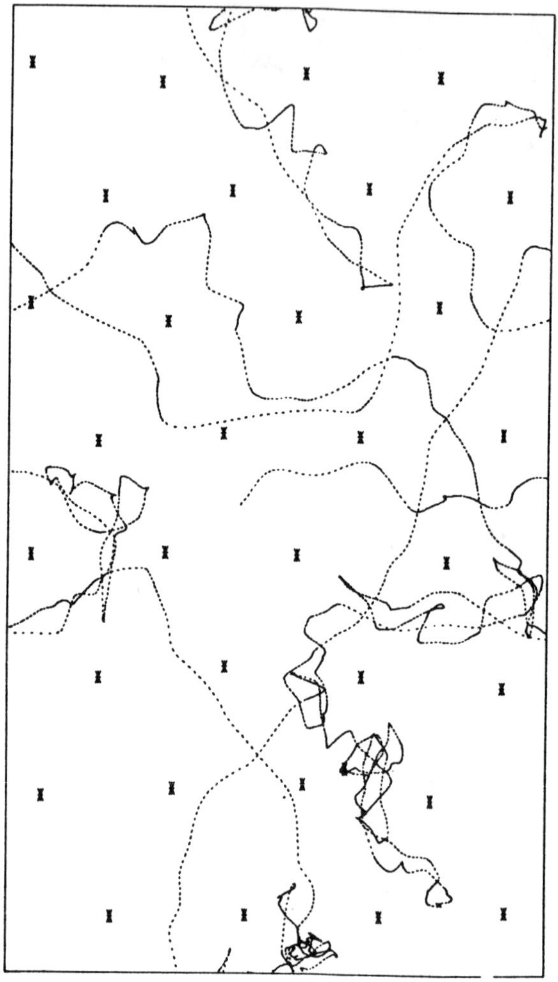

Fig. 2: Fully delocalized trajectory of the adatom on the FCC (111) surface. The trajectory was obtained at $T = .43\ T_M$ during ten thousand time steps. Asterisks represent the surface atoms at the start of the run (from DeLorenzi, Jacucci, Pontikis (1982)[7].

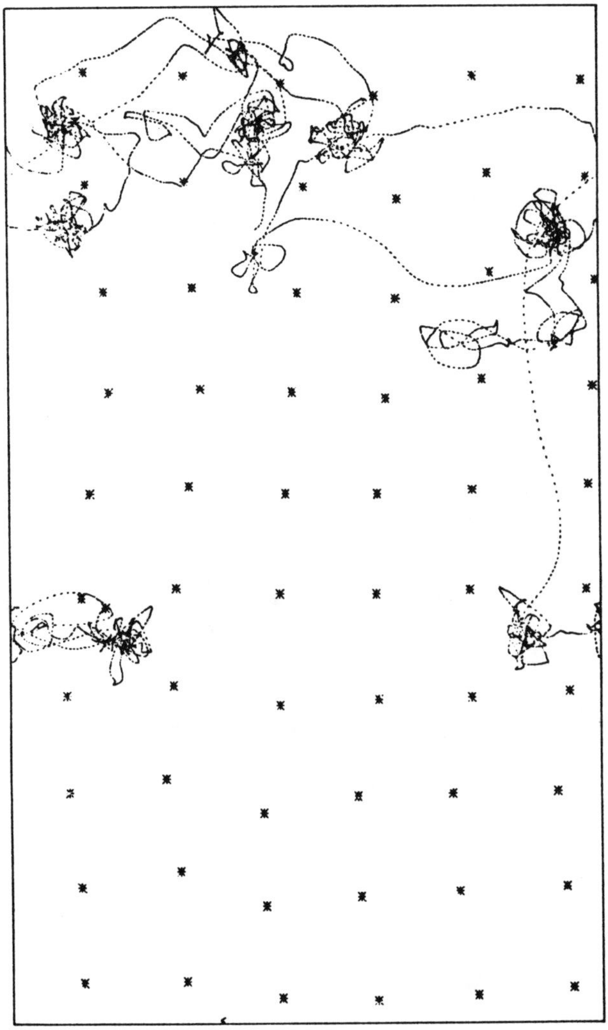

Fig. 3: Twenty thousand steps trajectory of the adatom on the FCC (100) surface showing non nearest neighbour jump diffusion at T = .49 T_M. The atoms of the surface are indicated by asterisks in their positions at the start of the run (from DeLorenzi, Jacucci, Pontikis (1982)[(7)]).

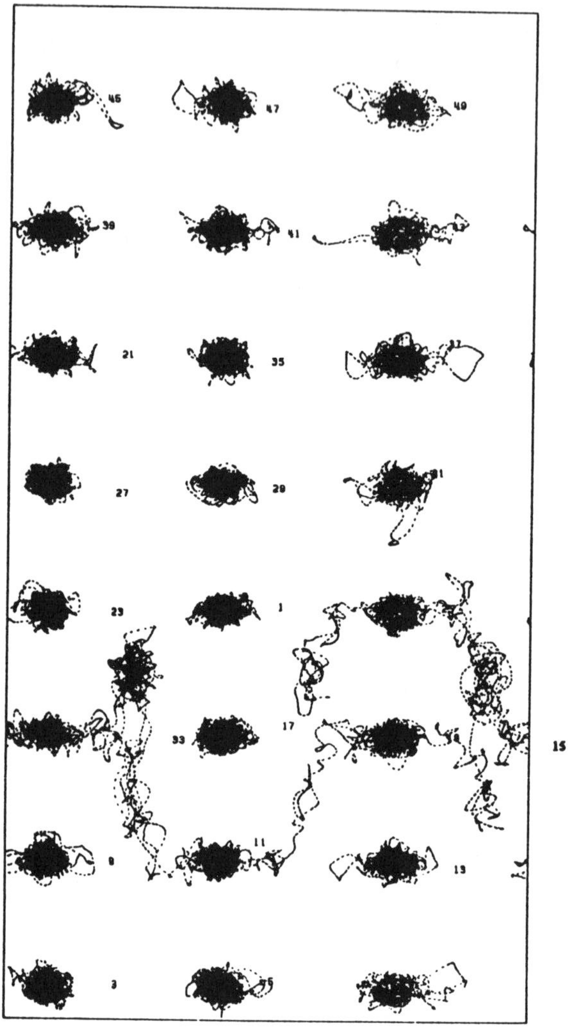

Fig. 4: Adatom diffusion on the channeled FCC (110) surface at T = .41 T_M.
a) Cascade of exchange events in a ten thousand step run.

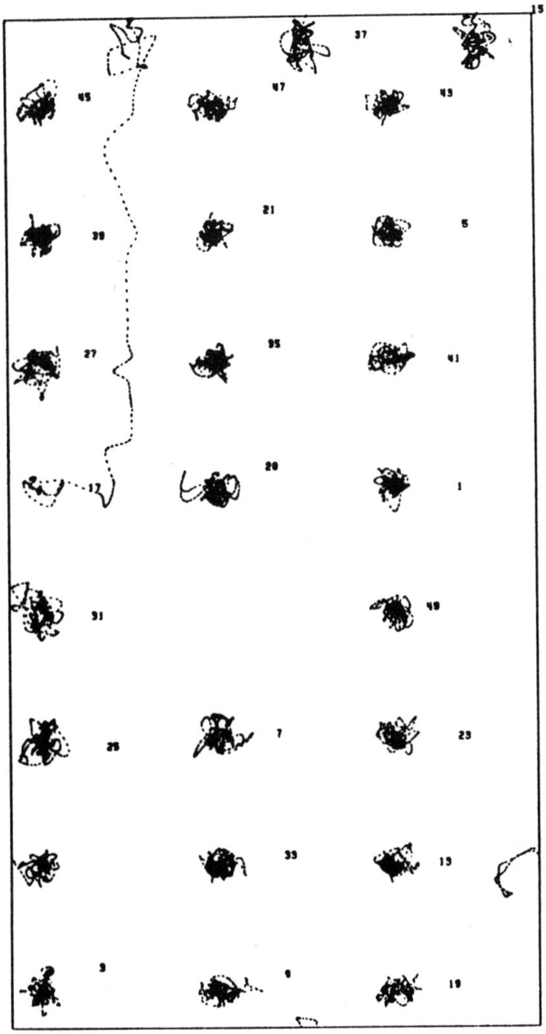

Fig. 4: Adatom diffusing on the channeled FCC (110) surface at T = .41 T_M.

b) Adatom multiple jumps along a channel during another thousand step run (from DeLorenzi, Jacucci, Pontikis (1982)[7].

IV. THE EXCHANGE MECHANISM

It was first observed by Bassett in 1979[15] that diffusion on the channelled Pt (110) surface did not only occur along the channels, but also across them, and with an activation energy that was much too low for atoms to jump over the ridges from one channel to the other. He soon suggested a possible mechanism involving an exchange between adatom and ridge atoms. Almost simultaneously MD simulations found exchange events occuring on FCC (110) surfaces of Lennard-Jones crystals.[7,16] Shortly after, Wrigley demonstrated experimentally the W atoms adsorbed on the Ir (110) surface quickly penetrate the substrate, while Ir atoms reside as new adatoms on the surface[6] (fig. 10).

The qualitative accord of these observations seems to say that the exchange mechanism is a general property of channelled surfaces. On the other hand, the Ir (110) surface reconstructs to a 2x1 structure and the same does Pt. Rh does not reconstruct and does not exhibit exchange events.[6] So the discussion is open on the relation between the exchange mechanism and the ability of a surface to reconstruct.

The model first proposed by Halicioglu and Pound[17] is presented in fig. 10c. There are four possible ways for the intermediate state to transform to a new stable configuration. With lattice statics it was possible to measure the energies involved, and it was soon clear that such a model already gives a migration energy lower than the energy necessary for direct jumps over the ridge. Large statistics of exchange events in MD simulations support the four escape possibilities with equal probability. It was not clear, however, what the intermediate configuration was: a saddle point or something else?

When studied in some more detail by combining lattice statics, lattice dynamics and refined techniques for the search of saddle points on an N-dimensional hypersurface, the exchange mechanism presents a number of interesting features. It is included in this lecture as an example of a complicated migration mechanism which deeply involves the many-body nature of the migration process and to show how MD may be usefully supported by complementary calculations.

If all the atoms of the surface were fixed and only the adatom could slide over it in a plane, the potential energy surface visited by the representative point in configuration space would appear as a sequence of identical wells separated by saddle points, each corresponding to the adatom jumping from one equilibrium site to the other (fig. 11[18]). If the substrate atoms move, however, the potential energy surface instead of being a function of only two variables (x and y coordinates of the adatom) becomes a complicated function of 3N variables:

$$V(x_{ad}, y_{ad}, z_{ad}, x_1, y_1, z_1, x_2 y_2, z_2, \ldots)$$

This instability to shear is a property of all Mie (m-n) potentials for values of the inverse power repulsion exponent m larger than 7.[12] As a consequence, it is certainly inappropriate to model BCC metals with the Lennard-Jones potential, with any Mie potential with m larger than 7, or with any potential with analogous stiffness. Potentials exhibiting unstable BCC lattices are *ipso facto* inappropriate to model metallic crystals exhibiting the BCC structure. Their structure instability will certainly be paired in the behavior of other properties of interest.

The body centered cubic structure is the structure exhibited by the alkali metals. For these substances reliable *ab-initio* potentials are available from pseudopotential theory. They correctly reproduce structural and dynamical properties of liquids and solid bulk phases, including thermodynamic properties of point defects. This last fact is of relevance in the present case because of analogies between lattice defects and surfaces with regard to ion-ion interactions. The latter are not known close to, or at, crystal surfaces, but certainly include additional many-body contributions coming from rapid spatial variation of the electronic density. Because of the success of point defect calculations with these potentials one can expect that the change in surface electronic density will not produce unrealistic effects, at least at low temperatures, where only the low lying regions of the potential are visited.

Based on the above considerations an alkali metal (Na, as calculated by Price et al.[13]) (fig. 5). was the choice for the first realistic simulation of surface diffusion on BCC surfaces[14].

It is interesting to see the behavior of various empirical potentials, compared to the *ab-initio* Na potential, in structural relaxation experiments on BCC slabs.[14] The calculation was performed for various orientations of the slab: (211), (110) and (100), and also for various thicknesses. The BCC (211) slab is in a sense most spectacular in this regard. Lack of lateral symmetry of pairs of adjacent atomic planes parallel to the exposed surface immediately generates a sliding of these planes against each other during relaxation. More commonly on low index faces (e.g. the corresponding FCC (110) face) only collective displacements perpendicular to the surface are involved in the relaxation process and the final spacing of adjacent planes is a function of the distance z from the surface.

During parallel collective sliding of the atomic planes in BCC (211) relaxation the atoms of each layer tend to sit in more symmetrical positions with respect to cells of neighbours of adjacent layers. The extent of the relaxation will depend strongly on the potential chosen among the particles. For certain potentials (e.g. Lennard-Jones) the displacement takes the form of a uniform shearing strain that stops only when symmetrical structures are recovered.

Tables I and II summarize the results of the investigation of BCC (211) slab stability. All slabs have periodic boundary conditions along the x and y directions. Relaxation is allowed for every particle independently and in all directions. This is achieved by performing a Molecular Dynamics run under the action of damping forces. In this way the slab, originally a portion of perfect bulk crystal, evolves towards a minimum energy configuration. Small random displacements are applied to the particles in order to break eventual crystal symmetries which could prevent the system from escaping from a locally unstable equilibrium configuration.

First we observe that the Na potential produces stable BCC slabs. Both parallel and perpendicular relaxation displacements depend on z; in the slab of double thickness the structural deformation at the slabs center is minor. Second: the Lennard-Jones potential completely reconstructs the lattice passing from a repetition of planes of type abc abc characteristic of the BCC (211) slab, to a repetition of type ab ab. Doubling the thickness of the slab produces the same reconstruction all across the slab. Each half slab shears just as much as a corresponding isolated slab, neglecting

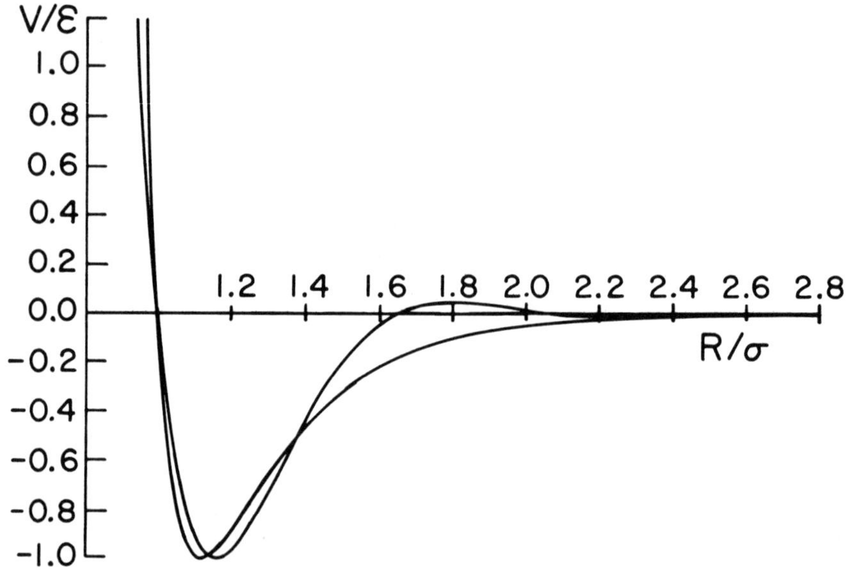

Fig. 5: Model potential for Na compared with the Lennard-Jones potential.

Table I

	LJ	Mie 6-3	W (Morse)		Na
density $[\rho/\sigma]$	1.00	1.00	1.10	1.05	.90
δz_1	-34.25	-22.98	14.45	3.89	-11.60
δz_2	22.41	20.94	7.12	3.60	8.19
δz_3	-36.80	-36.62	3.07	.90	-10.53
δz_4	22.24	18.01	1.70	1.62	7.04
δz_5	-37.30	-40.12	1.72	.59	- 9.00
δz_6	22.42	18.44	1.36	1.43	6.31
δy_1	16.67	16.80	13.33	12.76	9.96
δy_2	-33.33	-33.89	6.81	5.71	-9.96
δy_3	16.67	17.13	1.59	3.09	4.74
δy_4	-33.33	-33.09	1.07	.99	-6.29
δy_5	16.67	17.09	.52	1.88	3.70
δy_6	-33.33	-33.92	.80	.84	-5.62

Table I: Static relaxation of interplanar distances (above) and interplanar registry (below) from the perfect crystal, of a slab of 12 bcc (211) planes. Calculations for different empirical potentials and for the ab-initio metallic potential for Na are shown. The change in the distance between plane i and i+1 beginning from the outermost plane, denominated δz_i is expressed in percentage of the perfect crystal value. The relative parallel shift of the planes in the y direction (channel direction) δy_i is expressed in percentage of the interparticle distance along a channel. Relative shifts of 16.67 or -33.33 recover a repetition of planes of the type ab ab, as opposed to the abc abc repetition of the bcc (211) slab.

Table II

	LJ		Na	
	12 plane	24 planes	12 planes	24 planes
δz_1	-34.25	-34.27	-11.60	- 8.86
δz_2	22.41	22.43	8.19	6.00
δz_3	-36.80	-36.83	-10.53	- 7.40
δz_4	22.24	22.30	7.04	3.97
δz_5	-37.30	-37.40	- 9.00	- 5.01
δz_6	22.42	22.57	6.31	2.16
δz_7		-37.60		- 3.54
δz_8		22.68		1.10
δz_9		-37.66		- 2.73
δz_{10}		22.71		.55
δz_{11}		-37.68		- 2.37
δz_{12}		22.77		.38
δy_1	16.67	16.67	9.36	9.28
δy_2	-33.33	-33.33	- 9.90	- 7.22
δy_3	16.67	16.67	4.74	3.16
δy_4	-33.33	-33.33	- 6.29	- 3.71
δy_5	16.67	16.67	3.70	1.76
δy_6	-33.33	-33.33	- 5.62	- 2.43
δy_7		16.67		.99
δy_8		-33.33		- 1.73
δy_9		16.67		.55
δy_{10}		-33.33		- 1.37
δy_{11}		16.67		.34
δy_{12}		-33.33		- 1.26

Table II. Dependence of static relaxations (see Table I) from the number of planes in the slab, for Lennard-Jones and Na potentials.

small effects at the outermost planes. Third: A Morse potential fitted to W behaves in a way very similar to Na, albeit with somewhat different relaxation displacements. Similarly decreasing the value of m and n from 12-6 to 6-3 also produces behaviour similar to Na. The important last finding was that the density is a relevant parameter: the Mie 6-3 potential gives an unstable BCC slab for 1.0, and a stable one for 1.0.

The BCC (110) and (100) slabs behave similarly, with the noticeable difference that the starting configuration is laterally symmetric; the Na potential instead presents only a relaxation in the interplanar distance, preserving the lateral symmetry.

III.2 FCC and BCC surface diffusion.

Some information about the dependence of surface diffusion on the potential, as opposed to its dependence on structure surface, can be obtained by comparing MD simulations on Na-BCC surfaces and similar LJ-FCC surfaces. The following surfaces will be considered "similar":

i) FCC (110) and BCC (211); they are both channeled surfaces with the highest density of channels possible in their respective lattices;

ii) FCC (111) and BCC (110); they are both the most densely packed surfaces of their respective lattices;

iii) FCC (100) and BCC (100); they are both the lowest index surfaces of their respective lattices, with simple square structures.

The channelled structure allows two types of migration mechanisms for the adatom: the exchange mechanism and jumps of the adatom along the channels (parallel jumps). On both LJ-FCC (fig. 4) and Na-BCC (fig. 6) surfaces the mechanisms are the same: the exchange mechanism presents an intermediate dumbbell configuration while parallel jumps are mostly mutliple jumps. On the other hand, on the BCC (211) face, where the channels are wider (in terms of atomic radii 1.78 compared to 1.59 for the FCC (110)), parallel jumps dominate over exchanges, while they are only half as frequent as exchanges on the FCC (110). The specific structure of the surface (channels) is seen here to play the determining role.

On the densely packed surfaces, where atoms sit close to each other, one finds the effects of the repulsion which is harder in the Lennard-Jones potential and softer in the Na potential. In the LJ case, the hard compact FCC (111) surface (fig. 2) does not allow enough space for the adatom to penetrate, so that it remains well above the substrate, flying nearly free over it. The Na potential, on the other hand, offers a softer surface and the adatom is seen to be accommodated for long intervals in regular surface sites (fig. 7). Note that the surface density of these two faces is almost identical in terms of atomic radii, as the surface density for BCC (110) face is only 3% smaller than for the FCC (111) face. The same effect of the soft repulsion can be deduced from the behaviour of the vacancy.

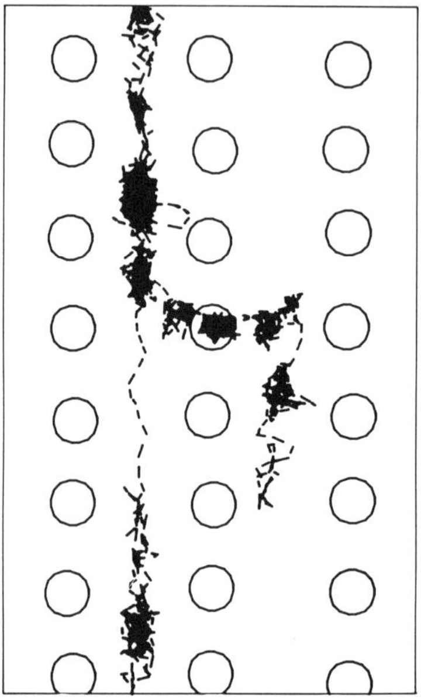

Fig. 6: Adatom migrating on the BCC (211) surface during a 50 psec interval at T=.40 T_M, showing quick diffusion along a channel and also one exchange event. During the exchange, both the atoms involved spend a long time simultaneously sitting in the two symmetrical positions near the ridge site ("dumb-bell configuration") showing more clearly than in the FCC (110) (fig. 1a) the local stability of this intermediate configuration. The positions of the surface atoms at the beginning of the represented interval are indicated with open circles.

In the hard LJ-FCC (111) face, the atoms around the vacancy relax towards the empty site, but the still compact packing does not allow any atomic displacement. The softer repulsion between neighbouring atoms in the Na-BCC (110) surface instead allows freer motion of the atoms within the surface. Several jumps of the vacancy are observed, even multiple jumps, (fig. 8), so that the mobility of the vacancy is comparable to that of the adatom on the same face.

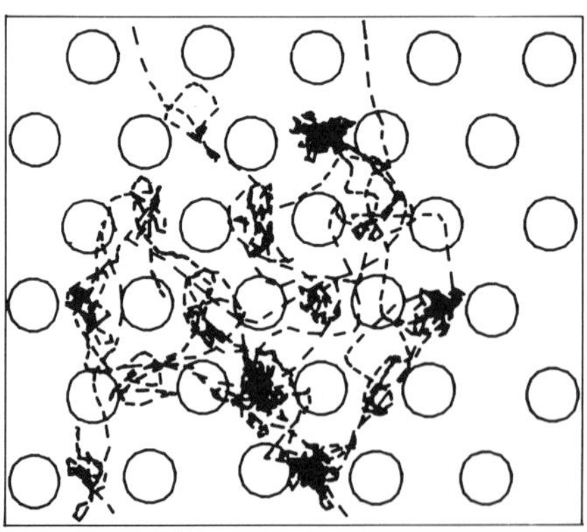

Fig. 7: Adatom migration on the BCC (110) surface, during a 2000 psec run, at T=.40 T_M[14]. Although quite complicated, the trajectory clearly shows randomization sites and multiple jumps, reminiscent of the diffusion on the FCC (100) surface (fig. 3).

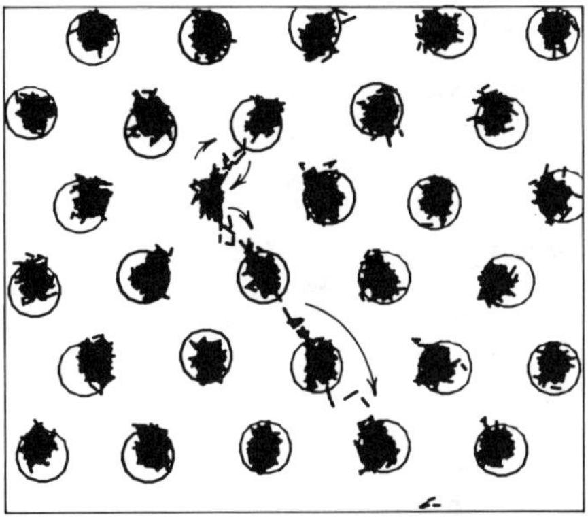

Fig. 8: Vacancy migration on the BCC (110) surface, during a 50 psec interval, at T=.40 T_M. The trajectories of all the atoms are shown. Their positions at the beginning of the interval are represented with open circles; the absence of a circle in one of the surface sites indicates the initial position of the vacancy. Arrows indicate jumps of the vacancy: a long arrow finally indicates a double jump of the vacancy, i.e, the simultaneous jump of two atoms. Jumps of the vacancy are performed very quickly.

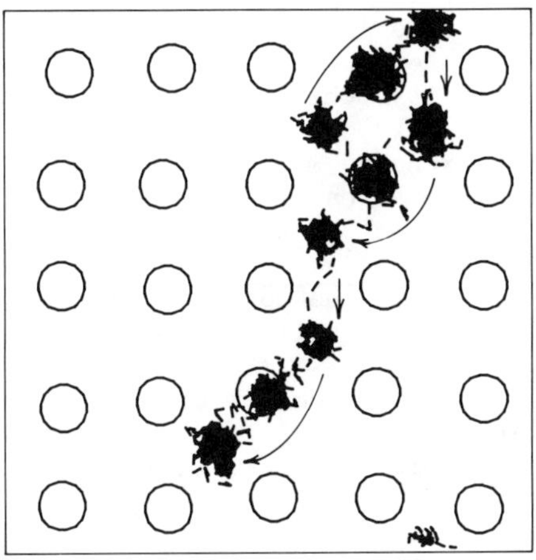

Fig. 9: Adatom migration on the BCC (100) surface, during a run of 100 psec at T=.40 T_M. The positions of the surface atoms at the beginning of the run are represented with open circles; then only the trajectory of the adatom originally present on the surface and of the surface atoms directly involved in the migration process are represented. Short straight arrows indicate direct jumps of the adatom from one equilibrium site of the surface to another. Long curved arrows indicate exchange events, in which the adatom replaces one of the substrate atoms (figs. 6,7,8, 9 from G. DeLorenzi, G. Jacucci[14]).

Another effect of the softness of the metallic potential is found on the BCC (100) surface (fig. 9). In this case however the softness of the potential is coupled to a lower surface density (26% smaller for the BCC (100) respect to the FCC (100)). On the LJ-FCC (100) (fig. 3) both the adatom and the vacancy are quite mobile, and behave in a way very similar to the BCC (110) face, that is, they perform single and double jumps between equilibrium sites, where they spend long intervals randomizing. On the looser Na-BCC (100) face, the large space of the equilibrium sites traps the adatom and renders direct jumps from site to site very difficult. On the other hand, the softness of the potential allows broad oscillations of the substrate atoms, which are often seen to perform large excursions out of their lattice sites, thus generating temporary spontaneous defects. It is then easy for the adatom to find sometimes an empty substrate site into which it may sink. These exchange processes are different from those on channeled surfaces in that no intermediate dumbbell configuration is ever formed.

III.3 Discussion

At the end of such a list of "experimental" observations, we are tempted to classify the reported properties in the following way:

a) properties determined by the structure of the surface:
- i) close packing hinders vacancy mobility, while it enhances adatom mobility;
- ii) channelled structures favour the exchange mechanism with an intermediate stable dumbbell configuration;
- iii) large channels favour the motion of adatoms along them as compared to narrow channels;

b) properties determined by the softness of the potential repulsion:
- i) soft repulsion favours vacancy mobility, while it lowers adatom mobility;
- ii) soft repulsion favours direct exchange processes (no intermediate stable configuration).

In spite of a classification between properties of FCC and BCC structures, we are led to a classification between properties determined either by the surface structure, or by the softness of the potential. Although the information is still very limited and only qualitative, it suggests that indeed the two determining factors in surface diffusion processes are the surface structure, more than the underlying bulk, and the form of the potential, in particular, the softness of the repulsion between nearest neighbours.

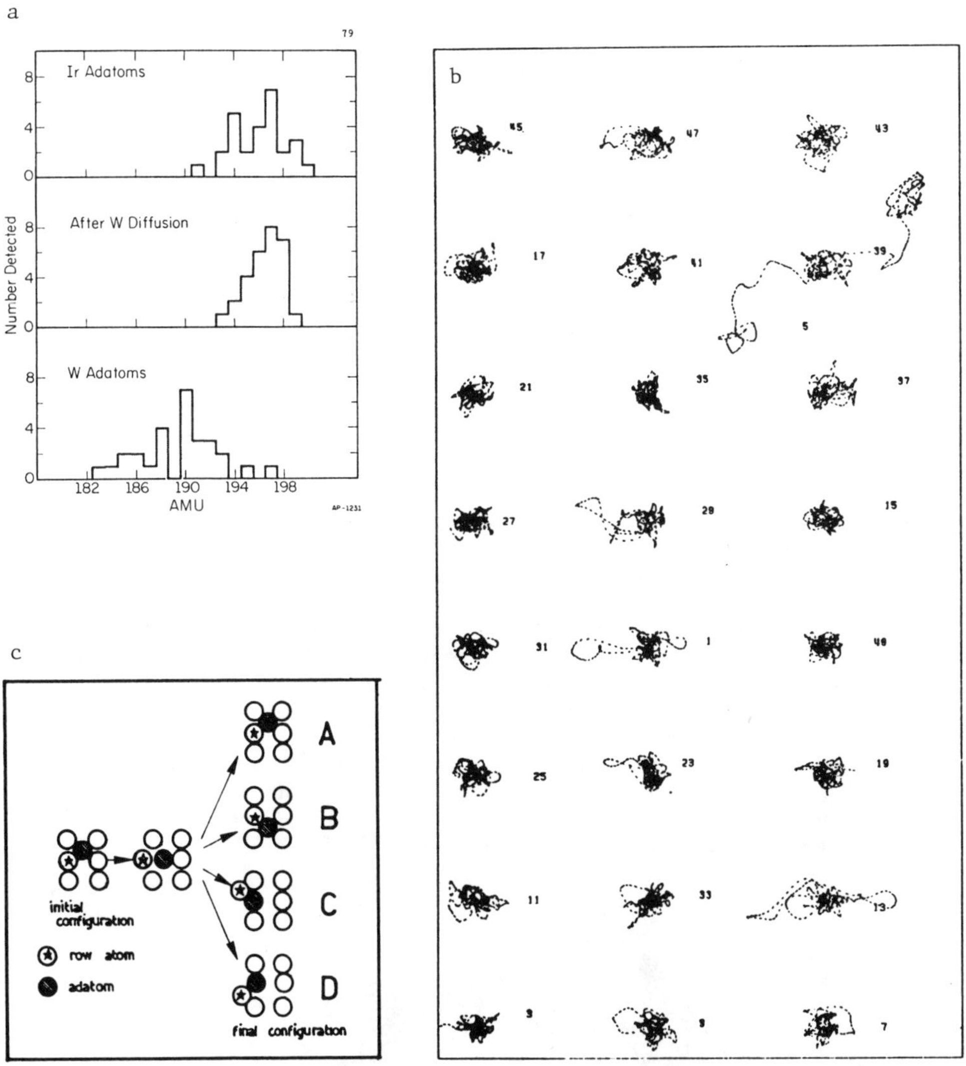

Fig. 10: Exchange mechanism for adatom migration on the channelled FCC (100) surface: a) Experimental evidence, for W diffusion on Ir, as shown by the "atom-probe" mass-spectrometer before and after the migration event, from Wrigley and Ehrlich (1980)(6); b) MD trajectories, showing the details in the dynamics of the exchange event (from DeLorenzi, Jacucci, Poutikis (1982)(7)); c) model for the mechanism as first proposed by Halicioglu and Pound.

Fig. 10 could then be a section of the hypersurface in the plane of the x and y coordinates of the adatom, for a fixed value of all the others:

$$V(x_{ad},y_{ad},z^0{}_{ad},x^0{}_1,y^0{}_1,z^0{}_1,x^0{}_2,y^0{}_2,z^0{}_2,\ldots)$$

For a different choice of the other coordinates, or, for a different choice of the section plane, this picture can be very different and show new minima, perhaps lower saddle points and easier migration paths. This is the essence of the many-body nature of our problem.

It is often difficult to find a section of the potential energy surface exhibiting the path of lowest energy for a specific transition (in other words, to find the appropriate reaction coordinate).

Returning to the exchange mechanism, the intermediate configuration (dumbbell) turns out to be a minimum of the potential instead of a saddle point. A search of extrema in the potential energy surface near the dumbbell configuration then reveals the existence of six distinct saddle points, four of which are simple saddles (only one unstable mode). These correspond to the lowest escape paths from the minimum and thus to the four transitions of the Haligioglu model. The other two are double saddles (two unstable modes) situated symmetrically with respect to the ridge. We call these crossroads. When plotted along these two direction, i.e. along the unstable mode eigenvectors in a crossroad, the potential energy surface looks as shown in fig. 12, in which the properties we just mentioned are nicely visible. From the picture it becomes clear how the intermediate configuration can evolve into four distinct but equivalent final configurations. Also the role of parallel jumps, as opposed to exchange jumps, becomes clear.

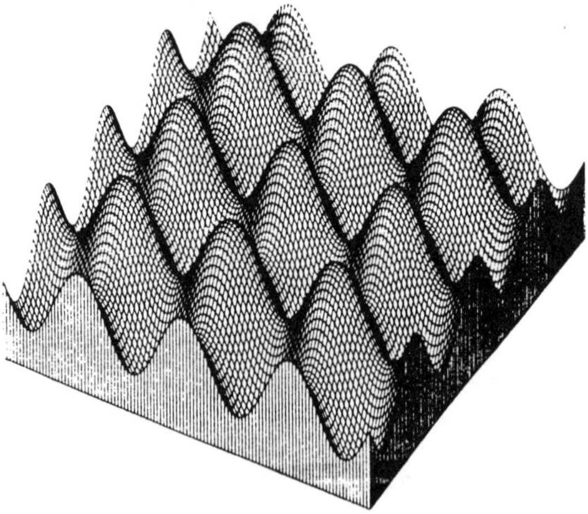

Fig. 11

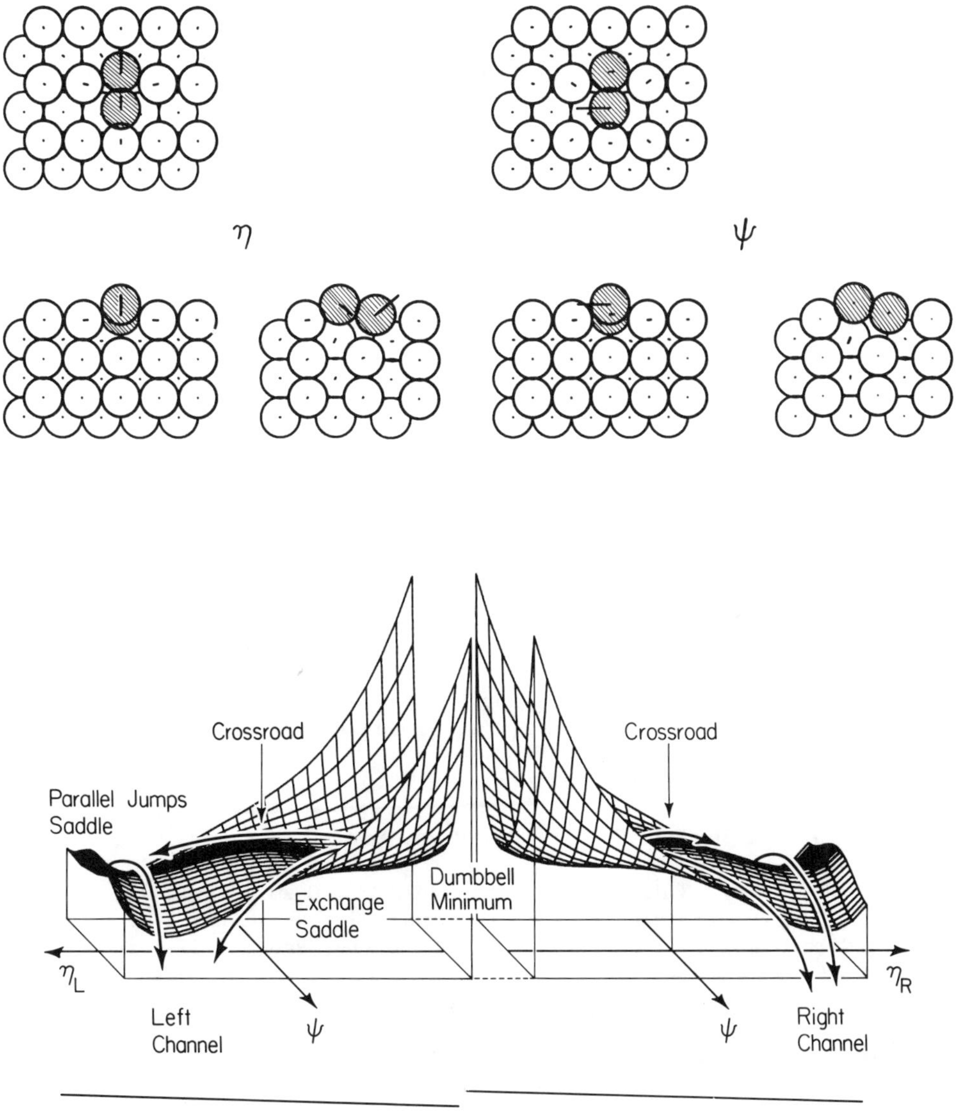

Fig. 12: Topography of the potential energy surface in the region near the saddle points for the exchange mechanism,[19] showing the paths of escape from the dumbbell minimum. V is plotted along the two unstable mode directions departing from the crossroad (see text); the atomic displacements involved are shown above, projected onto the surface plane and onto planes normal to the surface.

Finally we note that Morse potentials, with a softer repulsion than the Lennard-Jones potential, give essentially the same results: the topography of the potential surface is the same, but the relative heights change, thereby inducing different roles for the parallel jumps and for the exchange mechanism respectively.

ACKNOWLEDGEMENTS

The work presented in this lesson has its origin in a thesis at the University of Trento, with advisor Gianni Jacucci. Its later development was possible thanks to a continuing joint effort and collaboration. IBM fellowship is gratefully acknowledged. This work was partially supported by the National Science Foundation, MRL grant DMR-83-16981.

REFERENCES

[1] E.W. Muller and T.T. Tsong, Field Ion Microscopy, Principles and Applications, American Elsevier, New York (1969).

[2] G. Ehrlich, K. Stolt, Ann. Rev. Phys. Chem. 31, 603 (1980).

[3] D.W. Bassett, in: Surface Mobilities on Solid Materials. ed. Vu Thien Binh, Plenum (1981) NATO-ASI Series, Vol. B86

[4] G. Ehrlich, Diffusion in surface layers, CRC Crit. Rev. Solid State Mat. Sci., May 1982, p. 391 and ref. therein.

[5] G. Ehrlich, Physics Today, June 1981, p. 2.

[6] J. D. Wrigley and G. Ehrlich, Phys. Rev. Lett., 44, 661 (1980).

[7] J.C. Tully, G.H. Gilmer, M. Shugard, J. Chem. Phys., 71, 1630 (1979). M.R. Mruzik and G.M. Pound, J. Phys. F11, 1403 (1979). G.DeLorenzi, G. Jacucci and V. Pontikis, Proc. ICSS-4 and ECOSS-3 Cannes (1980), vol. I, D.A. Degra and M. Costa eds., p. 54 and Surf. Sci., 116 391 (1982).

[8] A. DaFano, G. Jacucci, Phys. Rev. Lett. 39, 950 (1977).

[9] A. Rahman, G. Jacucci, submitted to J. Chem. Phys.

[10] M. Parrinello and A. Rahman, Phys. Rev. Lett. 45, 1196 (1980).

[11] J. D. Doll and H. K. Dowell, Surf. Sci., 123, 99 (1982).

[12] W. G. Hoover, D. A. Young, R. Grover, J. Chem. Phys. 56, 2207 (1972).

[13] D. L. Price, K. S. Singwi and M. P. Tosi, Phys. Rev. B2, 2983 (1970).

[14] G. DeLorenzi and G. Jacucci, submitted to Surf. Sci.

[15] D. W. Bassett, P. R. Webber, Surf. Sci., 70, 520 (1978).

[16] S. M. Garofalini and T. Haliciogu, Surf. Sci., 104, 199 (1981).

[17] T. Halicioglu and G. M. Pound, Thin Solid Films, 57, 241 (1979).

[18] H. P. Bonzel, in: Surface Mobilities on Solid Materials. op. cit.

[19] G. DeLorenzi and G. Jacucci, in preparation.

DIFFUSION IN GRAIN BOUNDARIES

G. CICCOTTI
Istituto di Fisica G. Marconi
Universita' degli Studi di Roma I
I - 00185 Roma, Italy

The problem of resolving the self diffusion mechanism for grain boundaries in connection with their specific structure has received considerable attention during the last years, due to the importance of these defects in materials properties (1). However, the scale on which experiments can actually be performed does not permit the direct investigation of grain boundaries at the atomic level and therefore much work is currently devoted to computer simulation studies of intergranular diffusion and structure.

The most important contributions of recent computer simulations are: (a) the direct verification of the vacancy mechanism predominance for grain boundary self diffusion (2); (b) the proof of the existence of a structural transition in grain boundaries in the high temperature range (3).

In the present work (4) we focus our attention more specifically on the temperature dependence of intergranular diffusion mechanism studied by the standard molecular dynamics technique (5). In the high temperature range, our model is the previously studied fcc (310) [001]$\Sigma = 5$ coincidence grain boundary, for which a structural transition occurs accompanied by high diffusivity values (3). The trajectories of the atoms inside the grain boundary have been reanalyzed by plotting their mean squared displacement component along the tilt axis, say $<[z(t)-z(0)]>$, as a function of the similar quantity, $<[y(t)-y(0)]>$, where y designates the direction parallel to the boundary plane and normal to the tilt axis.

The qualitative behaviour of intergranular diffusion, illustrated by such a plot, is in excellent agreement with our earlier conclusion that the high temperature structure of the grain boundary remains crystalline despite the high disorder in it.

G. Jacucci (ed.), Computer Simulation in Physical Metallurgy, 67–68.

In the low temperature range ($T<0.5T_m$), diffusion has been studied by following the migration of a vecancy created in the otherwise perfect fcc (210)[001]$\Sigma=5$ coincidence grain boundary.
The analysis of the observed vacancy jumps permitted the description of the detailed mechanism for vacancy diffusion as well as the evaluation of the associated jump frequencies. The latter were used as the input data for a random walk simulation of the vacancy diffusion which provided the tracer correlation factor and an evaluation of the intergranular diffusion anisotropy.
Moreover, simulation gives evidence of a spontaneous generation of Frenkel pairs inside the boundary. The dissociation of these defects and the absence of any migration for the interstitials during the simulation, indicates the predominancy of the vacancy mechanism in intergranular diffusion. This result is in agreement with experimental and theoretical work for fcc metals, indicating that the vacancy mechanism is responsible for grain boundary diffusion and with conclusion of Kwok et al. (2) obtained by MD simulations in bcc bicrystals.

REFERENCES

1. Grain Boundary Structure and Kinetics (American Society for Metals, Metals Park, OH, (1980)

2. T. Kwok, P.S. Ho, S. Yip, R.W. Balluffi, P.D. Bristowe and A. Brokman, Phys. Rev. Letters 47 (1981) 1148

3. G. Ciccotti, M. Guillope and V. Pontikis, Phys. Rev. B27 (1983) 5576

4. M. Guillope, G. Ciccotti and V. Pontikis, Surface Science 144 (1984) 67

5. G.L. Pollak, Rev. mod. Phys. 36 (1964) 748

DYNAMIC COMPACTION OF POWDERS

WILLIAM G. HOOVER

1. INTRODUCTION

Powder metallurgy and explosive forming are both well-established technologies. With powder metallurgy, relatively complex parts, requiring little machining, can be made by casting or by static compression of powdered metals or ceramics. The voids and irregularities in the powders can be reduced by exposure to high-temperature, "sintering", to strengthen the finished product. With explosive forming, large, relatively simple shapes can be made. Costly large-scale static presses can be replaced by transient pressure waves created by detonating high explosives. The powder-metallurgy and explosive-forming methods, were first combined, in dynamic compaction, about 25 years ago.[1,2] The purpose was to make parts from nonequilibrium powders which were either hard to compact or which might lose their unique properties in the high-temperature sintering operation. Just as in ordinary explosive forming, dynamic compaction has the advantage of lighter-weight equipment. The main disadvantages are, first, the noise and vibration associated with high explosives and, second, the material inhomogeneities produced by dynamic loading.

There are many methods for making nonequilibrium powdered materials. The distribution of powder particle shapes, sizes, grain structure, and hardness can be varied widely. Most methods use high-speed cooling. The particles

G. Jacucci (ed.), Computer Simulation in Physical Metallurgy, 69–85.

themselves can be made from bulk materials using mechanical, inertial, electrical, or thermal separation methods. Because the time required to cool a conducting object varies as the square of its size, separated powder particles can be cooled much more rapidly than bulk matter. Thus the most interesting nonequilibrium materials are available in the form of powders. Such powders can be consolidated into technologically useful bulk materials either by static presses or by dynamic compaction. Dynamic processing enhances locked-in elastic strains and lattice defects, often resulting in a stronger material. Because the dynamic high-pressure state is transient, lasting only nanoseconds or microseconds, the whole compacting process can be completed before thermodynamically immiscible components of the powder particles have time to separate. Thus dynamic compaction can lead to nonequilibrium materials with unique mechanical, electromagnetic, or chemical properties.

The dynamic methods involve sufficiently complicated and inhomogeneous material behavior and geometrical wave-propagation effects, that numerical simulation is a relatively inexpensive and attractive way to predict the effects of changing experimental parameters. The dynamical process of powder compaction, using explosively-driven shock waves, is well-suited to numerical simulation.

On the most fundamental, atomic, level, nonequilibrium high-strain-rate simulations[3] should prove useful in understanding the compaction process. Presently, computer simulations are restricted to a few thousand particles. Simulations involving more particles but with much less realistic interactions can be carried out with bubbles. Figure 1 shows two stages in a "bubble-raft" simulation of the sintering process. In the movie version of this simulation, nucleation and motion of dislocations are readily followed.[4]

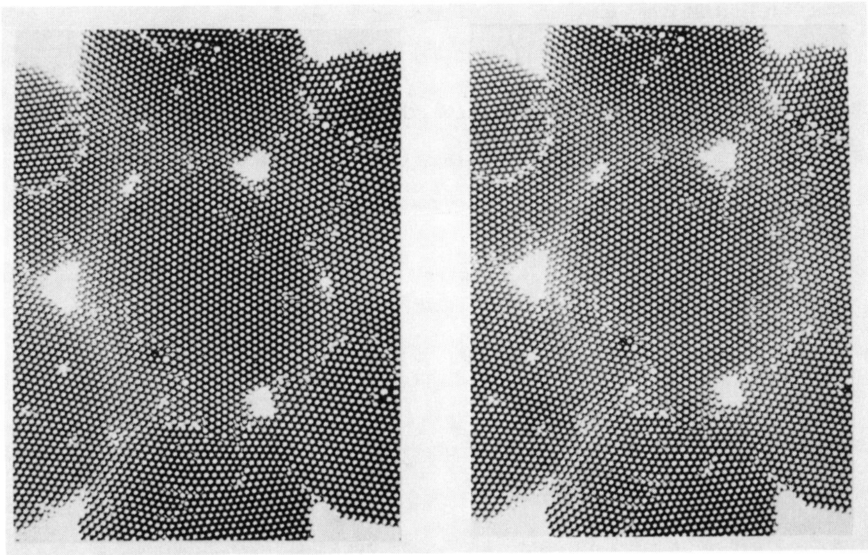

Figure 1. Bubble-raft model of copper-powder sintering from reference 4.

On the macroscopic scale, computer solution of the partial differential equations of motion can be used to provide a realistic model of compaction. To do this, equations of state for metals, powders, and high explosives all must be included. Here I describe some of the simulation work carried out by Mark Wilkins, at the Lawrence Livermore National Laboratory, modeling dynamic compaction. His finite-difference simulations of dynamic compaction were carried out in parallel with corresponding laboratory experiments. He will present a comprehensive report at the San Antonio (Texas) High-Energy-Rate Fabrication meeting next month.[5]

2. NONEQUILIBRIUM POWDERS

Whether a state of matter is "equilibrium" or "nonequilibrium" depends upon the timescale of observation. From the standpoint of a nuclear physicist, the stable form of matter is iron. From the standpoint of thermodynamic stability, graphite is stable under normal conditions while diamond is not. From the standpoint of a physicist at dinner, oil-and-vinegar salad dressing is sufficiently stable for prompt use. Under ordinary conditions, diamond has a reasonable lifetime, much longer than that of an oil-and-vinegar emulsion. Diamond powder can be produced by compressing graphite. Because the economic value of diamonds increases in a nonlinear way with size, it is commercially feasible to compact diamond powder into macroscopic pieces useful for machinery, and possibly, for jewelry.

Diamond is only one example of the many metastable forms of matter with desirable properties. Mechanical and electrical properties can often be improved by using nonequilibrium compositions. For instance, amorphous metals, made by rapid quenching of a melt on the surface of a cold substrate, can have much higher mechanical strengths than do their stable parent ingredients. This is because the principal mechanism for plastic flow in metals, dislocation motion, is not available to a material lacking crystalline order. Flow can also be inhibited by an admixture of atoms of different size.

For a list of more than 80 materials, metals, alloys, ceramics, and composites, all of which have been compacted dynamically, see page 55 of reference 2.

3. SHOCK AND DETONATION WAVES

Typical solid yield strengths can be estimated by noting that Mount Everest must exert a pressure of a few kilobars on the supporting earth beneath. Failure of the earth to flow, in the presence of this stress, suggests that solid yield strengths are also a few kilobars. Static apparatus designed to compact powders must, accordingly, be capable of exerting such pressures over areas of several square centimeters. The need for a supporting rigid structure can be eliminated if the high pressure is produced within a transient pressure wave that need not be supported by material boundaries. Such waves can be generated by using high explosives.

The energy-density of typical explosives, relative to their reaction products, can accelerate metals to velocities of a few kilometers per second. This speed corresponds to an energy density of hundreds of kilobars, fully adequate to provide the pressures necessary for the irreversible deformation and compaction of solids, even diamond.

Steady shock or detonation waves convert one equilibrium state of matter into another through an adiabatic process conserving mass, momentum, and energy. In the case of detonation waves, which drive shockwaves, the explosive's potential energy is divided between external work done and heat of its reaction products.

In a monatomic fluid a typical shockwave has a thickness on the order of an atomic diameter.[6] Figure 2 shows the variation in density, pressure, energy, and temperature in such a fluid-phase shockwave. The distance scale is measured in atomic diameters. The curves were calculated using the Lennard-Jones equation of state and state-dependent transport coefficients in the Navier-Stokes equations. The circles come from molecular dynamics simulations.[6] The strength of the shockwave in these calculations carried

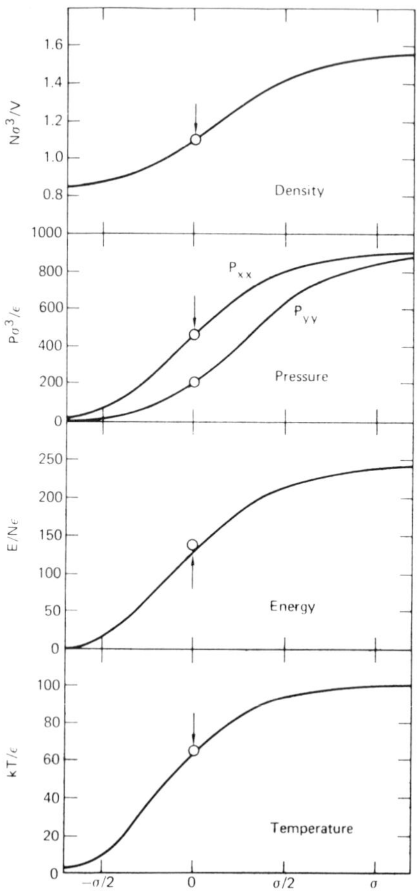

Figure 2. Dense-fluid shockwave.

triple-point argon to about 400 kilobars, with a final temperature of about one electron volt.

Because shockwaves move somewhat more rapidly than sound, the typical time for conversion from initial to final state, in a monatomic fluid, is only about a picosecond.

If relaxation processes must occur from excited states (vibrational or electronic, for example), then the shock process takes more time. In solids, deformation involves cooperative nonlocalized motion. This lack of localization serves to spread out the macroscopic shock zone. Nevertheless the shock process can be completed before structural or chemical changes occur. Typical powders are made up of micron-sized particles, a few hundred atoms in diameter. A typical shockwave travels at speeds of order one kilometer per second. Such a micron-sized particle will therefore be traversed in about one nanosecond. Thus the deformation rates near voids, at which the gaps between powder grains are filled in by deforming material, must be 10 gigahertz or more. Friction and stress-wave reflections at these gaps can cause intense local heating, melting, and jet formation. On the scale of individual grains, the compaction process is violent and inhomogeneous.

The time required for thermal equilibration of a micron-sized metal particle is about a nanosecond (assuming a thermal diffusivity of about one square centimeter per second). Thus localized melting, vaporization, and light emission can be expected to occur during shock compaction.

Computer simulations of homogeneous high-strain-rate deformation in solids are consistent with the limited information available from plastic wave-shape experiments. The "nonequilibrium molecular dynamics" simulations are carried out with shearing boundary conditions, changing with time. Typically, a few hundred atoms are used and the deformation proceeds to a total strain of order 100. As the system is sheared, plastic work is done. To avoid the irreversible heating associated with this work, the equations of motion are additionally constrained to preserve the temperature or the energy unchanged. Thus nonequilibrium simulations of high-strain-rate flow can be carried out in the desired steady state.[7]

Figure 3 is a corresponding-states description of the dependence of yield strength on strain rate. The figure includes both atomistic simulation data

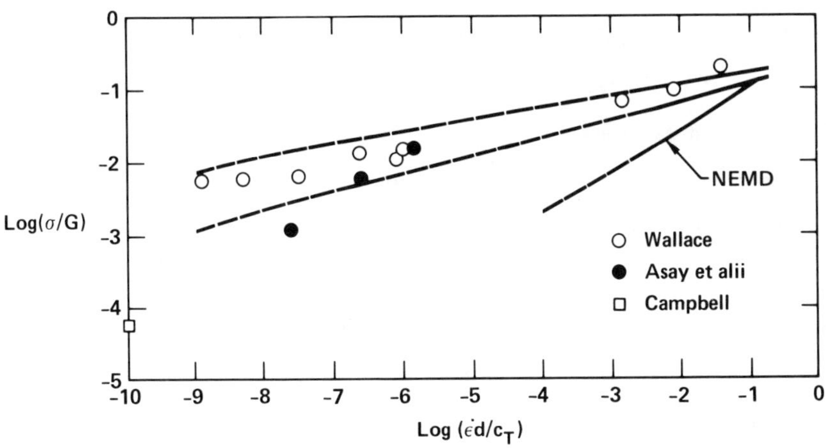

Figure 3. Solid-phase shear stress σ as a function of strain rate $\dot{\varepsilon}$.

(solid lines on the righthand side of the figure) and experimentally-based estimates (points on the lefthand side) for the yield strength as a function of strain rate. G is the shear modulus, d is the interparticle spacing, and c is the transverse sound speed. The maximum strain rate shown in the figure exceeds a terahertz and corresponds to neighboring atoms shearing past one another at approximately the velocity of sound.

Together, the simulation results, and the experimental evidence[1], provide ample reason to believe that the yield strength during dynamic shear deformation substantially exceeds its static value. For a yield strength of 10 kilobars and a total plastic strain of unity, the work of deformation would provide a temperature increase of several hundred kelvins. Much higher temperatures would be expected near the surfaces of non-spherical powders, or for powders with higher yield strengths.

4. LAGRANGIAN SIMULATIONS[8]

In a typical "finite-difference" (as opposed to differential) computer program, a structure is divided up into comoving "Lagrangian zones." These zones are just line segments in one-dimensional simulations. Zones are usually triangular or four-sided in two dimensions, and six-sided polyhedral "bricks" in three dimensions.

These continuum simulations are more complicated than atomistic ones because each zone has <u>many</u> properties (the stress and strain-rate tensors, for instance) which must be followed as functions of time, not just location and velocity. Just as in the atomistic case[9], simulations involving thousands of zones are possible. For a typical calculational grid, see figure 4. In the figure each rectangular zone represents a ring of material (powdered metal, confining plastic tube, or high explosive, from inner radius to outer

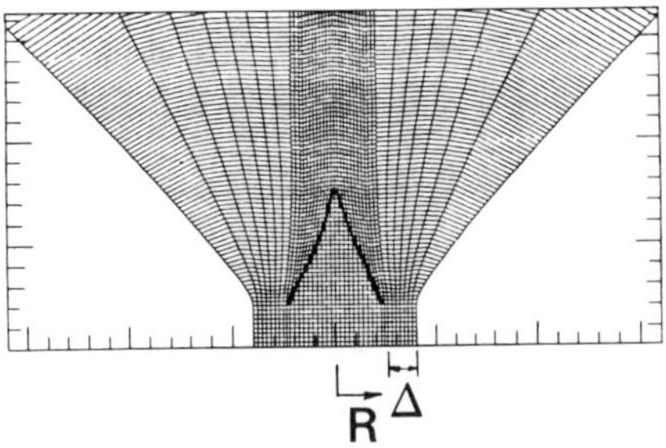

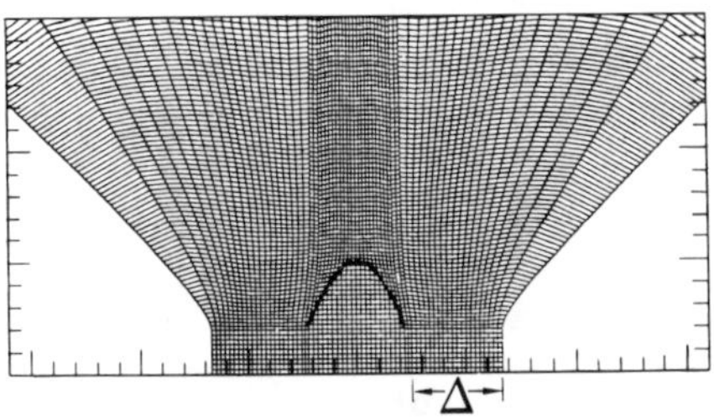

Figure 4. Explosive compaction. The maximum radius shown is 15 centimeters.

boundary) in a geometry with cylindrical symmetry. The cylindrical axis is vertical. The region originally filled with explosive is indicated by Δ. At the instant shown in the picture the detonation wave in the explosive has progressed from the top surface of the cylinder nearly to the base. The fanshaped region is made up of expanding Lagrangian zones filled with hot product gases from the detonating explosive. The motion of each of the materials shown (explosive, plastic, and powdered metal) obeys finite-difference approximations to the continuum equation of motion, $\rho\dot{u} = \nabla\cdot\sigma$. Stress σ and strain ε in each zone are approximated in terms of boundary "node" displacements, leading to ordinary differential equations for the motion of these nodes. How are the finite-difference equations converted into ordinary differential equations?

As a simple example, consider a meter stick divided into 100 1-cm zones by 101 equally spaced nodes. Now consider longitudinal deformations. In the elastic regime the longitudinal stress in each zone would be given by Young's modulus E times the strain, $\varepsilon_{xx} = (V - V_o)/V_o$. By considering the stress levels in the two zones adjacent to each node, the accelerations $\ddot{x}_J$ can be calculated. The differential-difference equation of motion for a typical node has the form

$$m\ddot{x}_J = E(x_{J+1}-2x_J+x_{J-1})/d \quad , \qquad (1)$$

where m = ρd is the mass of a zone and d is the unstrained zone length. If we fix the two ends of the meter stick, ($x_1 = 0$, $x_{101} = 100$) this problem can be solved analytically. With the nodal displacements given by $x_J = A\exp(i\omega t - ikx_J^0)$, the equations of motion lead to the eigenvalue solution for the vibration frequency ω, as a function of the wave-vector k:

$$\omega^2 = (E/\rho d^2)\ (2-2\cos kd) \quad . \qquad (2)$$

This solution corresponds to the exact continuum solution, $\omega^2 = (E/\rho)k^2$, in the limit that d vanishes. For nonzero d the error is of order d^2. An explicit finite-difference solution of equation (1) would typically be carried out by approximating the second-order time derivative,

$$\ddot{x}_J(t) \doteq [x_J(t+dt) - 2x_J(t) + x_J(t-dt)]/(dt)^2 \quad . \qquad (3)$$

Here again the errors in the finite-difference approach are second order, of order $(dt)^2$.

The set of 99 difference equations resulting from combining the spatial discretization in (1) with the temporal discretization in (3) can <u>still</u> be solved by the plane-wave solution appropriate to (1), but the solution is unstable, and diverges exponentially in time, unless dt is sufficiently small. The numerical instability can be analyzed by examining the high-frequency motion in which adjacent nodes move in opposite directions, $kd = \pi$. The quadratic equation for $\exp(i\omega t)$ produces unstable solutions unless dt is less than the time required for sound to cross the zone width d, $dt < d/c$, where the sound speed c is $(E/\rho)^{1/2}$. This condition on dt, called the Courant condition, is necessary for stability. The analysis is obviously more complicated by geometrical and constitutive nonlinearities, but the principle applies to any explicit finite-difference solution of the equations of motion.

In applying the numerical explicit finite-diffence approach to powder compaction, there are two additional complications. Both the equation of state of a porous (powder) material and the equation of state of the explosive driving the material must be specified. For powders, the simple interpolative equation of state

$$(V_o/V) = (1 + AP) / (1 + BP) \quad , \qquad (4)$$

where A and B are chosen to match the powder's initial compressibility and density, is sufficiently accurate for many purposes. An accurate version would have to include more parameters as well as an explicit dependence on energy and the size and surface finish of the powdered material.

For high explosives the equation of state is energy-dependent and *time*-dependent. The detonation energy (taken from experiment) in each zone is released at the time and with a rate consistent with the (experimental) detonation speed.

In dynamic compaction both the magnitude and duration of the stress are important. Because dynamic pressure waves move at approximately sonic velocities, the relative values of the sound speeds in the explosive and in the material being compacted are crucial to determining the stress history followed by the material. Figure 5 shows the difference between a fast and slow explosive (relative to the sound velocity in the steel) compressing a stack of four steel plates.

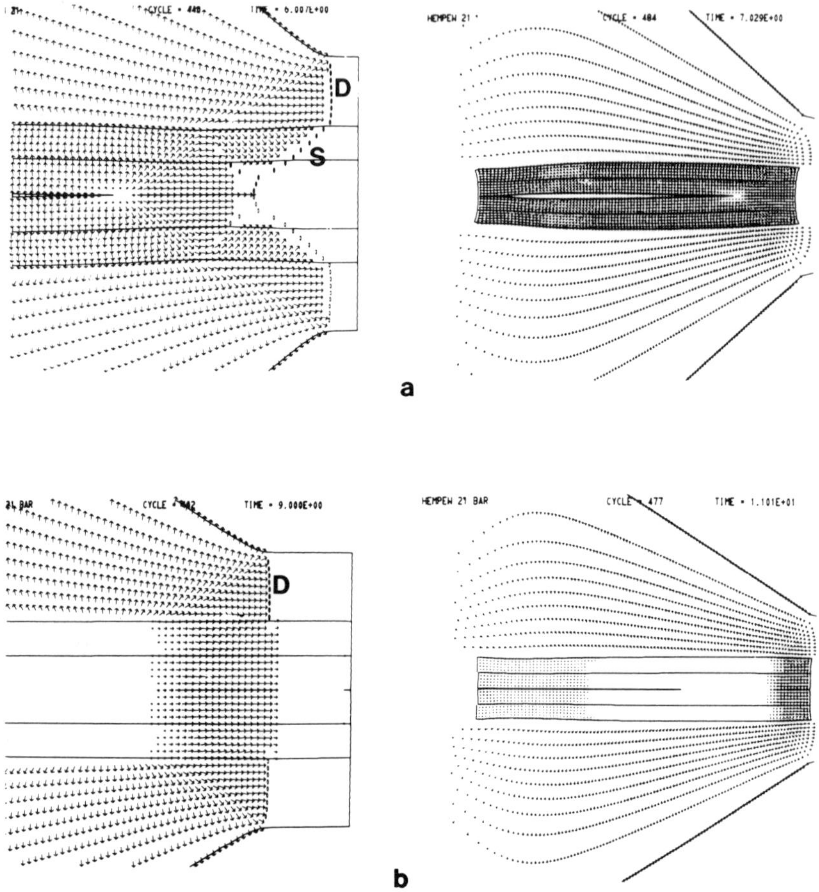

Figure 5. Fast (top) and slow (bottom) explosive compression processes.

There are explosive charges both above and below the stack of steel plates. In the fast-explosive case a strong shockwave is formed. At the horizontal centerline the high pressure generated is sufficient to force the plates apart, releasing the stored compression energy. The compression occurs too rapidly. This undesirable behavior can be avoided by using a slower explosive. In that case (baseof figure) the pressure waves in the steel can outrun the detonation wave (D) in the explosive, resulting in permanent plastic deformation.

Because a considerable part of the energy lost by compression waves in compressing and compacting powders, typically at only half normal density, offsets the gain from cylindrical stress convergence, it is possible to achieve a fairly uniform stress level. Computer simulations are particularly useful in studying the variation of results with changing geometrical features. If too little explosive is used, losses predominate and powder near the center of the cylinder is not compacted. If too much is used, the central region can be melted or vaporized by stress convergence. A method for moderating excessive compressive stresses (which can cause melting) or tensile stresses (which can cause cracking) is to place a massive "mandrel" along the cylinder axis. This geometry is shown in figure 6 for two different thicknesses (Δ) of explosive.

The dynamical properties of the mandrel influence the quality of the compacted powder. The effect of an aluminum core within a steel mandrel is illustrated in figure 7, in which copper powder was compressed. In the upper photograph a pure steel mandrel was used. In the lower photograph the steel contained a central aluminum bar. The softer aluminum cushioned the compacting powder better the compacting powder better than did the steel, which exhibits the multiple fractures characteristic of excessively-strong rarefaction waves. The calculations shown in the figures all correspond to real laboratory experiments.

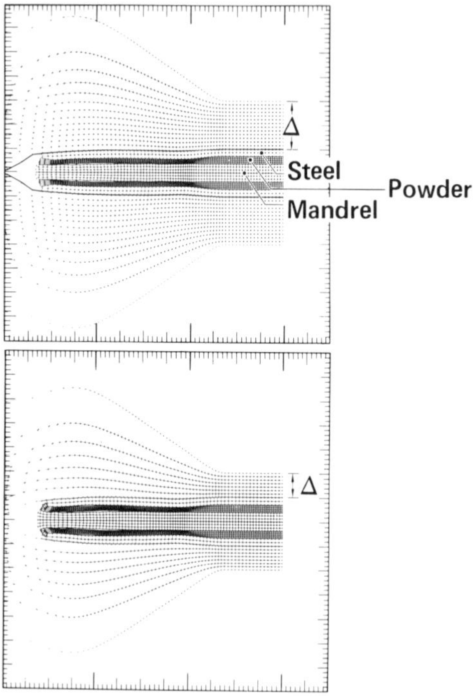

Figure 6. Cylindrical powder compaction with central mandrel. Two explosive thicknesses (Δ).

Figure 7. Cylindrical compaction of copper powder. Steel mandrel with and without aluminum filler.

Figure 8. Aluminum nitride and metallic glass cylinder.

The sections of aluminum nitride (top) and metallic glass (bottom) cylinders shown in figure 8 were both made using aluminum mandrels. Because the equation of state of powders is complicated by the effects of grain size, hardness, shape, and surface oxidation, and is not simply related to the equation of state of the parent materials, series of simulations must be carried out to make the parameters of the equation-of-state models fit experimental results. Once this fitting procedure has been accomplished, the resulting model is capable of predicting the results of further experiments.

6. OUTLOOK

The potential importance of dynamic compaction, coupled with the sophistication required for effective modeling make dynamic compaction an exciting area of physics right now. In particular, microscopic modeling of surface-surface interactions, and mesoscopic modeling of sintering, deformation, and cold welding should have useful impact on the macroscopic models required to help design physics experiments and refine commercial powder processes.

7. PROBLEMS

A. Calculate the minimum radial strain required to change a disk to a hexagon at fixed density. If only the outer half of the disk deforms, what is the minimum strain required?

B. Show that the time required for a grain to reach uniform temperature varies as the two-thirds power of its volume. Assume that heat flow follows Fourier's law $\dot{T} = D\nabla^2 T$, where the thermal diffusivity D is constant.

C. Show that the finite-difference solution for the vibrations of a meter stick fails to converge if the Courant condition $dt < d/c$ is violated.

ACKNOWLEDGMENT

I am particularly grateful to Mark Wilkins for making his unpublished work available to me before his presentation at the High-Energy-Rate-Fabrication Conference in San Antonio.

I acknowledge the support of the Army Research Office and the Air Force Office of Scientific Research. Work performed at the Lawrence Livermore National Laboratory, under contract W-7405-ENG, is carried out under the auspices of the United States Department of Energy.

REFERENCES

1. Explosive Welding, Forming and Compaction, T. Z. Blazynski, Ed. (Applied Science Publishers, New York, 1983).

2. Dynamic Compaction of Metal and Ceramic Powders (Report of the National Materials Advisory Board, National Academy Press, Washington, 1983).

3. Physics Today Issue on "Nonequilibrium Fluids", January 1984.

4. F. V. Lenel, G. S. Ansell, and R. C. Morris, in Advanced Experimental Techniques in Powder Metallurgy, J. S. Hirschhorn and K. H. Roll, Eds. (Plenum, New York, 1970), p. 61.

5. M. L. Wilkins, "Dynamic Powder Compaction", Lawrence Livermore National Laboratory Report UCRL-90142 (1983).

6. B. L. Holian, W. G. Hoover, B. Moran, and G. K. Straub, Phys. Rev. A **22**, 2798 (1980).

7. W. G. Hoover, A. J. C. Ladd, and B. Moran, Phys. Rev. Letters **48**, 1818 (1982).

8. M. L. Wilkins, in Methods in Computational Physics, **3**, B. J. Alder, S. Fernbach, and M. Rotenburg, Eds. (Academic, New York, 1964). p. 211.

9. F. F. Abraham, W. E. Rudge, D. J. Auerbach, and S. W. Koch, Phys. Rev. Letters **52**, 445 (1984).

FLOW AND PLASTICITY VIA NONEQUILIBRIUM MOLECULAR DYNAMICS

WILLIAM G. HOOVER

1. SYNOPSIS

The viscous flow of fluids and the plastic flow of solids, such as metals, are interesting from both the practical and the theoretical points of view. Atomistic "molecular dynamics" simulations provide a way of visualizing and understanding these flows in a detailed microscopic way. Simulations are necessarily carried out at relatively high rates of strain. For this reason they are ideally suited to the study of nonlinear flow phenomena: normal stresses induced by shear deformation, stress rotation, and the coupling of stress with heat flow, for instance. The simulations require appropriate boundary conditions, forces, and equations of motion. Newtonian mechanics is relatively inefficient for this simulation task. A modification, "Nonequilibrium Molecular Dynamics," has been developed to simulate nonequilibrium flows.[1] By now, many high-strain-rate rheological studies of flowing (viscous) fluids and (plastic) solids have been carried out. Here I describe the new methods used in the simulations and some results obtained in this way. A three-body shear-flow exercise is appended to make these ideas more concrete.

G. Jacucci (ed.), Computer Simulation in Physical Metallurgy, 87–106.

2. INTRODUCTION

Molecular dynamics is the numerical solution of the equations of motion of N interacting particles. The original "equilibrium" version of molecular dynamics solved Newton's equations of motion. The volume V was kept fixed, so that the energy E was a constant of the motion. This early work was mainly devoted to determining the thermodynamic "equation of state", relating the pressure P and temperature T to E and V. A few studies of linear transport phenomena were carried out, using expressions linking equilibrium fluctuations to linear transport properties. Velocity fluctuations, stress fluctuations, and heat current fluctuations respectively were used to find the diffusion coefficient, the viscosity coefficients, and the thermal conductivity. In addition, the structure of fluids, including gas-liquid mixtures, and the dynamic approach to equilibrium were studied. The equilibrium results led to theoretical advances in estimating the equation of state. Results for simple systems were used as a basis for a perturbation treatment of more complicated cases. This approach was successful. The equation-of-state problem can now be considered solved, at least for simple, pairwise-additive short-ranged interactions.

Progress on nonequilibrium problems has been slower due to the lack of a suitable perturbation theory. For this reason there has been considerable activity in simulating, as opposed to calculating from theory, nonequilibrium flows. Shockwaves, surface collisions, rapid expansions, heat flows, and the like have all been studied. Here I will describe only the shear deformation of simple materials (either fluid or solid), a path leading to viscosity for fluids[2] and to yield strength for solids.[3]

These simulations use a modified "nonequilibrium" molecular dynamics which incorporates the deformation being studied as well as a steady-state thermostat to increase computational efficiency. I describe the thermostatting and deformation techniques first. Then I outline the applications of these nonequilibrium techniques to particular fluid and solid systems.

I believe there is great potential for combining the nonequilibrium methods with models for metals to study problems of metallurgical interest. This application of nonequilibrium molecular dynamics is still in its infancy.

3. NONEQUILIBRIUM MOLECULAR DYNAMICS

Irreversible flows invariably involve the conversion of work into heat. The irreversible heat then acts to change material properties in a very nonlinear way. Thus transient irreversible problems are much more complicated than steady flows. The irreversible heating is particularly severe in microscopic computer simulations because very high strain rates must be used. (Otherwise the simulations would require too much time.) Gigahertz to terahertz strain rates are necessary if the induced stresses are to exceed the natural thermal fluctuations present in small systems. Thus the time required for a significant change in shape in a computer simulation lies between a picosecond and a nanosecond.

A nanosecond of real time corresponds to about 100,000 time steps in a computer simulation. Terahertz strain rates exceed the rates in rapid Hopkinson-bar experiments by about seven orders of magnitude. Because the irreversible heat varies as the square of the deformation rate, the heating rate in a straightforward computer simulation would be of order 10^{12} kelvins/second.

To avoid the irreversible heating, steady states can be simulated by forcing the microscopic temperature, energy, or enthalpy to remain fixed.[4]

From the microscopic viewpoint temperature is simply a measure of mean-squared particle velocity. This quantity, when summed over all particles, can be held constant by a frictional damping force $-\zeta\dot{r}$. That additional force, added to the Newtonian Force F, can be expressed directly in terms of the interparticle forces and velocities $\dot{r}$:

$$m\ddot{r} = F - \zeta\dot{r} \quad ;$$

$$\zeta \equiv m(\sum\sum F\cdot\dot{r})/(\sum m\dot{r}^2) \quad ;$$

$$(d/dt)(\sum m\dot{r}^2) \equiv 0 \quad . \tag{1}$$

The isothermal equations of motion (1) conserve kinetic, rather than total, energy. They can be derived directly from Gauss' "Principle of Least Constraint". This Principle is a generalization of Newtonian mechanics which makes it possible to satisfy constraints (constant temperature, stress, heat flux, ...) typical of nonequilibrium systems. These constrained systems make possible the study of systems with large gradients or flows. For example, under shockwave conditions, "temperature" can be very anisotropic. The temperature parallel to the direction of shockwave propagation can exceed the transverse temperature by a factor of two. Such nonequilibrium temperature distributions can have profound effects on chemical reaction rates. These interesting problems can be treated by a tensor generalization of equation (1), but a thorough discussion would take us too far afield.

With the thermostat provided by equations of motion (1) we can study flow problems involving deformation. The irreversible heat which deformation provides--viscous heating in fluids, plastic work in solids--can than be absorbed as rapidly as it is produced.

Deformation can be induced through boundary forces or displacements, or by imposing a macroscopic deformation throughout the system. Three types of macroscopic shear deformations are indicated in figure 1:

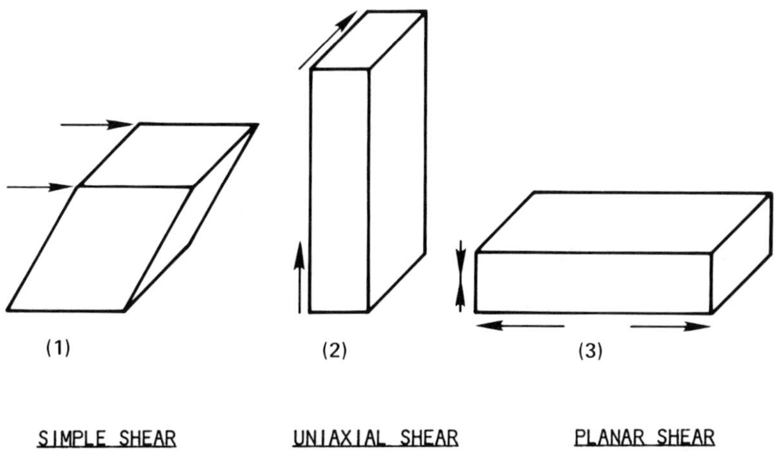

Figure 1. Three Macroscopic Shear Deformations.

In each case the figure shows the effect of deformation on an initial cube. Simple shear (or "plane couette flow") is a <u>rotational</u> flow in which the fluid elements rotate clockwise with an angular velocity equal to half the shear strain rate $\dot{\varepsilon} = du_x/dy$. Uniaxial shear (with $\dot{\varepsilon}_{xx}=-2\dot{\varepsilon}_{yy}=-2\dot{\varepsilon}_{zz}$) is typical of the shape change undergone during shockwave compression. This flow, as well as the two-dimensional "planar shear" version shown last (with $\dot{\varepsilon}_{xx}=-\dot{\varepsilon}_{yy};\dot{\varepsilon}_{zz}=0$), is irrotational. The first of these flows has been extensively studied, for it can be simulated as a true steady state.

For small strain rates, all three shear flows can be described by the same phenomenological coefficient: shear viscosity in the case of fluids; yield strength in the case of solids. When nonlinear effects must be included, all three flows are fundamentally different. Here we will study in more detail the first flow shown above, simple shear, in the steady state at constant temperature. We will consider a range of strain rates for which nonlinear effects are important.

We begin by distinguishing the systematic macroscopic velocity, $u = (\dot{\varepsilon}y,0,0)$, from the local microscopic velocity $p/m = (p_x,p_y,p_z)/m$. We expect that the microscopic velocities p/m will have nearly an equilibrium (Maxwell-Boltzmann) distribution at small enough rates of strain $\dot{\varepsilon}$. Because the systematic velocity varies from point to point, there is a Coriolis' acceleration affecting the momenta p. In the case of simple shear, a free particle moving in the y direction with x fixed would undergo an effective acceleration:

$$\dot{x} = 0 = (p_x/m) + \dot{\varepsilon}y \quad \text{---------->} \quad \dot{p}_x = -\dot{\varepsilon}p_y \quad . \tag{2}$$

Adding this effective acceleration to the Newtonian forces gives the set of equations describing adiabatic simple shear:

$$\begin{aligned} \dot{x} &= (p_x/m) + \dot{\varepsilon}y, \qquad \dot{y} = (p_y/m), \qquad \dot{z} = (p_z/m), \\ \dot{p}_x &= F_x - \dot{\varepsilon}p_y, \qquad \dot{p}_y = F_y, \qquad \dot{p}_z = F_z \quad . \end{aligned} \tag{3}$$

Temperature is $\langle p^2/3mk \rangle$. By generalizing the friction coefficient ζ from (1) to include the Coriolis contribution, we find the equations describing <u>isothermal</u> simple shear:

$$\dot{x} = (p_x/m) + \dot{\varepsilon} y \quad , \quad \dot{y} = (p_y/m), \; \dot{z} = (p_z/m) \quad ,$$
$$\dot{p}_x = F_x - \dot{\varepsilon} p_y - \zeta p_x \; , \; \dot{p}_y = F_y - \zeta p_y \; , \; \dot{p}_z = F_z - \zeta p_z \quad . \qquad (4)$$

The boundaries of the system are usually taken to be periodic, as shown[5] in Figure 2, to eliminate the edge effects associated with small systems.

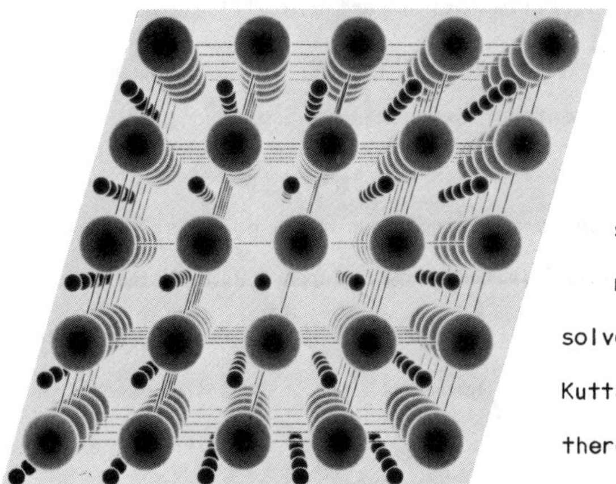

Figure 2. Two-Dimensional Shear Flow.

Shear flow simulations have been carried out for a wide range of thermodynamic states and strain rates, for two- and three-dimensional fluids and solids. In three dimensions there are 6N equations of motion. These can conveniently be solved using the fourth-order Runge-Kutta method. In two dimensions there are 4N equations of motion. The initial velocities are usually chosen from a Maxwell-Boltzmann distribution. The momenta $\{p\}$ are then scaled to the desired temperature. That temperature is subsequently maintained by the equations of motion (4). The trajectories which result can then be analyzed for structure.

Thermodynamic functions, such as the pressure tensor, can be obtained by time averaging the corresponding phase functions.

The pressure tensor P is the same as the momentum flux, a measure of the flow of momentum, per unit area and time. Because the flow direction (normal to the area) and the momentum direction are both _vector_ quantities, pressure is a _second-rank_ (two-index) tensor, 2x2 in two dimensions and 3x3 in three dimensions. The pressure tensor has two kinds of contributions in a system characterized by pairwise-additive forces. The single-particle convective contributions (pp/m)/V give the flow of momenta p across comoving planes oriented normal to the velocities p/m. In a stationary system with sides L_x, L_y, and L_z, for instance, the probability that a particle will cross a plane of area dA perpendicular to the x axis, during the time interval dt, is $(p_x/m)dtdA/V$. Thus the averaged momentum flow, per unit area and time, is the sum, over all particles, of (pp/m)/V, with $V = L_xL_yL_z$.

There is a second contribution to momentum flux which does not depend upon particle motion, the action-at-a-distance contribution due to interparticle forces. If particle i, at x_i, interacts with particle j, at x_j, with the force F, the probability that a plane normal to the x axis intersects that interaction is $\pm(x_i-x_j)dA/V$ and the corresponding flow of momentum per unit time and area is $\pm F(x_i-x_j)/rdA$, where r is the distance separating the two particles. Thus the flow of x momentum in the x direction, per unit time and area, has the form $(x_i-x_j)F(x_i-x_j)/rV$. The complete pressure tensor P is the sum of both terms, convective and action-at-a-distance:

$$PV = \sum(pp/m) + \sum\sum rF \quad , \tag{5}$$

where the single sum runs over all particles and the double sum runs over all distinct pairs of particles. If we specify a strain rate $\dot{\varepsilon}$ describing the plane couette flow of Figure 1, the corresponding pressure-tensor element P_{xy} can then be averaged as a function of time. At sufficiently high rates, in fluids, or even at low rates, in solids, P_{xx}, P_{yy}, and P_{zz} can also be affected by the shear. These "normal stress effects" are responsible for a host of nonlinear phenomena[6], two of which are indicated in Figure 3.

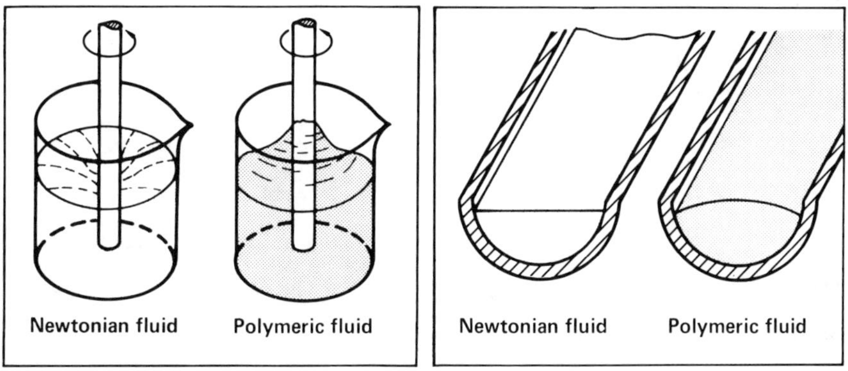

Figure 3. Two Non-Newtonian Flows exhibiting normal-stress effects.

Fluid <u>viscosity</u> is obtained by dividing the shear stress $-P_{xy}$ by the strain rate $\dot{\varepsilon} = du_x/dy$. Solid <u>yield strength</u> is simply $-P_{xy}$. Both kinds of simulations have been carried out. Some results are described in sections 4 and 6.

4. APPLICATIONS OF NONEQUILIBRIUM MOLECULAR DYNAMICS TO FLUID FLOW

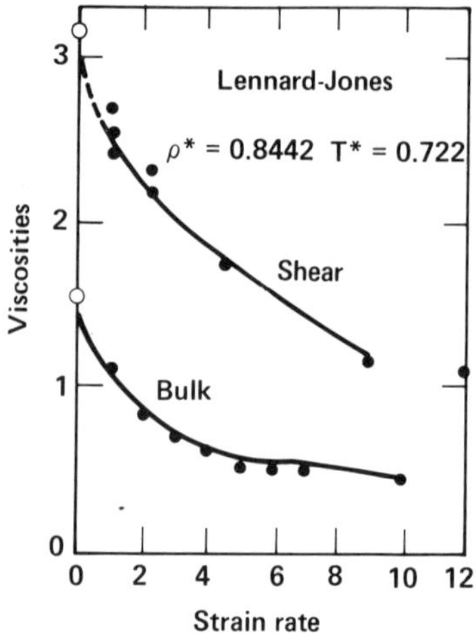

Figure 4. Variation of Viscosities of Liquid Argon with Strain Rate.

A thorough study of the dependence of shear viscosity on thermodynamic state and strain rate has been carried out, using the Lennard-Jones 6-12 pair potential. Near the freezing line the data depend relatively strongly on strain rate, so that it is necessary to extrapolate from the high rates of the computer simulation to obtain the zero-rate macroscopic value. Typical data[7] are shown in Figure 4. The viscosities and strain rate are given in reduced units: the atomic mass, well depth, and collision diameter are all equal to unity. The extrapolated zero-strain-rate viscosities (open circles) are in good agreement with laboratory measurements.

Laboratory viscosity measurements are relatively straightforward because long-time steady-flow studies can be carried out. Fluid can be forced through a constriction, as in Poiseuille flow, or stirred by a submerged disk. In either case the viscous power loss can be easily measured.

5. SHOCKWAVE COMPRESSION

Much larger shear stresses, and higher strain rates, are found in the shockwaves used to compress fluids and solids to high pressures, pressures which can be orders of magnitude higher than the pressure at the center of the earth. The basic shockwave geometry is shown in Figure 5.:

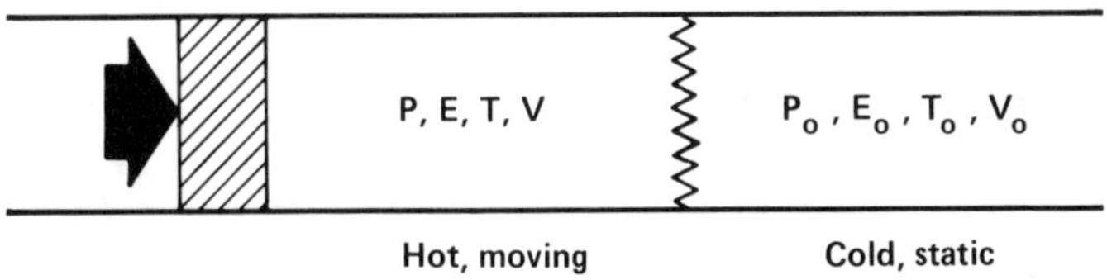

Figure 5. Steady Shockwave Generated by a Piston Compressing a Fluid

In a steady one-dimensional (planar) shockwave cold material is transformed by the shock process into steadily moving hot material with higher pressure, energy, density, and entropy. By measuring the velocities of the moving hot material and the shockwave, it is possible to calculate the final pressure P_{xx} and energy. For fluids P_{xx} and the final energy are equilibrium values (because a fluid cannot support shear) and interpretation of the shockwave velocity data is straightforward. For solids P_{xx} differs from P_{yy} (the difference is twice the yield strength) and the energy slightly exceeds the value it would have under isotropic conditions. Despite these difficulties, estimates of the solid's shear stress can be made.

6. APPLICATIONS OF NONEQUILIBRIUM MOLECULAR DYNAMICS TO PLASTIC FLOW

Shear flow simulations using the equations of motion (4) have also been carried out for solids, both in two and in three dimensions.[8] Although the details of such simulations certainly depend upon the interatomic forces, only one (very simple) force law has been used so far, the piecewise-linear force law shown in Figure 6. This force varies linearly from the force-free separation d up to a maximum attractive value d + w. With further stretching the force again decreases linearly to zero at r = d + 2w.

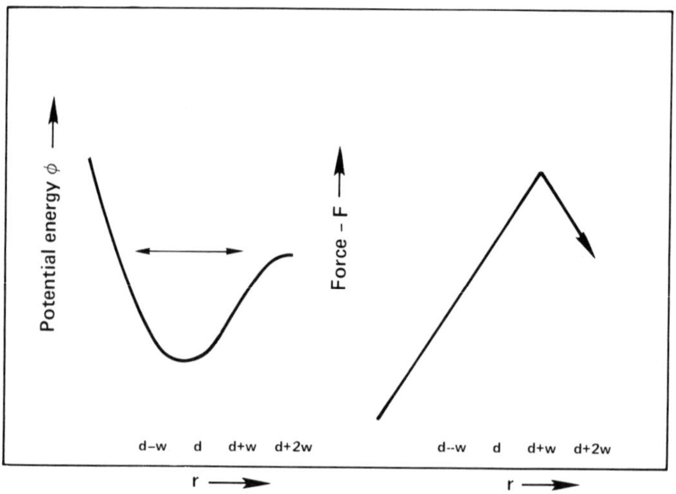

Figure 6. Piecewise-Linear Force Law used in Plasticity Simulations.

In the vicinity of the potential minimum d, Hooke's Law is obeyed so that crystal properties can be calculated using quasiharmonic lattice dynamics.

That is the main motivation for studying this system. The thermodynamic equation of state, frequency spectrum, and properties of various lattice defects including dislocations, vacancies, and cracks, have all been worked out.

In the case of isothermal steady shear, simulations were carried out for a variety of system sizes, and over about two decades of strain rate, from ten gigahertz to one terahertz. At the low-frequency end of this range, the deforming crystal typically contained only a single pair of dislocations. At the high-frequency end, dislocations were as numerous as particles, so that individual lattice defects could not be identified. The measured dependence of stress (yield strength) on strain rate is summarized in Figure 7. G is the shear modulus; d the interparticle spacing; and c_T is the transverse sound speed.

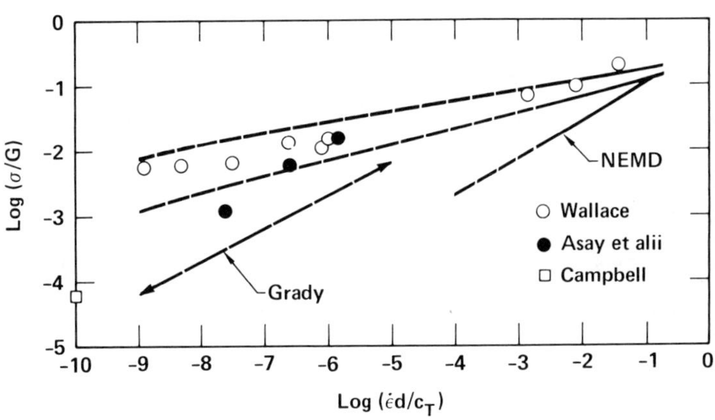

Figure 7. Yield Strength σ (Flow Stress) as a Function of Strain-Rate $\dot{\varepsilon}$.

The molecular dynamics data lie at the righthand side of the figure. Experimental data, estimates from plastic shockwaves in real metals, appear near the lefthand side of the figure.

The log-log slopes (for three different temperatures) lie between the limits of perfect plasticity (zero slope) and Newtonian Viscosity (unit slope), with the slope increasing linearly with temperature. The correspondence between the computer and laboratory experiments suggests that the same physical mechanism is responsible for both sets of data in the high-strain-rate region.

It seems likely that an understanding of rapid deformation would be useful in interpreting dynamic compaction experiments carried out on powdered amorphous metals. The rapid frictional interactions among the compacting grains certainly lead to terahertz strain rates. The detailed mechanism of the compaction process may thus prove sensitive to rate-dependent plasticity.

7. DISLOCATION THEORY OF PLASTIC FLOW

The textbook explanation of plastic flow ascribes that deformation to lattice defects, dislocations, that move through the lattice at speeds somewhat below the transverse sound speed c_T. Series of simulations of disclocations have been carried out to determine their static and dynamic structure, speed as a function of shear stress, and interaction energy.[3,9] These studies show that the energies of dislocation structures can be adequately described by a pairwise-additive interaction energy of the form predicted by elastic theory. According to the theory, the interaction energy of two dislocations, separated by a distance r and in the presence of a shear stress σ, is of the form $A\ln r - Br\sigma$, within an additive constant. Because the interaction energy has a maximum, Arrhenius kinetics can be used to estimate the rate at which pairs of dislocations are produced, varying as $\exp(-\Delta E/kT)$. In the steady state, this rate must equal the annihilation rate, which in turns varies as the square of the strain rate. This equivalence leads to the result that stress must vary as a power of the strain rate:

$$\sigma/\sigma_0 = (\dot{\varepsilon}/\dot{\varepsilon}_0)^{2kT/Db^2} \quad , \tag{8}$$

where D is an elastic constant and b is the interatomic spacing. On the other hand, the molecular dynamics data in figure 7 show conclusively that the 2 in (8) should be replaced by 3, a result impossible to extract from the theory.

8. WHAT REMAINS TO BE DONE?

It is clear that the nonequilibrium simulation methods will prove useful in understanding both experimental data and in advancing the theory of plastic flow in metals. The main difficulties, estimating forces in metals near defects, and understanding the properties of defects when the forces are known (as in computer simulations), still need considerable attention.

9. PROBLEM: SAMPLE SIMULATION OF STEADY SHEAR

To illustrate the application of the nonequilibrium equations of motion, we consider an example which is nearly the simplest possible, namely the shear deformation of three particles in two dimensions. Periodic boundaries are used to eliminate edge effects. See Fig. 8.

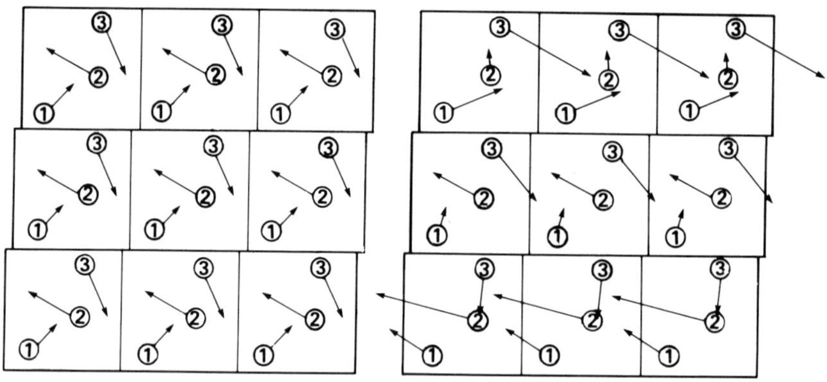

Figure 8. Arrows indicate (p/m) [left] and $\dot{r}$ [right].

For convenience we use a truncated Hooke's law force with the force constant, mass, and force-free spring length all set equal to unity. Thus the (purely-repulsive) force has a magnitude $F = 1 - r$ for $r < 1$ and $F = 0$ for $r > 1$. Longer-range forces, even forces extending to infinity, can be treated with Ewald's Fourier-transform method, with a moderate increase in computer time. In our sample calculation we choose fixed values of both temperature and strain rate. To start out we specify initial conditions. We choose the time $t_0 = 0.6$ and the strain rate $\dot{\varepsilon} = 0.1$, so that the initial shear strain is

0.06, as indicated in Fig. 8. We arbitrarily choose a basic cell of side-length L = 2. As the calculation proceeds the particle coordinates will eventually leave the central 2x2 box. Provided that the accelerations of each particle are based on the nearest images of its two neighbors, as calculated in the three-step process below, it is not necessary to replace the coordinates in the central box. Let us start with the arbitrary set of coordinates x, y and thermal velocities p_x, p_y shown in the Table.

Table I. Initial configuration with L = 2 and ε = 0.06 at t_o = 0.6.

i	1	2	3
x	-0.55	+0.21	+0.34
y	-0.66	-0.09	+0.75
p_x	+0.10	-0.20	+0.10
p_y	+0.10	+0.10	-0.20
F_x	-0.064	+0.017	+0.047
F_y	-0.012	-0.118	+0.130
i,j	**1,2**	**2,3**	**3,1**
y_i-y_j	-0.57	-0.84	+1.41
n	0	0	-1
y_{ij}	-0.57	-0.84	-0.59
$x_i-x_j+2n\varepsilon$	-0.76	-0.13	+0.77
r_{ij}	0.95	0.85	0.97
F	+0.05	+0.15	+0.03

To solve the equations of motion by difference-equation methods we need to calculate the forces on each of the particles. The sheared periodic boundaries must be taken into account in calculating the nearest-image separation of each of the three particle pairs (1,2; 2,3; 3,1). For each (i,j) pair the forces are calculated in three steps:

1. Calculate y(i) - y(j) and add to it nL, where n is the integer chosen to make the final "nearest-image" separation y_{ij} lie between -L/2 and +L/2.

2. Calculate $x(i) - x(j) + nL\dot{\varepsilon}t$, to take into account the offset of periodic images separated by nL in the y direction at time t. Add to this the integer multiple of the box length L which causes the final nearest-image separation x_{ij} to lie between -L/2 and +L/2.

3. Accumulate the x and y forces for the nearest-image separation just computed into the forces on i and the forces on j. The same numbers are used, with opposite signs.

This force calculation is illustrated in the lower half of the Table. From the forces and thermal velocities the friction coefficient ζ can be calculated. For the configuration in the Table, the "friction" coefficient ζ, calculated from the forces and momenta according to Gauss' principle:

$$\zeta = m\sum\sum[(F\bullet p/m) - \dot{\varepsilon}(p_x p_y/m)]/\sum(p^2/m),$$

is -0.343. In this nonequilibrium problem, where velocities play a direct role in the equations of motion, it is computationally convenient to solve _first-order_ differential equations for particle coordinates and momenta rather than second-order coordinate equations. With the forces and friction coefficient known we have sufficient information to advance the time using a standard first-order differential equation solver. The set of 12 equations corresponding to the homogeneous deformation in Fig. 8 is the following:

$$\dot{x}(i) = [p_x(i)/m] + \dot{\varepsilon}y(i) \quad , \quad \dot{y}(i) = [p_y(i)/m] \quad ,$$
$$\dot{p}_x(i) = F_x(i) - \zeta p_x(i) - \dot{\varepsilon}p_y(i) \quad , \quad \dot{p}_y(i) = F_y(i) - \zeta p_y(i) \quad ;$$

for i = 1, 2, and 3. By taking into account the fixed center of mass and fixed kinetic energy these 12 equations could be reduced to seven, but it is simpler to integrate the full set. The inevitable drift in kinetic energy due to finite computational accuracy, can be offset by occasional rescaling of the velocity. Simultaneously, we accumulate the energy and pressure tensor, and the probability distributions in space and velocity. The resulting _new_ configuration, at $t_o + dt$, then replaces the initial condition. We repeat the integration process by returning to step 1 above. Fluctuations in cumulative averages fall off as $(t - t_o)^{-1/2}$ as the averages stabilize. The shear viscosity is obtained by dividing the long-time steady value of the shear stress $-P_{xy}$, by the strain rate $\dot{\varepsilon}$: $\eta = -P_{xy}/\dot{\varepsilon}$.

REFERENCES

1. W. G. Hoover, Annual Rev. of Phys. Chem. 34, 103 (1983).

2. D. J. Evans, H. J. M. Hanley, and S. Hess, Physics Today 37 (1), 26 (1984).

3. A. J. C. Ladd and W. G. Hoover, Phys. Rev. B 28, 1756 (1983).

4. D. J. Evans, W. G. Hoover, B. H. Failor, B. Moran, and A. J. C. Ladd, Phys. Rev. A 28, 1016 (1983).

5. W. G. Hoover, K. A. Winer, and A. J. C. Ladd, "Irreversibility in the Two-Body Hard-Disk Lorentz Gas", Int. J. of Eng. Sci. (in press).

6. R. B. Bird and C. F. Curtiss, Physics Today 37 (1), 36 (1984).

7. W. G. Hoover, D. J. Evans, R. B. Hickman, A. J. C. Ladd, W. T. Ashurst, and B. Moran, Phys. Rev. A 22, 1690 (1980).

8. W. G. Hoover, A. J. C. Ladd, and B. Moran, Phys. Rev. Letters 48, 1818 (1982).

9. W. G. Hoover, N. E. Hoover, and W. C. Moss, J. Appl. Phys. 50, 829 (1979).

ATOMISTIC EQUILIBRIUM MODELS OF CRACK TIP BEHAVIOUR

J E Sinclair
Theoretical Physics Division
AERE Harwell, Oxfordshire
OX11 0RA
U.K.

ABSTRACT. A review is given of atomistic treatment of crystal defects using static models. The emphasis is on applicability to crack tip modelling, where long ranged mechanical interactions between the modelled tip region and the surrounding body are vital to the behaviour. Several modelling methods are explained, and a new one put forward. Current application to the problem of competition between dislocation generation/motion and cleavage is described.

1. ATOMISTIC MODELLING – IN CONTEXT

Treatments of the behaviour of cracks can be classified according to the underlying representation of the nature and behaviour of the material (Figure 1). This lecture is concerned with atomistic rather than continuum models, and with static models in particular. One particular continuum treatment — using linear elasticity with sheets of dislocations — will be covered by me in another lecture, while other authors will describe molecular dynamics models.

Of course, to model the behaviour of a cracked body it may not be possible or appropriate to use the same type of representation throughout the body. There is, for instance, no question of modelling any cracked body by treating every atom individually. Elastically, a loaded crack is the longest ranged defect that a crystal can have (except perhaps a disclination), so that a crack tip can never be considered as an isolated object, as (say) a vacancy can. On the other hand, it is possible, even advantageous, to conceptually separate the crack tip, and the processes which occur there, from the rest of the body, and its mechanics.

This separation is advantageous, because in many cases the two are in a sense independent. Fracture mechanics is a discipline which has flourished by shutting its eyes to the details of the fracture process zone, and which maintains its form and results under a variety of different tip processes. Likewise, it is probable that the properties of a certain type of crack tip remain the same under a change of the mechanics of the surrounding body. To take an example, in brittle intergranular cleavage a certain amount of plasticity

G. Jacucci (ed.), Computer Simulation in Physical Metallurgy, 107–128.

accompanies the cracking, and the dislocations no doubt strongly affect the state at the crack tip. But it is possible to lump them in with the elastic portion of the body and the loading system, as a complicated mechanical system which provides boundary conditions for the tip processes of bond rupture. These tip processes might have just the same critical points (limits of stability, activation barriers for dislocation generation, etc.) even when some variable such as temperature alters the mechanics by affecting flow properties in the bulk.

So we see that provision for mechanical interaction between an atomistic model of a crack tip and the surrounding crystal and loading system is a more vital requirement than when modelling other defects.

In setting up an atomistic model of a crack (or indeed any crystal defect), the main choices to be made are—

- the problem to be studied and quantities to be determined;
- the interatomic potential;
- the modelling method.

These are now considered in turn, with emphasis on the methods available.

2. WHAT CAN BE MODELLED?

Some of the output quantities and topics which might be tackled by atomistic modelling of cracks include—

- the near-tip structure and distribution of cohesive forces;
- the limits for mechanical equilibrium (e.g. critical stress intensity factors);
- energy barriers for propagation & closure — with straight crack front, with kinks, ...
- dependence of above on mixed mode loading (due to dislocations, grain-boundary orientation ...)
- repeat above for blunted tip, or even a corner (can cleavage re-initiate after blunting?);
- extrinsic effects: atoms/molecules on surfaces, free in opening, in bulk, ...
- effect of GB structure, as function of orientation;
- effect of jogs in crack plane, of dislocations threading tip, ...

Dynamic effects are also obviously of great interest, but the above list concentrates on simpler questions for which static models might be used.

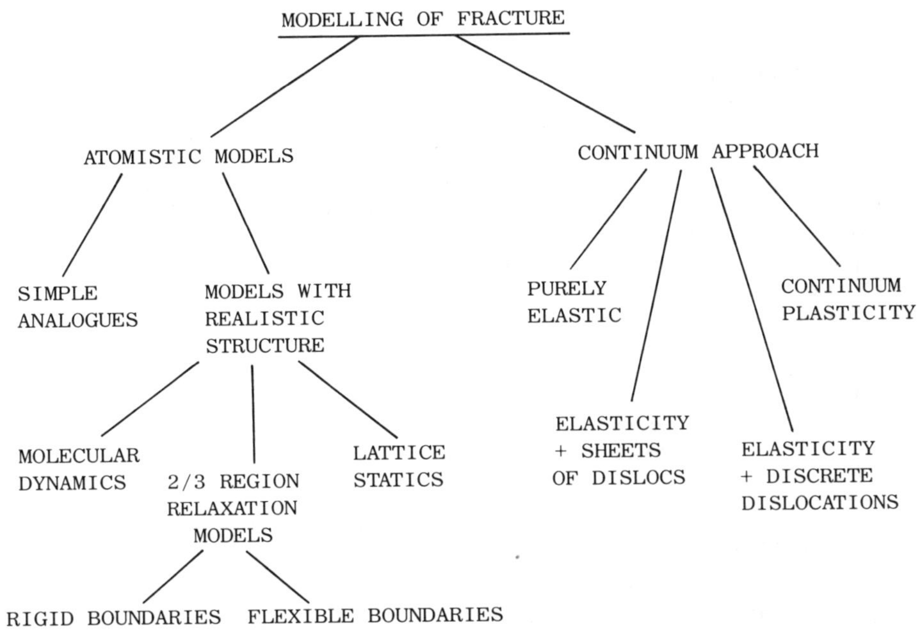

Figure 1. Various approaches to fracture modelling.

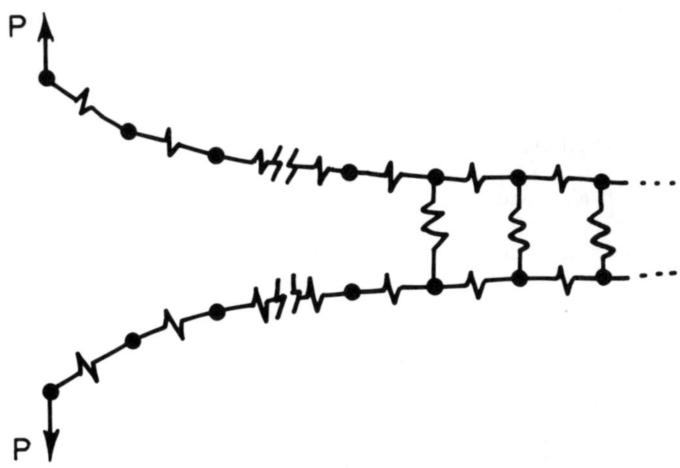

Figure 2. One dimensional model of Thomson Hsieh & Rana.

3. INTERATOMIC POTENTIALS

Choice of an interatomic potential is always a vexed question. The level of realism required depends upon the problem. With simple defects, for which the range of possible structures and processes is now well understood, the emphasis is on accurate representation of a specific material. In such cases, sophisticated electronic calculations may be in order. By contrast, in the field of crack modelling, the classes of atomic structures and re-arrangements which are possible are not known, not experimentally accessible, so there is still great scope for exploratory calculations using fairly simple empirical potentials. Even so, it is widely felt that to restrict such explorations to the regime of central pair potentials may be inadequate to discover the full range of possible types of behaviour, and there has been some recent effort to develop simple potentials with more complex, many-body properties (Finnis & Sinclair 1984, Daw & Baskes 1983). Essential properties of a potential for crack modelling include—

- it must give equilibrium at the correct lattice parameter, and stability with respect to alternative structures;
- the harmonic small-strain behaviour must match the elastic constants;
- the cohesion must fall off at large strains;
- the work of separation must match the surface energy.

Further desirable features include—

- a change in form from bulk to surface (many-body effect);
- correct prediction of stability and mobility of relevant dislocations;
- potentials for 'foreign' atoms should be chemically sensible (bond saturation effects, etc.)

4. MODELLING METHODS

The question of methodology is perhaps more important for crack modelling than for any other type of defect. The following review therefore covers a wide range of approaches.

4.1 Simple Analogue Models

The Peierls-Nabarro and the Frenkel-Kontorova models of a dislocation (e.g. Hirth & Lothe (1968)) are well known one-dimensional models for the interaction of a dislocation with the periodicity of the crystal structure. In a similar spirit, Thomson et al. (1971) constructed a one-dimensional analogue model of a crack (Figure 2), further developed by Fuller & Thomson (1977). This consists of a twin string of atoms, coupled laterally by tensile springs, the last of which is non-linear and breakable, and longitudinally by bending springs, representing the elastic coupling provided by the bulk of the body.

Fuller et al. (1980) have used this simple model to make a surprisingly far-reaching analysis of the energetics of activated crack-tip processes with and without chemical reactions at the tip. Actually, the arguments of Fuller et al. cover any situation where loads act through a harmonic lattice (system of linear springs) on a single non-linear interatomic bond. As in electrical systems, so in mechanical ones, a complex network of linear devices can be reduced to a simple 'equivalent circuit' (Figure 3). The equilibrium conditions are then obtained from the intersection of a linear and a non-linear graph of force against bond length (Figure 4).

Taking the example of chemical attack by a diatomic molecule, Fuller et al. deduce how the non-linear crack-tip bond's effective force law would be altered. They find the external load for thermodynamic equilibrium, which they interpret as K_{Iscc}, to be considerably lowered from the intrinsic value, while the load for propagation (K_{Ic}) is lowered only a little.

While this kind of modelling depends for its success on the validity of treating only one bond non-linearly, it has been described here in some detail since it suggests how more complex situations might be modelled most effectively (see later).

4.2 Lattice Statics

An approach devised by Kanzaki (1957) and developed by Hardy & Bullough (1967) for modelling point defects used the zero frequency limit of the equations of harmonic lattice dynamics, and thus became known as lattice statics. An equivalent formulation in terms of a Greens function in spatial rather than wave-number coordinates was developed by Tewary (1973). These works on point defects all retained the harmonic approximation, although the force-constant changes introduced by the defect were taken into account. A similar approach to modelling cracks has been used by Fuller & Thomson (1978), Hsieh & Thomson (1973) and Esterling (1976). At least one interatomic bond at each crack tip is now considered to be completely non-linear (Figure 5).

It should be noted that in the works just cited, the whole crack, and not just one crack tip, is treated as the crystal defect. This is because it is the perfect lattice Greens function which is used to build up the displacements of the cracked lattice. Thus, the free surfaces have to be treated in terms of inter-atomic force constant changes for the atoms adjoining the crack faces. True, most of the equations determining the equilibrium are linear — all except those involving the single non-linear bond at each crack tip. Linear algebra can therefore be used to eliminate all variables except those for the non-linear bond. One is therefore left with a single non-linear equation to solve, just as in the linear chain model described in the previous section. However, the derivation of this equation involves the force-constant changes (bond breaking) all along the crack interface. Thus, long cracks are expensive to model, and it is not

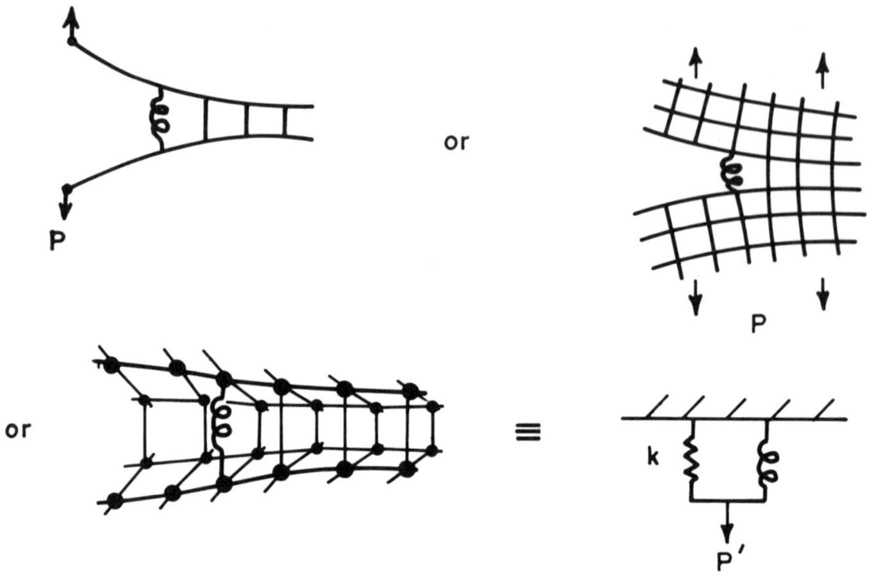

Figure 3. Any loaded network of linear springs is seen as a single spring, k, by a non-linear element.

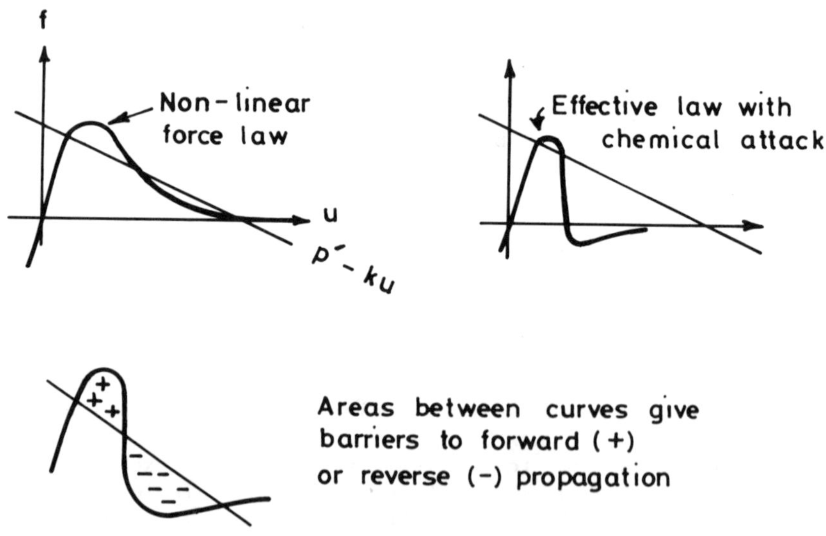

Figure 4. Equilibrium of non-linear bond with linear loading system.

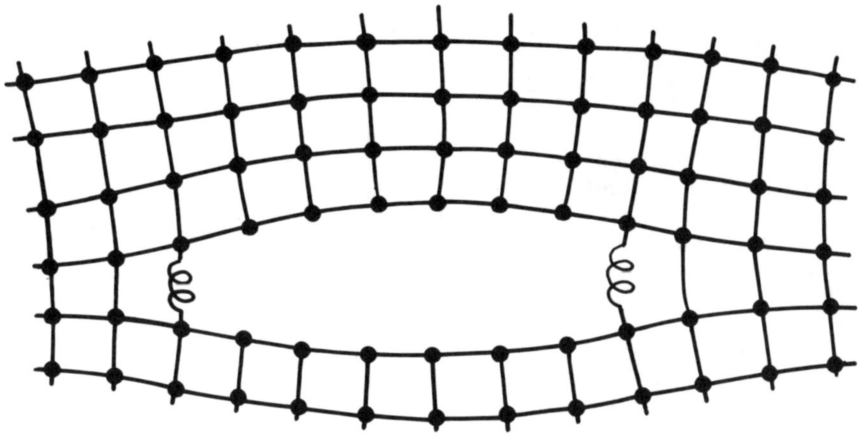

Figure 5. Lattice Statics model of crack. Displacements are generated from the Greens function of the perfect lattice, with forces at atoms with broken or non-linear bonds in model.

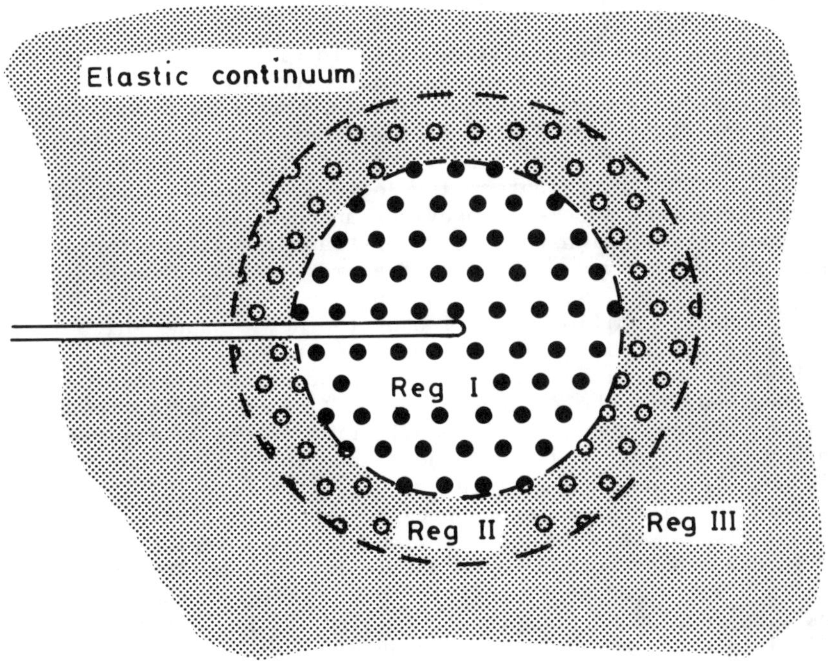

Figure 6. Three region atomic/continuum model

possible to study a single crack tip whose properties could be incorporated into other calculations.

4.3 Two or Three Region Atomic/Continuum Models

The most commonly used atomistic modelling approach for all kinds of crystal defects has been to treat independently the atoms in a small region containing the defect, while using displacement fields obtained from linear elasticity to position the remainder of the crystal. Let us first establish a uniform notation for discussing models of this type. The inner region containing the defect is region I. Surrounding this, let all atoms close enough to interact, via the chosen interatomic potential, with at least one region–I atom, be called region II. The remainder of the crystal is region III (Figure 6). In the simplest relaxation model, the postions of the region–I atoms are adjusted to equilibrium while those in region II are fixed in positions derived from an elastic displacement field appropriate to the defect. Region III atom positions are not required. The equilibrium of region I is determined from an assumed interatomic potential, and several computational procedures for the relaxation have been used. This model has been used in studies of dislocation structures and energies too numerous to cite here, and for crack tips (Sinclair & Lawn 1972, Kanninen & Gehlen 1971). There is no fundamental way to determine how large region I must be to give accurate results. The assumption is simply that at some sufficiently large distance from the defect the elastic displacement field will to sufficient accuracy give equilibrium under the interatomic potential.

An improved scheme (Sinclair 1971, Sinclair 1975) was introduced, in which the elastic displacement field determining the atom positions in region II and III is allowed to vary during the relaxation along with the region–I coordinates. The 'standard' field — that used in the fixed-boundary scheme — provides an unchanging contribution, but additional terms, which fall away more rapidly with distance from the defect, can be derived within the framework of linear elasticity theory, provided allowance is made for the existence of (unknown) inner boundary conditions supplied by the defect zone which is not linear-elastic. The coefficients of these terms are now considered as variables in the relaxation, their corresponding forces at any stage being derived from the forces on region II atoms. After relaxation, all three regions should be in equilibrium, whereas with rigid boundary conditions, regions I and II are in equilibrium, but region II is not. This new scheme later acquired the name 'Flex–S'.

This and other similar schemes are known as flexible boundary methods. Flex–S was shown to allow much smaller region II sizes than the rigid boundary method for a particular dislocation (Sinclair 1971), and was also used (Sinclair 1972, Sinclair 1975) in modelling crack tips in diamond-structure crystals. In the latter study, the use of flexible boundaries was particularly necessary for accurate determination of the 'lattice trapping' effect. Rigid boundary conditions can introduce constraint on the motion of dislocations or crack tips (a kind of image force), and can thus frustrate attempts to determine the stability against motion of the defect under various applied stresses. It is clear also, that if dislocation generation were possible from the tip of a loaded crack, then this might not be correctly exhibited in a model with rigid boundaries.

An alternative flexible boundary method, which goes under the title 'Flex–II', was developed at the Battelle Columbus laboratory (Hoagland 1973, Sinclair et al. 1978). In this method, an approximate Greens function (lattice response to a point force) is used to remove the forces on region II atoms which develop when region I is relaxed with fixed boundaries. Since the Greens function cannot be expected to 'work' for the non-linear and defected region I, relaxation of region I and the boundary fix-up must be repeated alternately until convergence.

In comparative trials on a dislocation model, Flex–II was found to converge considerably faster than Flex–S (Sinclair et al. 1978). The method is somewhat harder, however, to apply to crack modelling. This is chiefly because of the difficulty of generating a good enough approximation to the Greens function for a cracked lattice. The *continuum* Greens function for a cracked body is available (Sinclair & Hirth 1975), at least in 2 dimensions, but for nearby source and field positions, the *lattice* response must be estimated, and dependence upon both source and field positions relative to the crack tip makes this a much harder task than in a lattice with no free surfaces. A second problem with Flex–II, which applies to any defect system, is that it is not cast overall as a variational problem. Location of saddle-point configurations, required for activation energy calculations, therefore cannot be pursued with algorithms such as that of Sinclair & Fletcher (1974).

4.4 Flex–G — a New Modelling Procedure

This new method aims to combine the best features of lattice statics and the flexible-boundary atomic/continuum models, while culminating in a set of non-linear equations with as few variables as possible. The simple model of Fuller et al. (1980) discussed earlier managed to treat bond breaking at a crack tip in terms of a single non-linear equation, by eliminating all the internal variables of the surrounding lattice. It seems desirable to do the same thing for situations with more complicated crystal structure and more extensive non-linear zone. To achieve such a model requires a rather complex set of preparatory calculations, computing approximate Greens functions, and assembling coupling matrices. One point of view may certainly be that it is much simpler to use a crude rigid boundary model, employing a large region I of several thousand atoms to avoid undue constraint from the boundaries. The advantages of reducing the system to a few tens of coordinates, at the cost of considerable analytic and computational overheads, probably only appear when it is desired to study a system with a number of metastable states, and to find the energetics of transitions from one to another. Such a system is provided by the problem of crack tip behaviour where alternatives such as blunting and cleavage are possible.

The name coined for the method indicates the use of Greens functions, and there is a very strong connection with the lattice statics method. As with all modelling methods, approximations are involved. Firstly, there is an idealisation of the defective crystal as having anharmonicity in its potential energy function only through non-linear interatomic bonds lying in a limited zone. Secondly, the so-called Greens functions used are recognised not to relate to any explicitly identifiable linear lattice, but are assumed to be usable (in a manner to be explained) for generating equilibrium configurations of that part of the crystal which is harmonic.

The method consists of two parts: a general formalism applicable to all defect systems, and approximate Greens function calculation specific to each system. To describe the general formalism, we use the same notation for regions as before, and suppose that region I (the only one containing any anharmonic interactions) is described by coordinates x_1 to x_{N_1}. Region II will extend to x_{N_2}, while the remainder will have a nominally infinite number of coordinates. In accordance with the hypothesis of spatially limited anharmonicity, the potential energy of the crystal will take the form

$$V(x_1,..x_\infty) = V^{\mathrm{I}}(x_1,..x_{N_1}) + \tfrac{1}{2}\sum_{i,j=1}^{\infty} u_i \Phi_{ij} u_j, \tag{1}$$

where

$$u_i = x_i - x_i^o,$$

and x_i^o are 'initial' coordinates which give equilibrium in regions II and III. These would include fixed displacements appropriate to the defect, such as the basic '$K\sqrt{r}$' field for a loaded crack, and would probably be supplied from continuum theory. Corresponding to (1), the forces in any configuration are given by

$$f_i \equiv -\partial V/\partial x_i = f_i^{\mathrm{I}}(x_1,..x_{N_1}) - \sum_{j=1}^{\infty} \Phi_{ij} u_j, \tag{2}$$

where

$$f_i^{\mathrm{I}} = -\partial V^{\mathrm{I}}/\partial x_i, \tag{3}$$

and is therefore zero for $i > N_1$.

Now suppose that some quantities G_{ij} are available, which we shall refer to as 'approximate Greens functions'. The idea is that these provide a set of displacement fields (for fixed j) which produce equilibrium in the harmonic regions, and roughly so in region I, except at the 'source' position (where $i = j$). Actually, we shall not even require exact equilibrium in regions II and III. Formally, suppose

$$\sum_{j=1}^{\infty} \Phi_{ij} G_{jk} = \delta_{ik} + \varepsilon_{ik}, \tag{4}$$

where δ_{ik} is the Kronecker delta, and

$$\varepsilon_{ik} \ll 1 \quad \text{for } i > N_1. \tag{5}$$

A pictorial way to express this is to imagine a fully harmonic reference lattice with interatomic force constants Φ^o_{ij} such that the G_{jk} were exact Greens functions (Figure 7). Any Φ^o_{ij} involving atoms in regions II or III must be very similar to the corresponding Φ_{ij}. Then one could apply a set of forces f^*_i to region I of the reference lattice and the G_{jk} would then supply the displacements everywhere to give equilibrium:

$$u_i = \sum_{j=1}^{N_1} G_{ij} f^*_j \, . \tag{6}$$

Because of (4) and (5), this same set of displacements would produce near equilibrium in regions II and III of the 'real' lattice, but not in region I for general $\mathbf{f}^*$. Using the potential (1), one could adjust the $\mathbf{f}^*$ until equilibrium was achieved in region I of the 'real' lattice, and regions II and III would automatically also be in equilibrium. More convenient, however, is to take the region I coordinates as the independent variables, not the $\mathbf{f}^*$. To do this, define $\mathbf{H}$ as the inverse of the first $N_1 \times N_1$ block of $\mathbf{G}$:

$$\sum_{j=1}^{N_1} H_{ij} G_{jk} = \delta_{ik} : \qquad i \leqslant N_1, \;\; k \leqslant N_1. \tag{7}$$

Then

$$f^*_i = \sum_{j=1}^{N_1} H_{ij} u_j , \tag{8}$$

and (6) and (8) can be combined to give the displacements in regions II and III which will give equilibrium there for an arbitrarily chosen set of region I positions. Combining with (2), the resultant forces appearing in region I are

$$f_i = f^{\mathrm{I}}_i (x_1 , ... x_{N_1}) - \sum_{j=1}^{N_1} \Psi_{ij} u_j : \quad i \leqslant N_1, \tag{9}$$

where

$$\Psi_{ij} = \Phi_{ij} + \sum_{k=N_1+1}^{N_2} \Phi_{ik} \sum_{m=1}^{N_1} G_{km} H_{mj} \, . \tag{10}$$

The first sum in (10) extends only to N_2 because of the finite range of the Φ_{ik} (which defines the extent of region II). The forces in regions II and III are not precisely zero:

$$f_i = -\sum_{k=1}^{N_1} \varepsilon_{ik} \sum_{j=1}^{N_1} H_{kj} u_j: \quad i > N_1, \tag{11}$$

but are, with the ε_{ik}, small by supposition.

Equation (9) expresses the desired result: we now have a small set of coordinates x_1 to x_{N_1} implicitly determining the configuration of the whole crystal, and the forces f_i relate to the x_i via the non-linear functions f_i^{I}, and a set of effective spring constants Ψ_{ij}. The latter express the interactions mediated by a compliant lattice which moves around (maintaining internal equilibrium) as the trial configuration of the region I is adjusted to bring the forces f_i to zero. It can be shown that the Ψ_{ij} as defined do not form an exactly symmetric matrix, but that small asymmetries exist proportional to the ε_{ik}. At a consistent level of approximation, then, $\boldsymbol{\Psi}$ can be forcibly symmetrized, and an effective potential obtained which relates to (9) just as (1) does to (2), but with a finite $(N_1 \times N_1)$ sum. The equilibration is then a variational problem, and energy minima or saddle points may be sought conventionally, using for instance conjugate gradient methods.

The task of generating approximate Greens functions G_{ij} will have strong dependence upon the problem being studied. With point defects, the perfect lattice Greens function could be used, and the method then becomes a particular formulation of Lattice Statics. Dislocations, which introduce ambiguities in the relative indices of two atoms, even at long range, prevent the lattice Greens function being used directly. However, continuum solutions are generally useful for generating equilibrium displacements of a lattice, provided the field is slowly varying and small. Thus, the continuum elastic Greens function for an infinite body will be adequate for a dislocated crystal, except near the dislocation core, and for very close source and field positions. The Flex–G model does not require the G_{ij} to be even approximate equilibrium fields in region I, so any kind of interpolation of the continuum functions will do there. It is only for field locations in region II which are close to a source position in region I that accurate, short-range lattice Greens function values are required, and in these circumstances, the distortions of the initially estimated dislocated configuration are sufficiently small that a local correspondence to a perfect lattice can unambiguously be made, and the perfect lattice Greens function components used.

For a crack model, the same kind of approach can be used, but there are more complications. In a two dimensional model, periodicity is maintained in one direction. The elementary source is a series of forces on equivalent atoms, and the continuum approximation is a line of body force, lying parallel to the crack front. The solution for this problem (Sinclair & Hirth 1975) can be written as the Greens function for the infinite, uncracked body, plus 'image' terms. When the source and field positions are far apart, and each is far from the crack surfaces, this continuum solutions can be used direct. For many

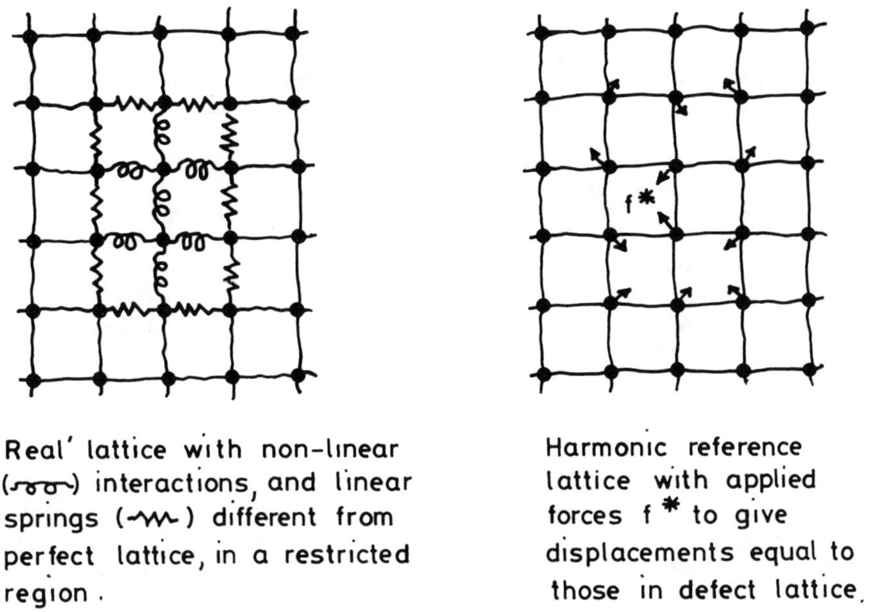

Figure 7. Real and reference lattices in Flex–G method.

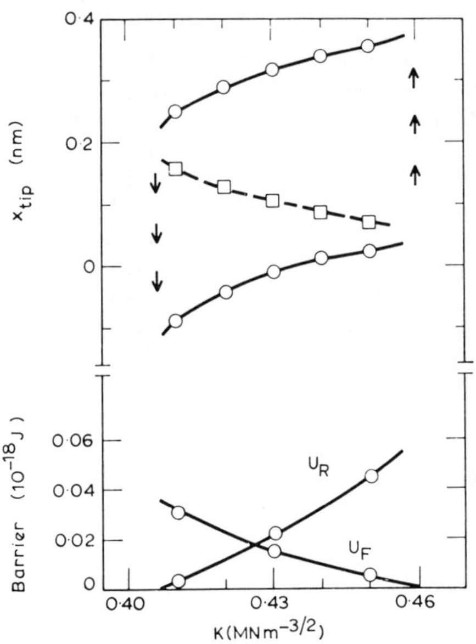

Figure 8. Lattice trapping study in silicon (Sinclair 1975).

combinations of nearby source and field positions, the infinite-body part can be replaced by the perfect lattice Greens function, just as in the dislocation case, while the image term is a slowly varying field, and the continuum version is adequate. If the image term is rapidly varying, either the source or field position must be close to a surface, and it is now possible to use an alternative splitting of the continuum result into the half-space Greens function plus correction terms. Now, a lattice version of the half-space part can be substituted, and the correction terms give no problems. Thus, although a quite complicated steps is required to form the required G_{ij} values, the result can be stored and used for a number of related problems. In particular, a range of interatomic potentials can be studied which differ only in their anharmonic parts.

5. ATOMISTIC CRACK MODELS – RESULTS

Space is insufficient for a complete review of past work in this field. In the realm of simple analogue models, the paper by Fuller et al. (1980) already mentioned is worth reading. Smith (1977) has also used similar models. These, and the works using lattice statics already cited, have concentrated on the subject of lattice trapping, analogous to the Peierls–Nabarro barrier for a dislocation. When an unrealistic force law is used, such as a suddenly snapping linear spring, gross overestimates of the trapping effect result. Whether the use of a simple square lattice misrepresents the effect in realistic three-dimensional lattices is not yet known. The model of Sinclair (1975) combines realistic structure (in the diamond class) with plausible interatomic potential forms (a variety is tested). This work found the lattice trapping effect to be modest: the ratio of maximum to minimum stress intensity factor within which the crack was meta-stable against propagation ranged from 1.1 to 1.4, depending upon the potential. This was for a straight crack front aligned with a low-index crystallographic direction. When the front was forced to lie in a high-index direction, a series of like-signed kinks formed, and the lattice trapping was greatly reduced. The calculations of Sinclair (1975) were unique in using saddle point location to determine the energy barriers for crack propagation or closure as a function of load (Figure 8). The use of a flexible boundary method (Flex–S) was essential for this study. The importance of kinks is confirmed by Esterling (1976), who used a version of lattice statics in 3 dimensions.

Models of cracks in metallic lattices have mostly used 2 or 3 region atomic/continuum models. For some reason, these have been dominated by representations of bcc iron, using the empirical pair potential of Johnson (1964). See for instance Gehlen (1973), Markworth et al. (1981) and Mullins & Dokainish (1982). With plastic deformation now a distinct possibility, the range of possible tip structures and processes is potentially much more complex. Shear deformations have certainly been observed, but not systematically studied. The Johnson potential is not really very suitable for this kind of study. It is very short ranged (1st and 2nd neighbours only), to the point that in some circumstances (Mullins & Dokainish 1982) the state of bi-axial tension around the tip is large enough to cause the second neighbour interactions to be completely lost, and a lattice structural transformation occurs, the bcc lattice having lost stability (Figure 9).

Although such exploratory modelling in large atomistic models has its value, it would seem to be much more attractive to study a system with rather fewer degrees of freedom, so that the competition between different processes, such as shear vs. tensile deformation steps, could be more clearly studied. It is for such an approach that the Flex–G model was developed, and work is in progress using it to find the conditions which determine whether sliding (with dislocation formation), or bond breaking in cleavage mode, occur in metal-like lattices. So far, only the bcc lattice with (001) crack plane has been studied, using short ranged pair potentials similar to that of Johnson. Crack front orientations of [010] and [110] have been used.

A first step is to test the stability of the perfectly sharp crack. This is a function of applied load (expressed as up to three stress intensity factors K_M for the three modes of loading). Continuum theory (anisotropic, in 2 dimensions) is used to generate the initial positions from which the u's in eq. (9) are defined, and can be continued right to the tip to provide a starting point for the relaxation. Figure 10 shows a relaxed sharp tip configuration.

If the flexible boundary system worked perfectly, the crack tip could reside at any of a number of positions related by the lattice repeat distance, and the atomic configurations would be exactly equivalent by translation. To test this in practice, it is useful to have the facility of artificially forcing selected atom pairs to have zero interaction. If this is used to cut the last unbroken bond, and the relaxation repeated, the next equivalent configuration should result, so that at the end, the artificial bond breaking is not required. Having thus found two adjacent stable tip positions, an intermediate saddle point can be sought using the Sinclair & Fletcher (1974) conjugate gradient algorithm. The energy difference between the two stable configurations should give the energy release rate, and the differences between the saddle point and the two minima give the barriers to forward and reverse propagation (Figure 11). To test the effectiveness of the flexible boundary system, this process may be repeated for the series of tip positions across the model. If the departure from a straight lines connecting the minima and the saddle points is small compared with the space between the two lines, then the interference is small compared with the lattice trapping effect. Such was found to be the case with rather small models (Figure 10 is the whole of region I).

Note that it is *not* possible to produce cleavage extension through one atomic spacing to the right simply by gradual increase of the applied K_I and re-relaxation. At any K_I within the lattice trapping range the tip should be 'happy' to reside anywhere across the model. Above the upper limit of the trapping range, there should be no equilibrium possible. Likewise, it is not possible to explore the

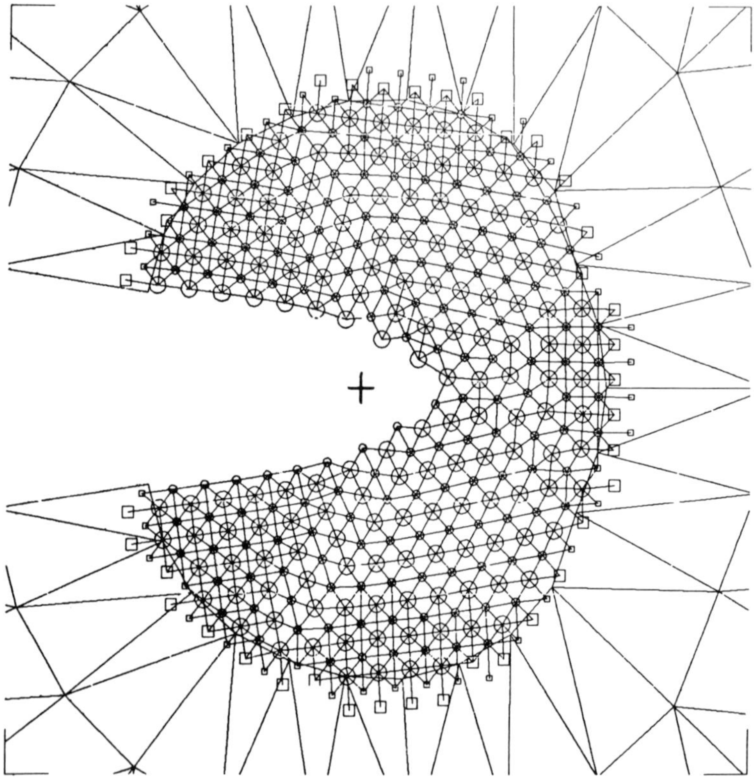

Figure 9. After Mullins & Dokainish (1982).

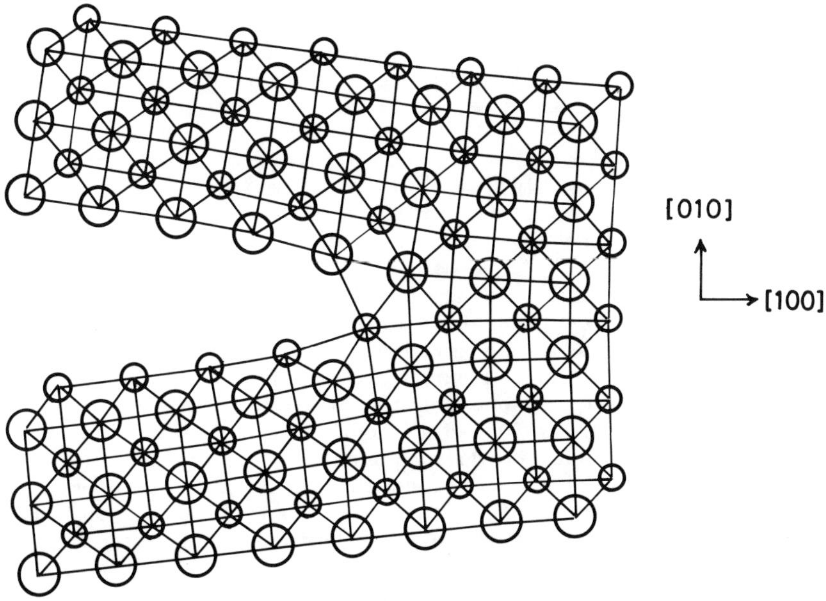

Figure 10. Stable sharp tip configuration.

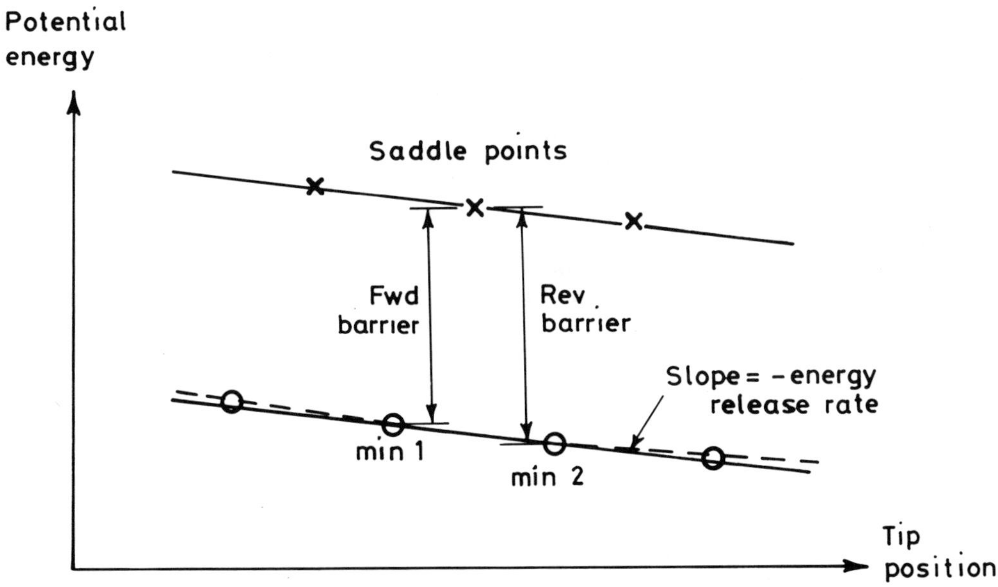

Figure 11. Illustrating derivation of lattice trapping barriers from energy minima and saddle points.

possibilities for shear deformation (dislocation generation) simply by trying a range of K's (single or mixed mode) in the initial conditions. Ideally, something akin to the trick of artificial bond breaking to force cleavage extension is required to force dislocations to appear. So far, however, such a scheme has not been devised. Rather, the continuum solution for the displacements around a dislocation near a crack tip (Sinclair & Hirth 1975) have used to generate initial atom positions. Then, trial and error was used to find combinations of the K_M and initial dislocation position such that the dislocation remained stable after relaxation. This is illustrated in Figures 12 to 15. With initial positions as in Figure 12, relaxation led to the configuration of Figure 10 again. That is, the dislocation annihilated at the surface. With the initial state of Figure 13, the dislocation ran during relaxation away from the tip (Figure 14), to be arrested only by the model boundary, beyond which perfect-lattice bonding is forced (linear springs). Note that the adjustments during relaxation do *not* correspond to any physical motion, and the occurrence of such changes as those just described does not conclusively prove that stable dislocation configurations do not exist. Figure 15 shows a stable relaxed dislocation position found for a particular choice of mixed mode loading.

What is called for, but has not yet been completed, is a thorough search over (K_I,K_{II},K_{III}) space for stable dislocation positions as close as possible to the tip, and for each to find (by saddle point location) the energy barrier between this state and that with a sharp tip. Comparison with the barrier for cleavage extension without blunting could then be made. A further test could be for the barrier to cleavage propagation from the blunted state. Figure 16 illustrates that from the loading and blunting state of Figure 15, cleavage through a single bond is not possible: artificial breaking of the bond leads to a relaxed state in which the two atoms have not moved apart beyond the range of the potential. This means that cleavage re-initiation is now harder after blunting (the activation is via breakage of more that one bond). What is now needed is a comparison of the relative ease of such a process and the alternative that the dislocation moves further from the tip, possibly to be followed by further blunting dislocations. By such investigations, it is hoped to address the fascinating question whether a mixture of alternating blunting and cleavage steps can be predicted, and thus to throw light on the process of brittle cleavage in metals, which occurs to the accompaniment of considerable dislocation activity.

REFERENCES

Daw, M S, & Baskes, M I, 1983, Phys. Rev. Letters **50,** 1285.

Esterling, D M, 1976, J. Appl. Phys. **47,** 486.

Finnis, M W, & Sinclair, J E, 1984, Phil. Mag. in press (AERE Harwell report TP1030).

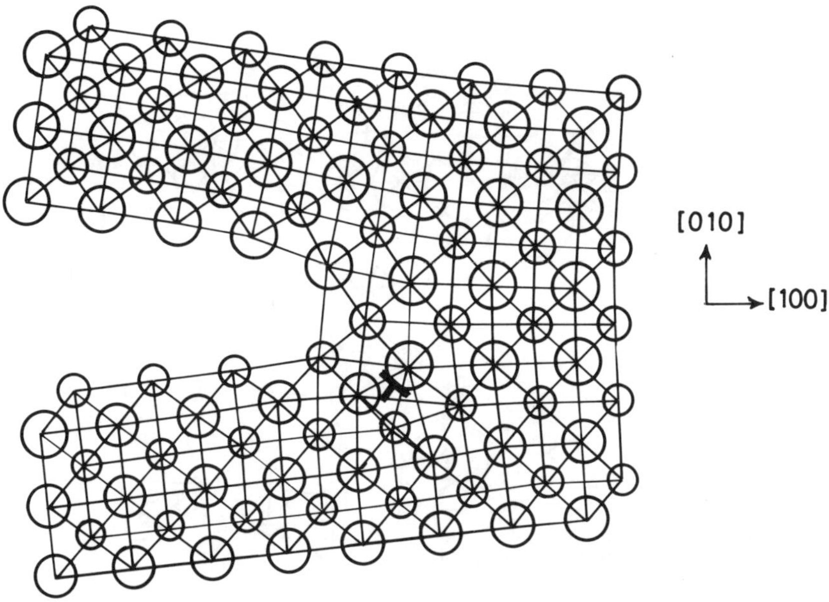

Figure 12. A trial dislocated configuration before relaxation.

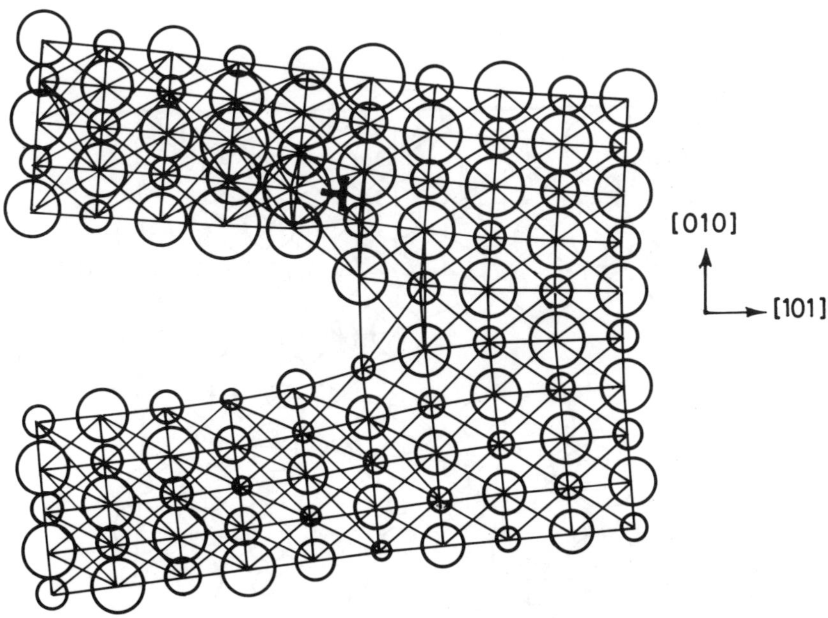

Figure 13. Similar to figure 12, with different crystallographic orientation.

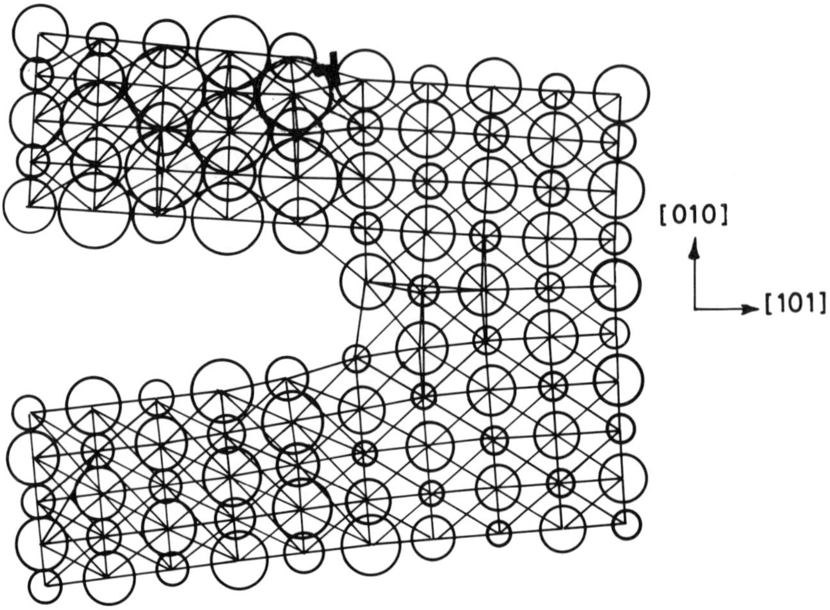

Figure 14. Result of relaxing configuration of figure 13.

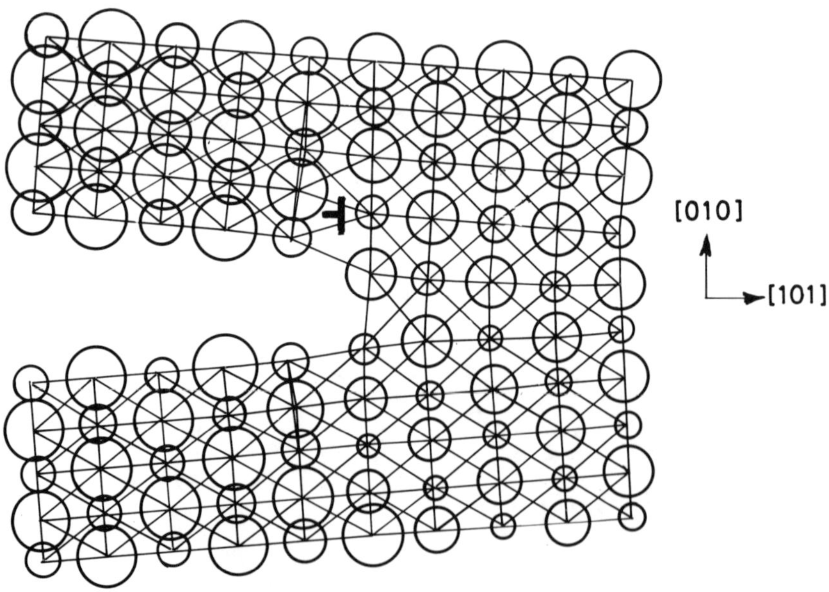

Figure 15. A relaxed configuration with dislocation stable.

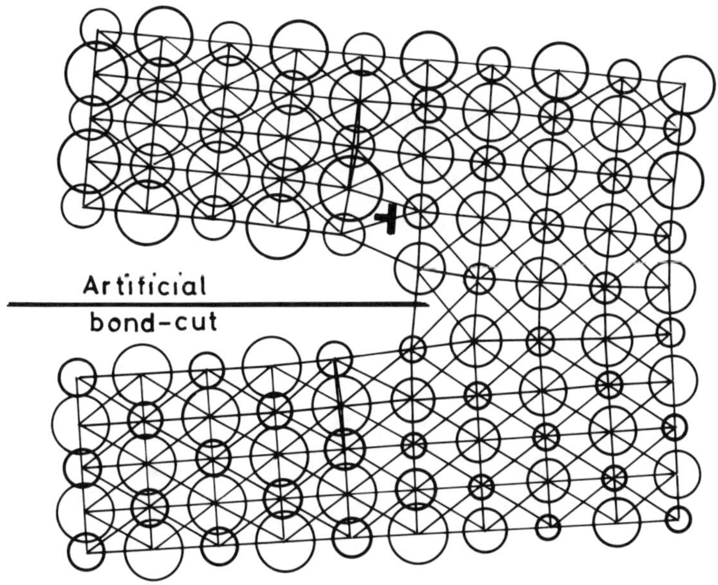

Figure 16. With the applied loads as in figure 15, no stable configuration can be found with just one further bond broken.

Fuller, E R, & Thomson, R M, 1977, in *Fracture 1977,* vol. 3, ed. D M R Taplin, Univ. Waterloo Press, p. 387.

Fuller, E R, & Thomson, R M, 1978, in *Fracture Mechanics of Ceramics,* vol. 4, ed. R Bradt, D P H Hasselman & F F Lange, Plenum, New York, p. 507.

Fuller, E R, Lawn, B R, & Thomson, R M, 1980, Acta Met. **28,** 1407.

Gehlen, P C, 1973, Scripta metall. **7,** 1115.

Hirth, J P, & Lothe, J, 1968, *Theory of Dislocations,* McGraw Hill, New York.

Hardy, J R, & Bullough, R, 1967, Phil. Mag. **15,** 237; Phil. Mag. **16,** 405.

Hoagland, R G, 1973, PhD Thesis, Ohio State University.

Hsieh, C, & Thomson, R M, 1973, J. Appl. Phys. **44,** 2051.

Johnson, R A, 1964, Phys. Rev. **134,** A1329.

Kanninen, M F, & Gehlen, P C, 1971, Int. J. Fracture Mech. **7,** 471.

Kanzaki, H, 1957, Phys. Chem. Solids **2,** 24.

Markworth, A J, Kahn, L R, Gehlen, P C, & Hahn, G T, 1981, Res Mechanica **2,** 141.

Mullins, M, & Dokainish, M A, 1982, Phil. Mag. **A46,** 771.

Sinclair, J E, 1971, J. Appl. Phys. **42,** 5321.

Sinclair, J E, 1972, J. Phys. C: Solid State Phys. **5,** L271.

Sinclair, J E, 1975, Phil. Mag. **31,** 647.

Sinclair, J E, & Fletcher, R, 1974, J. Phys. C: Solid State Phys. **7,** 864.

Sinclair, J E, Gehlen, P C, Hoagland R G, & Hirth, J P, 1978, J. Appl. Phys. **49,** 3890.

Sinclair, J E, & Hirth, J P, 1975, J. Phys. F: Metal Phys. **5,** 236.

Sinclair, J E, & Lawn, B R, 1972, Proc. R. Soc. **A329,** 83.

Smith, E, 1977, in *Fracture 1977,* ed. D M R Taplin, University of Waterloo Press, vol. 4, p. 65.

Tewary, V K, 1973, Adv. Phys. **22,** 757.

Thomson, R M, Hsieh, C, & Rana, V, 1971, J. Appl. Phys. **42,** 3154.

DEFECTS IN AMORPHOUS SOLIDS: A POSSIBLE APPROACH

Marco Ronchetti
Dipartimento di Fisica
Universita' Degli Studi Di Trento
38050 Povo di Trento - Italy

ABSTRACT. Are amorphous systems disordered by definition? The idea of some kind of order in these systems is here discussed, and evidence of orientational order in supercooled liquids is presented. The notion of defects in amorphous systems is then introduced in two possible ways: the first one is to define them as deviations from the order. The second one is to try to extend to amorphous systems the idea of line and point defects of crystals. An investigation by computer simulation of the existence of vacancies and dislocations is reported. Possibility and limits of the application of computer simulation (Molecular Dynamics) are also discussed.

1. Introduction.

Equilibrium solid state of the matter has the structure of nearly-perfect crystal, tought we know many examples of non-crystalline amorphous solids, also called "glasses". The familiar glasses are silicates (compounds of silicon and oxigen) that did not crystallize when cooled from the liquid state to room temperature. For a long time it was tought that most of the other substancies could only solidify in their ordered crystalline state. Cohen and Turnbull (1959) predicted that liquids of even the simpest structure would became glasses if sufficiently undercooled. Later on experiments of very fast cooling expanded the class of glass-forming substancies. Today many classes of substances are known to fall in a glassy state when suitable conditions do occour: among these oxydes ($SI-O_2$, B_2-O_3,...), chalcogenide (As-S, Ge-Se,...), halogenures ($Be-F_2$...), Metal-Metalloide alloys (Pd-Si, Fe-B...), Metal-Metal alloys (Ni-Nb, Cu-Zn...), organic polymers and many others. Also some elements are known to vitrify: P, S, Se. Interest in the glassy state of the matter has been growing in the last two decades, mainly because of the (also economically) interesting properties of the metallic glasses. Non-crystalline solids can be obtained following three different main routes:

G. Jacucci (ed.), Computer Simulation in Physical Metallurgy, 129–143.

1) freezing the structural disorder of the liquid state (e.g. splat cooling)
2) using the absence of order of the gases (e.g. by vapour deposition)
3) disordering the solid crystalline state (e.g. by radiation damage).

It is not obvious, and probably it is not true, that the properties of the amorphous states produced in so different ways are the same, even thought many important similarities do exist. The principal characteristic of amorphous systems is the absence of long range order, and the existence of short range order. This property makes the amorphous solids structurally similar to the liquids, even tought important differences do exist, the most striking one being the split of the second peak of the pair correlation function g(r). Many efforts have been devoted to modelling the atomic structure of glasses. This issue (limited to metallic glasses) has been recently reviewed by Gaskell (1983).

2. Computer simulation of glasses.

The earliest simulations were mechanical rather than numerical: a large number of hard spheres (e.g. iron balls) were assembled in an irregular shaped container, and the system was shaked and compressed to achieve maximum density (random close packing). Density and local coordination were studied. Later on a number of algorithms were developed and tested by computer to investigate the properties of dense random packed structures. A review of these computer models is due to Hoare (1978).

A direct procedure to create computer amorphous systems imitates the most diffused way to obtain a glass in real world: fast cooling of a liquid. This can be done via Molecular Dynamics: one starts from an equilibrium liquid configuration, and suddenly drops the temperature. Since temperature is the same as kinetik energy, a system can be cooled (or heated) by rescaling the velocities of the particles (see Appendix). To our knowledge the first reports of "Molecular Dynamics glasses" are those of Rahman et al. (1976) and Damgaard Kristensen (1976). Since then many other works based on this method have been published. A review of the field ("Computer simulation of freezeng and supercooled liquids") has been presented by Frenkel and McTague (1980). An important question can be raised: to what degree do Molecular Dynamics glasses represent real world glasses? An important limitation of computer simulation is the fact that only relatively small systems can be handled. This problem is commonly solved by the use of periodic boundary conditions. Although in this way surface effects are eliminated, some other problems are introduced. A periodic fluid has the symmetry of the boundary conditions (usually cubic). In addition only fluctuation with wavevector $\underline{k}=(2\pi/L)(n_x,n_y,n_z)$ are allowed (L being the side of the cubic box). More, nucleation of crystals having lattice commensurated with the boundary condition are favored: Mandell et al. (1976) observed for a Lennard Jones fluid BCC nucleation instead of FCC. But a more subtle limitation is due to the very short times of the simulations, as we shall discuss in the next paragraph.

3. Observation time and relaxation time.

M.Reiner (1964) pointed out that our first understanding of the glassy state goes back to the Bible. We quote here, with minor changes, a few paragraphs from the 4th International Congress in Rheology. In Judges 5, verse 5, the Prophetess Deborah sang: "The mountains flowed before the Lord". (The authorized English version contains a translation error, and says 'melted' instead of 'flowed'.) Well, Deborah knew two things. First, the mountains do actually flow. Secondly, they flow before the Lord, and not before man, for the simple reason that the time of observation of God is infinitely long, while man's short lifetime does'nt allow him to observe this flow. Reiner therefore defined the dimensionless "Deborah Number" as the ratio between relaxation time and observation time. If your time of observation is very large compared to the time of relaxation of the material under observation (i.e. for little Deborah numbers), you see the material flowing. On the other hand if the Deborah number is very large then the material, for all practical purpooses, is a solid.
These considerations have a straightforward application to the studies of the amorphous state. Observation time during a computer experiment is at most a few nanoseconds, while in real world it ranges from one second to several days, i.e. at least 10^8 to 10^{13} times larger. Therefore what in computer simulation behaves like a solid has to be considered in real world as a viscous liquid. Other facts have to be remarked. Properties of real glasses are known to depend upon the history of the sample and in particular upon the cooling rate. (Real) metallic glasses are produced by cooling up to 10^6 K per second, while cooling rates in simulated systems range from 10^{12} to 10^{14} K/sec. Properties of computer glasses are therefore expected to differ from those of real glasses. Angell, Clarke and Woodcock (1980) observed that "vitrification in computer simulations occurs under conditions in which the usually sharp glass transition is smeared out almost beyond recognition". In fact it turns out to be very difficult to determine the glass transition temperature, and so different criteria were proposed, yelding different points (see Frenkel and McTague 1980).
So far we have underlined some of the difficulties found in computer simulation of the glassy state. From now on we shall be more positive, and show that simulation can be an useful tool indeed.

4. Transport properties and defects.

Transport in non-crystalline solids in not as well understood as in crystalline materials. The study of this process is complicated by the fact that glasses are not in thermodynamical equilibrium, and therefore are subjected to structural relaxation. In crystals atomic transport is allowed by the presence and motion of defects. The notion of defects is linked to the existence of a "perfect" reference system: in crystals this is obviously given by the regular lattice. A "perfect amorphous systems" is difficult to imagine. It is however possible to define structures which are amorphous (in the sense that they do not

present translational symmetry) but at the same time show a high degree of configurational order. One example would be a tetrahedrally coordinated random network. In such a system a defect may be individuated by a dangling bond. An example of a system showing orientational order will be discussed in the next paragraph.
After this, we shall describe one possible approach to the study of defects in amorphous solids. It is due to Praveen Chaudhari and co-workers, and is based on the idea to extend the notion of usual defects like vacancy and dislocation to the amorphous system, to introduce them in a sample and to study stability, motion and properties of these entities. This is neither the only nor the best possible approach, but we believe it is interesting and worth to be discussed.
We mention here two other ways to look at the problem of defects and diffusion. One is due to Ahmazadeh and Cantor (1981): it considers the interstitial diffusion of small particles in dense random packed metallic (DRP) glasses. The interstitial sites are defined in term of the Bernal holes and other polyhedra found in DRP structures. The other is due to Egami et al.(1980,1982): existing models are analyzed in term of internal stresses. A defect is then a region where either hydrostatic pressure, hydrostatic tension or internal stresses predominate.
A review of current understanding of atomic transport is due to Chaudhari, Spaepen and Steinhardt (1981).

5. Bond orientational order.

To show that also apparently disordered systems may present some degree of order, we report here the results of Steinhardt, Nelson and Ronchetti (1981,1983). Let's define a "bond" as the ideal joining between an atom and its nearest neighbours. We can associate with each bond a set of order parameters $Q_{lm}(r)$ given by the spherical harmonics associated with the orientation of the bond relative to a given coordinate system. In an isotropic liquid, one would expect that all the Q_{lm} (exept Q_{00}) vanish when averaged over r. Systems with cubic rotational simmetry are characterized by $\langle Q_{4m}\rangle=0$ (the brackets indicating average over the sample). In icosahedral clusters the first non-zero signal is observed for $\langle Q_{6m}\rangle$. Extended orientational symmetry can be examined by evaluating the bond angle correlation function

$$G_l(\underline{R}) = (1/(2l+1)) \sum_{m=-l}^{l} \langle Q^*_{lm}(\underline{R})\, Q_{lm}(\underline{0}) \rangle / \langle Q^*_{00}(\underline{R})\, Q_{00}(\underline{0}) \rangle$$

Non zero asymptotic values for some l show the existence of long range orientational order. For zero asymptotic values, a fit of the exponential decay gives an orientational correlation lenght ξ_l.
Bond orientational distibution has been analyzed in samples produced by Molecular Dynamics. Systems made of 600 and 864 particles subjected to periodic boundary conditions and interacting via a Lennard Jones potential have been used. Starting from an equilibrium liquid several models were obtained at various temperatures by rapid quenching.

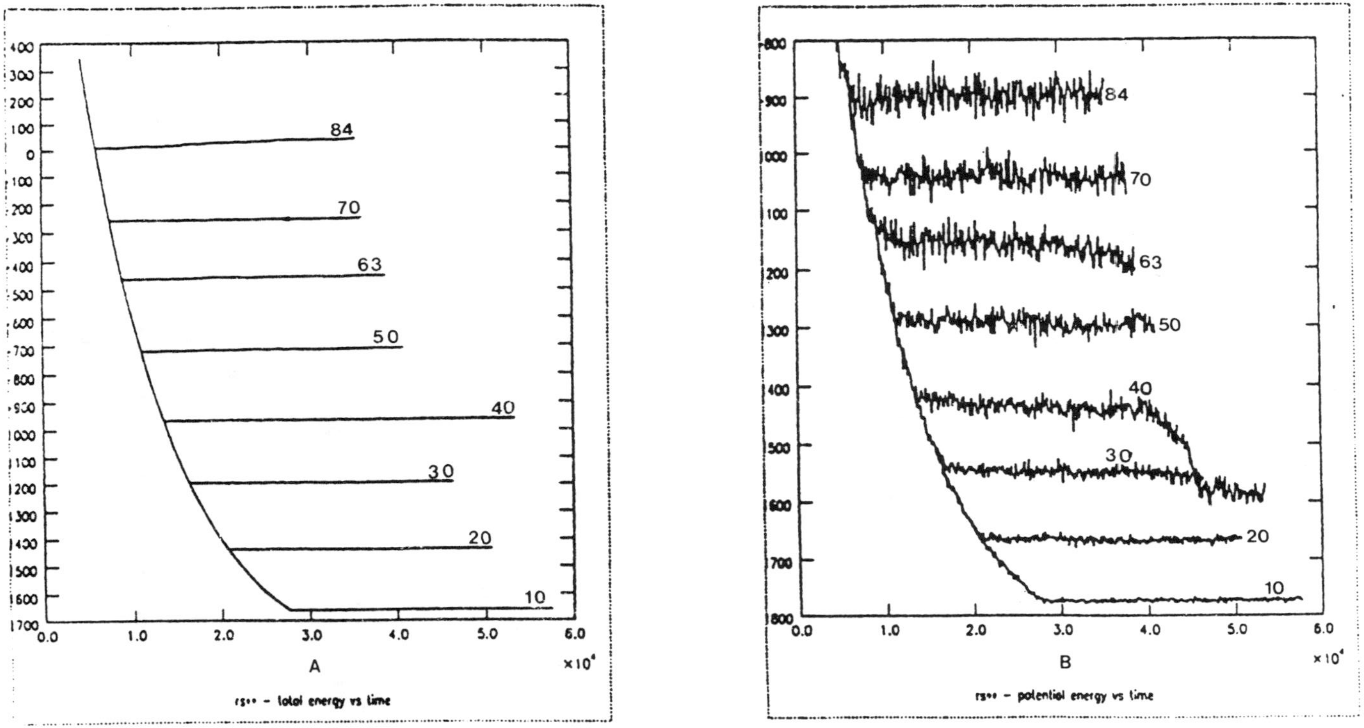

FIG 1. Molecular dynamics quenching at constant density of a Lennard Jones system of 864 particles, parameters for Argon and density 1.6 g/cm^3. Cooling rate was approximately 10^{12} K/sec. A) Total energy vs. time. Energy is subtracted until the desired temperature has been reached, then the system is relaxed for 30000 time steps. This operation has been performed for 8 different temperatures (10 to 84 K). The values of the temperature (in Kelvin) are indicated for every branch. B) Potential energy vs. time. Thermal fluctuations are clearly observable. The sudden drop in the 40 K branch is due to cristallisation of the sample. Energy in arbitrary units, time in program units: 10^{-14} sec

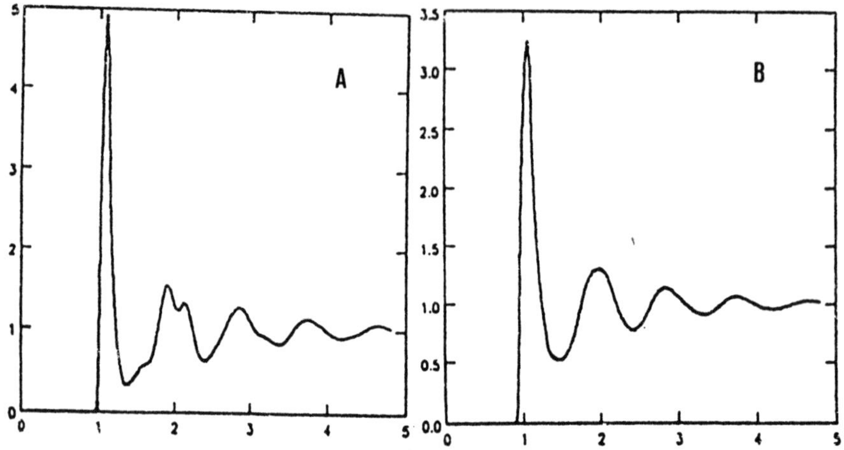

FIG 2. Pair correlation function g(r) of the system described in Fig.1:
A) quenched amorphous at 30 K. Note the splitting of the second peak: a signature of the amorphous state.
B) equilibrium liquid at 120 K (the melting point is at 84 K).

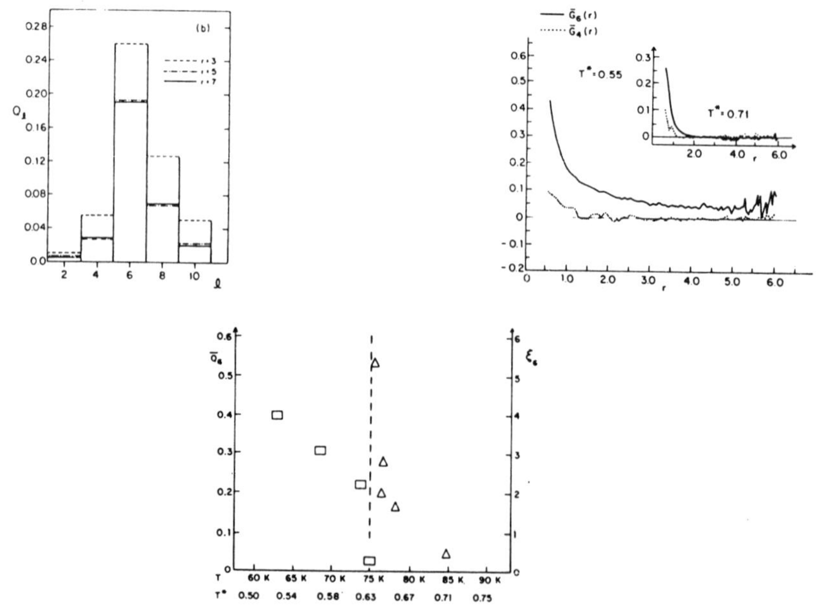

FIG 3. Evidence of bond orientational order (after Steinhardt, Nelson and Ronchetti, 1981)
A) Histograms of bond orientational order parameters.
B) Bond-angle correlation functions $G_l(r)$ at two temperatures.
C) Behaviour of asymptotic value Q of the $G_l(r)$, and of the correlation lenght ξ.

Each system was then evolved for 30000 time steps (300 picoseconds) (Fig. 1 and 2) and the final configurations were analyzed. As expected, equilibrium liquid presented all Q_{lm} equal to zero within the noise. Samples at lower temperature (supercooled systems) showed enhanced signals for l=6. Bond angle correlation function $G_l(r)$ approached non-zero constans for large r at every temperature below 75 K (melting point being 84 K) for l=6, while for values of l lower then 6 the asymptotic value was zero. Orientational correlation lenght was observed to increase by lowering the temperature, and to diverge at T=75 (see fig. 3).These facts suggested that at T=75 K the system drops in a state characterized by short range orientational order, broken icosahedral symmetry and extended correlations in the orientation of local icosahedral packing units. Some kind of order has therefore been discovered in an apparently disordered system.

6. Point defects.

At least three classes of point defects are known in crystals: the vacancy (i.e. an empty lattice position), the interstitial (i.e. an extra atom not linked to the lattice), and the substitutional atom (i.e. a particle of a different chemical specie). While the third kind of defect can (in principle) be eliminated by using more and more pure materials, the first two are thermodynamically favored in crystals at finite temperature. In fact the free energy of the crystal has a minimum in corrispondence to some concentration of vacancies and interstitials, due to their entropic contribution. The definition of these structural defects is linked to the lattice periodicity, or at least to some reference state. The absence of these conditions typical of the amorphous solids precludes the usual definitions. In spite of this fact, one can re-introduce the concept of vacancy referring to suitable sized empty regions (holes) in the materials. In computer models these region can be introduced by removing one or more particles, in the very same way used in crystals. Mechanic and thermodynamic stability of these holes can be then checked by rispectively static and dynamic relaxation, the first one consisting of some method to locate the nearest local minimum of potential energy, the second by suitable sampling the phase space via either Monte Carlo or Molecular Dynamics. This work has been done by Bennett, Chaudhari, Moruzzi and Steinhardt (1979) for different amorphous models: a Lennard Jones glass and a covalentely bonded system. We shall describe their methods and results.
The L.J. glass consisted of 500 atoms in a cubic box whithin periodic boundary conditions. It was obtained by freezing via Molecular Dynamics an equilibrium liquid. Static relaxation was imposed as follows: periodic boundary was removed, and the system relaxed toward the nearest energy minimum (at zero temperature) by a conjugate gradient method (Fletcher and Powell 1963). Relaxation was stopped when the difference in energy between two iteration was of 1 part in one million. Three particles were then removed, and the sample relaxed again.

Dynamic relaxation instead was obtained by molecular dynamics: one to five atoms were removed at finite temperature from the box. The evolution of the system (again within periodic boundary conditions) was followed for approximatly 2 picoseconds (roughly 200 integration steps) at various temperature (6, 12 and 24 K) and at two different densities (1.6 and 1.75 g/cm^3). Two methods to analise the vacant volume evolution were devised. In the first one the sample was sliced at various heights, and the particles within the slice were plotted at various times (dynamic model) or before and after relaxation (static model). In this way it was possible to follow the motion of the particles, and see how the hole evolved. The second method was less subjective: the whole box was subdivided in many equal cubic cells which side was much less then the atomic radius. For each cell the distance d from the nearest particle was computed. Obviously, a vacancy consist of a cluster of cells having a value of d near to an atomic diameter. The presence of a vacancy can therefore be checked by studing the distribution of cells having d larger then a critical distance. Both methods presented evidence of the instability of the vacancies in either static and dynamic relaxations (see figure 4). Molecular Dynamics showed that increasing the temperature lowers the anneal time, and the same happens decreasing the density (and therefore the pressure). At the lower tested density the pressure was negative (pV/NkT=-6.7 at T=12 K), but this was not sufficient to stabilize the vacancy.
The covalently bonded system was a Polk (1971) model: 519 atoms with tetrahedral coordination, relaxed under a Keating potential. This sample was subjected to only to static relaxation, but different conditions were tested: after removing three atoms, the system was relaxed again under Keating potential with and without bending force. Since the results of these relaxations are very similar, and the vacant volume survives in both cases, it is concluded that vacancies are stabilized in random network models by topological constraints. The same sample, relaxed under L.J. potential, collapsed to a structure characteristic of metallic packing, and the vacancy disappares. This feature looks important because materials forming covalently bonded glasses and metallic liquids, like Ge or Si, should present such a network collapse.

7. Line defects: dislocations.

The presence of a dislocation in a crystal can easily be detected by looking at the atomic positions. It would be misleading to deduce from this fact that dislocations are possible only with reference to a regular state. It is in fact sufficient to recall that dislocations are defined and used in elasticity theory of continous media. Therefore it should not be too surprising the extention of the dislocation concept to amorphous systems. We shall report here the work done by P.Chaudhari, P.J.Steinhardt and co-workers at the IBM T.J.Watson Laboratory. They investigated the static (with Levi) and dynamic (with Ronchetti) stability of dislocations in amorphous solids, and more recently the dislocation motion (with Shi).

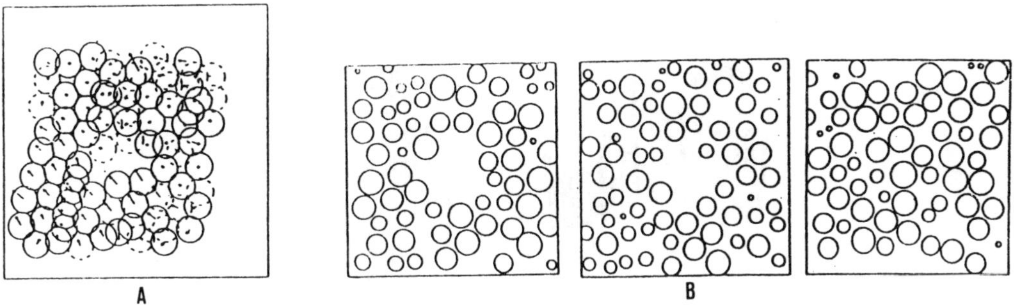

FIG 4. Relaxation of vacancy in an amorphous solid. (After Bennett et al. 1979)
A) Static relaxation. The slice of the sample containing the vacancy is considered. The lines show the motion of the atoms during the relaxation. Dashed circles indicate atoms which were outside the slice before the relaxation, and inside at the end.
B) Dynamic relaxation: plot of the atoms within a slice at three different times. The vacancy is present at the first time, and annealed at the end.

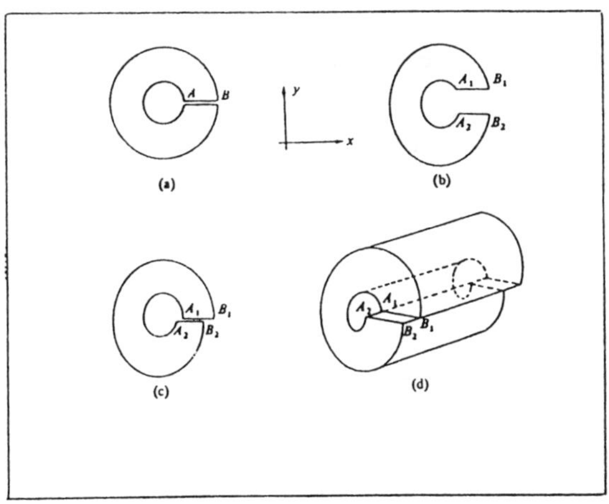

FIG 5. Introduction af a dislocation. Procedure B and C produce an edge dislocation, procedure D gives a screw dislocation.
A) A cylinder-shaped model is cut along the line A-B (top view).
B) Desplacement along the direction orthogonal to the plane defined by the cut.
C) Desplacement in the cut plane, along the radial direction.
D) Desplacement in the cut plane along the direction orthogonal to the cut.

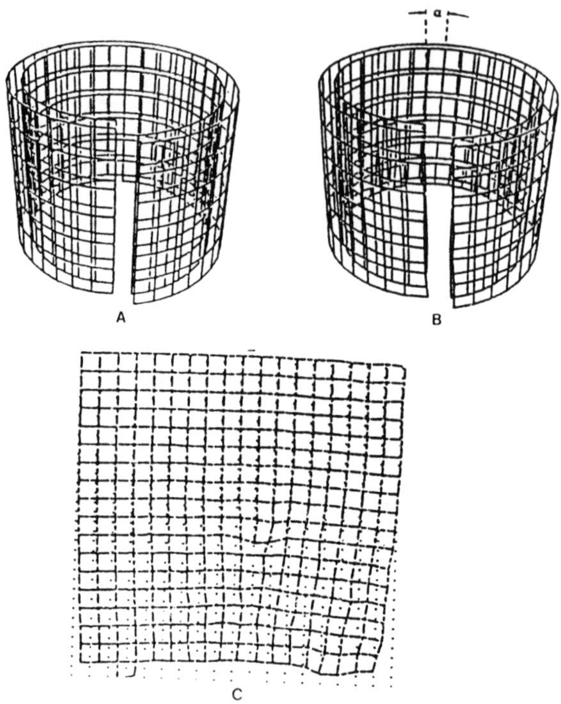

FIG 6. Reference net for a dislocation in an amorphous solid. (After Chaudhari, Levi and Steinhardt, 1979)
A) Screw dislocation before relaxation.
B) Screw dislocation after relaxation.
C) Edge dislocation: before (points) and after relaxation (lines).

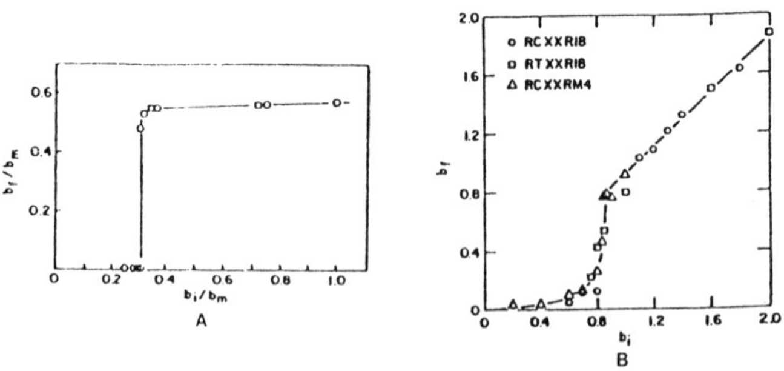

FIG 7. Burgers vector of a screw dislocation: final value (b_f) vs. initial value (b_i). (After Chaudhari and Steinhardt, 1982)
A) In a crystal.
B) In an amorphous.

The study of dislocations in amorphous solids breaks down in four steps:

1) introducing the dislocation
2) relaxing the system
3) detecting the dislocation after relaxation
4) quantifying the elastic properties of the dislocation.

Methods of introducing a dislocation in continous media has been described by Volterra, and they are reported in many treatises of elasticity. The same principles can be applied to amorphous sistems. Edge dislocation may be introduced by removing a strip of particles from the sample and melding.Screw dislocation is obtained by displacing two parts of the model on either side of a plane that cuts halfway through the solid. Introduction of edge and screw dislocation is shown in figure 5. The models used were a box of approximatly 34x69x69 Angstrom containing 4193 particles for the edge dislocation, and a cylinder of 5390 atoms for the screw. The samples were obtained using part of the Finney (1970) model. This is a mechanical dense random packing model composed by roughly 8000 particles.

Static relaxation was performed allowing displacement of the particles toward the nearest local energy minimum with the conjugate gradient method, as in the case of the vacancies. Dynamic relaxation was allowed by Molecular Dynamics, using a smaller sample: 3055 particles interacting via truncated L.J. potential, the cutoff being 1.2 times the L.J. radius.

The evolution of the dislocation has been tracked in several ways. One method made use of a reference net: a regular grid was superimposed on the sample before defect intoduction, and each node of the grid was related to the particles near to it. During the introduction of the dislocation and relaxation the displacement of the nodes was defined as the average displacement of the related particles (see figure 6) This method has the nice feature to visualize the dislocation, and to allow some measurement of interesting quantities, like Burgers vectors and Eshelby twist. This approach was found to be useful for the screw dislocation. Burgers vector was obtained by measuring relaxed displacements along a Burger circuit in the reference net and averaging along the lenght of the cut. The results for a crystal and for an amorphous solid were compared (fig. 7). In both cases exists a critical displacement above which the dislocation is stabilized. In the crystal an increasing screw displacement gives rise to a constant Burgers vector, as expected for a periodic arrangement of atoms. In the glass instead a linear relationship is observed between initial and final displacement if the initial one is larger then the critical value. This behaviour has been explained by observing that at the critical displacement many atoms change their nearest neighbours, and for higher displacements many shallow local minima associate with a less densely packed plane were found by the system. A slightly lower density along the slip plane has in fact been measured.

The reference grid method is not completely satisfactory because of the arbitraryness of the reference. It has also some ambiguity: for instance, in presence of diffusion the nodes may become related to particles that are actually in other regions. Therefore an other less subjective way to detect the presence of dislocations has been developed. It refers to the elastic properties of the defects. A solid in static equilibrium and having properties that are indipendent of one of the coordinates (say z) has only three indipendent components of the stress tensor, and these can be expressed as spatial derivatives of a single scalar potential called Airy stress function (ASF). For continous media the ASF can be found exactly in some special cases. F.C.Frank (1978) showed that also two-dimensional stress problems involving rods and pin-joints can be solved by using ASF. Steinhardt and Chaudhari (1981) applied this method to atomic models. The method was found to be very useful to describe the stress characterising the edge dislocation (fig 8). It was shown that the introduction of an edge dislocation in an amorphous system produces a stress similar to the one founded in dislocated crystals and expected by continuum theory. The edge dislocation was found to be stable under static relaxation.

Dynamic relaxation was allowed at various temperatures. At the highest one (36 K) the dislocation annealed out rapidly. At 12 K the system was evolved for 40 picoseconds, and the configuration were analyzed at every picosecond. This is a very heavy computational task, because thermal noise has to be subtracted in order to evaluate the ASF: each time the chosen configuration must be relaxed at zero temperature applying the conjugate gradient technique. At this temperature the dislocation did not anneal out up to the maximum running time.

Very recent work has been presented on dislocationlike motion in a glass (Shi and Chaudhari,1983). Although a computer experiment has been performed on both edge and screw dislocations, only the first one has been reported in some details. The sample used was a box of roughly 70x45x85 Angstrom containing 12274 particles, and it was derived again from the Finney model. Edge dislocation has not been introduced in the same way as in the stability studies: instead of cutting and removing a band of atoms from the center, the particles near to the left border in the upper half of the box were displaced to the right by a finite distance (1/6 to 1/3 of an atomic diameter). The system was then relaxed under special boundary conditions: in order to avoid back flow and to prevent the dislocation from being pinned down only in-plane freedom was given to the particles on the left, upper and lower faces of the box. Complete freedom was allowed on the right face and inside the box. In the remaining direction (front-back) the box was considered to be periodic. Under these conditions, a step appears after relaxation on the free surface opposite to the side where dislocation was introduced (see fig. 9). This step has been interpreted as being due to slip motion of the dislocation. The procedure has been repeated by introducing the dislocation at different distances from the right side of the box: the results do not show any significative difference in size and shape of the step obtained. Also different values of the height at which dislocation was introduced have been tested: no dependence of the results from this parameter has been observed . No measure of elastic properties has been reported.

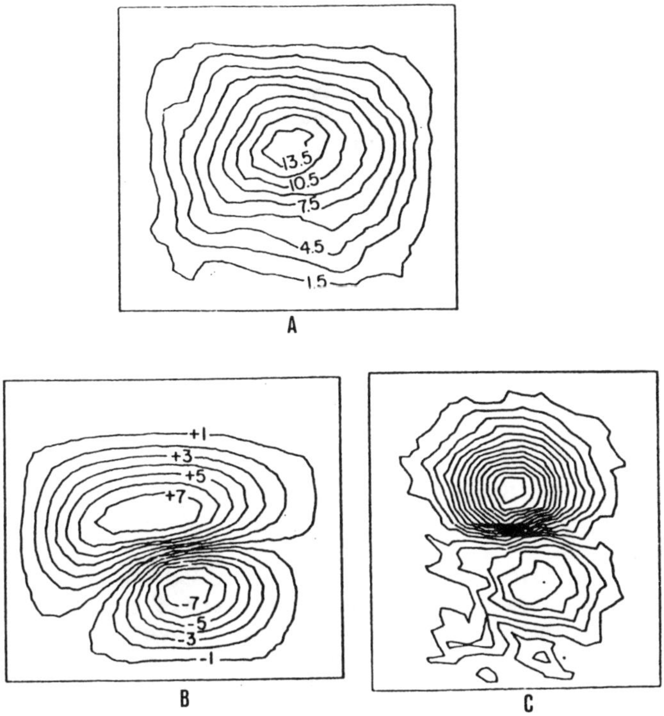

FIG 8. Airy stress function for an edge dislocation:
A) undislocated model.
B) (edge) dislocated crystal.
C) (edge) dislocated amorphous solid after dynamic relaxation at 6 K.
(A) and (B) after Chaudhari, Levi and Steinhardt (1979).

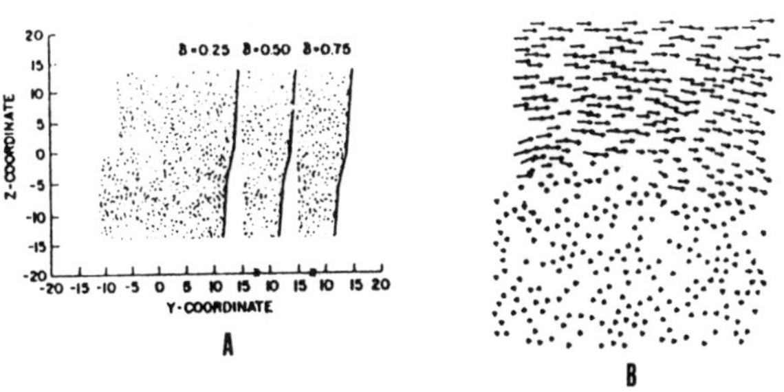

FIG 9. Motion of an edge dislocation in an amorphous solid (after Shi and Chaudhari, 1983)
A) Step on the free surface for various values of the distance at which the dislocation has been introduced.
B) Plot of the desplacement of the particles during relaxation.

8. Conclusions and summary.

The possibility to apply computer simulation to the glassy state, and some of the connected problems have been discussed. A possible definition of defects in amorphous systems has been described, and results of the study of these defects have been shown. The work is still in progress.

9. Appendix: quenching by Molecular Dynamics.

The Verlet algorithm for Molecular Dynamics obtains the new atomic coordinates as sum of Taylor expansion for t+dt and t-dt:

$$x(t+dt) = 2\, x(t) - x(t-dt) + \ddot{x}\, t^2$$

and the velocities as difference of the same expansions:

$$v(t) = [x(t+dt) - x(t)] \,/\, 2t$$

The velocities are therefore not used to predict the new coordinates, and in order to rescale the temperature one has to modify in some way the equations of motion. The method we used was based on the approximation vt=x(t)-x(t-dt) and therefore in order to modify the temperature by a factor f the following equation of motion was employed:

$$x(t+dt) = x(t) + f\,[\, x(t) - x(t-dt)\,] + \ddot{x}\, t^2$$

In any case, the particular algorithm to subtract energy from the system is probably not very important, as far as the ratio of energy energy decrease is the same.

REFERENCES

Ahmadzadeh M. and Cantor B. (1980) J.Non-Cryst.Solids **43**,189

Angell C.A., Clarke J.H.R., Woodcock L.V. (1981) Advances in Chemical Physics, vol XLVIII

Bennett C.H.,Chaudhari P., Moruzzi V., Steinhardt P. (1979) Philos.Mag. A40,485

Deborah, Holy Bible, Judges 5, verse 5.

Cohen M.H., Turnbull D. (1959) J.Chem.Phys. **31**,5

Chaudhari P., Levi A., Steinhardt P. (1979) Phys.Rev.Lett. **43**,1517

Chaudhari P., Steinhardt P.J. (1982) Philos.Mag.A **46**,25

Chaudhari P., Spaepen F., Steinhardt P.J. (1983) In "Glassy Metals II", ed. by H.J.Guentherodt, H.Beck, Topics Appl.Phys. Vol.**53** (Springer Berlin) p.127

Damgaard Kristensen W. (1976) J.Non-Cryst.Solids **21**,303

Egami T., Maeda K. and Vitek V. (1980) Philos.Mag.A 6,883

Egami T., Srolovitz D. (1982) J.Phys.F Met.Phys. **12**,2141

Finney J.L. (1970) Proc.Roy.Soc.London, Ser.A **319**,479
Fletcher and Powell (1963) Comput.J. **6**,163
Frank F.C. (1978) Phys.Educ. **13**,258
Frenkel D. and McTague J.P. (1980) Ann.Rev.Phys.Chem. **31**,491
Gaskell . (1983) In "Glassy Metals II", ed. by H.J.Guentherodt, H.Beck, Topics Appl.Phys. Vol.**53** (Springer Berlin) p.5
Hoare M.R. (1978) J.Non-Cryst.Solids **31**,157
Mandell M.J., McTague J.P., Rahman A. (1976) J.Chem.Phys. **64**,3699
Polk D. (1971) J.Non-Cryst.Solids **5**,365
Rahman A., Mandell M.J., McTague J.P. (1976) J.Chem.Phys. **64**,1564
Reiner M.(1964) Physics Today, **17**,62.
Ronchetti M., Chaudhari P., Steinhardt P.J., unpublished work
Shi L.T., Chaudhari P. (1983) Phys.Rev.Lett. **51**,1581
Steinhardt P.J., Chaudhari P.(1981) Philos.Mag.A **44**,1375
Steinhardt P.J., Chaudhari P.(1981) J.Comp.Physics **42**,266
Steinhardt P.J., Nelson D.R., Ronchetti M. (1981) Phys.Rev.Lett. **47**,1297
Steinhardt P.J., Nelson D.R., Ronchetti M. (1983) Phys.Rev.B **28**,784

SIMULATION OF BRITTLE FRACTURE via MOLECULAR DYNAMICS

WILLIAM G. HOOVER

1. SYNOPSIS

Fast fracture in brittle materials can be simulated with molecular dynamics. The results have many features in common with macroscopic experiments on real materials. These features include steady propagation of cracks at speeds comparable to the speed of sound; increasing crack velocity with increasing stress; crack "inertia", allowing the penetration of cracks into regions which would not fracture under static conditions; and crack-surface irregularity at moderate-to-high temperatures.

2. INTRODUCTION

The breaking of an object into two or more pieces is fracture. The economic consequences of fracture in jet aircraft, bridges, and oil shale, for instance, abundantly justify the study of fracture from a variety of viewpoints. These studies range from <u>microscopic</u> trajectory analysis of atoms near crack tips, through <u>mesoscopic</u> analyses of the formation and motion of dislocations and voids, to <u>macroscopic</u> finite-element or finite-difference simulations of failure for materials characterized by "damage functions" or other failure criteria describing susceptibility to fracture.

Fracture normally occurs through the enlargement of existing defects in solids--voids, grain broundaries, micro-cracks, or surface irregularities.

G. Jacucci (ed.), Computer Simulation in Physical Metallurgy, 145–158.

All such defects concentrate stress. The fact that failure stresses lie far below ideal crystal strengths shows that the stress concentration is large. It is not hard to see, from continuum elasticity theory, that a sharp crack has an infinite stress concentration. The crack shown in Figure 1 has a

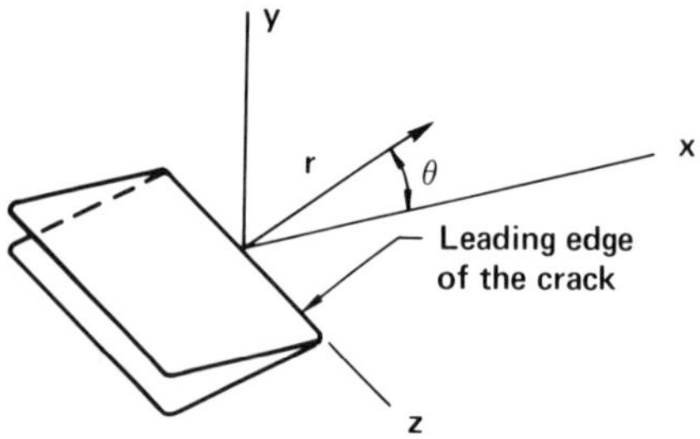

DISPLACEMENTS:

$$u_x = \cos(\theta/2)f(r)g(\theta) \;;$$
$$u_y = \sin(\theta/2)f(r)g(\theta) \;;$$
$$f(r) = (K/2G)(r/2\pi)^{1/2};$$
$$g(\theta) = 1 + 2\sin^2(\theta/2) \;,$$
$$\text{for } \lambda = \eta.$$

Figure 1. Crack in an elastic continuum[1]

displacement u_y in the vertical direction of order $(K/G)x^{1/2}$, where K is the "stress-intensity factor," G is the shear modulus, and x is the horizontal distance from the crack tip. The strain and stress both <u>diverge</u> as $r^{-1/2}$ at the crack tip. This singular behavior is unpleasant because many discrete zones are needed near the crack tip to give an accurate numerical description of the displacements and strains. Because interatomic forces are finite, it can be anticipated that the singularity changes form very near the crack tip. Atomistic simulations of fracture show how this comes about.

The classical Griffith model of fracture is based on energy conservation. Consider the application of the model to a strip undergoing fracture. The unbroken part of the elastic strip shown in Figure 2 has been stretched by an amount $2\varepsilon h$ vertically, so that its energy density e is $(2\eta+\lambda)\varepsilon^2/2$, where η and λ are the Lamé constants.

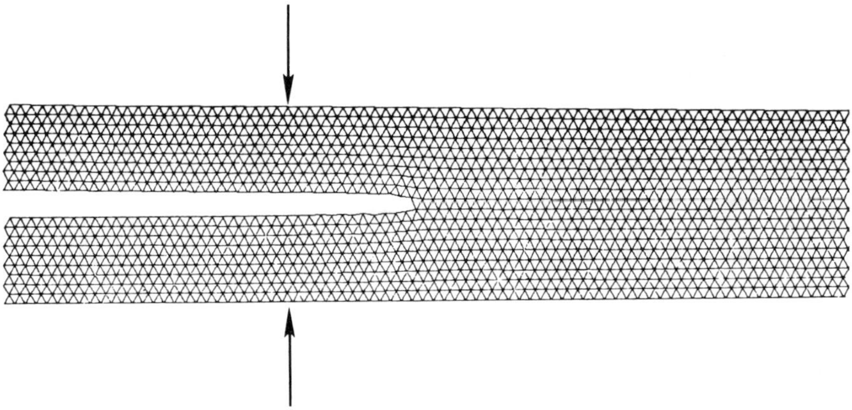

Figure 2. Elastic strip undergoing brittle fracture. Strip width is 2h.

The broken portion is stress free but has a surface energy, per unit length of surface γ, equivalent to an energy density γ/h. Griffith suggested that simple energy minimization causes fracture to propagate when the strain energy matches the surface energy:

$$\sigma_G = (2\eta+\lambda)^{1/2}(2\gamma/h)^{1/2} = (2\eta+\lambda)\varepsilon_G \quad . \tag{1}$$

This Griffith model is only an approximation. It ignores the irreversible nature of fracture, the sensitivity of fracture to temperature, and the dependence of crack-surface morphology on both temperature and velocity. Griffith's model always underestimates the failure strain, but it is still a useful semiquantitative guide, as we will see.

3. ATOMISTIC MODELS OF FRACTURE

A microscopic atomistic model for crack structure or growth requires (1) a force-law, (2) equations of motion, and (3) appropriate boundary conditions. For simple closed-shell atoms--argon is the prototype--the forces are well understood. For metals the situation is complicated, even for bulk matter. When empirical potential functions are fitted to bulk properties it is not unusual to find that defect properties are in error by as much as a factor of two.[2] Despite intense theoretical effort it appears highly unlikely that fundamental quantum-mechanical estimates of metal forces will prove useful soon. Instead, there is considerable room for judicious empirical work. An approach fitting <u>all</u> possible data (elastic contants, formation, fusion, surface, vacancy and stacking-fault energies, etc.) with short-ranged forces <u>might</u> work. At present it is unreasonable to expect quantitative agreement between computer simulations and experiment--the forces are simply not well enough known.

It is still possible to develop techniques for simulating brittle fracture by studying model systems with simple forces. This is the point of view we have adopted in studying fracture.[3,4] It is a particularly fruitful approach to follow in characterizing number-dependence, sensitivity to boundary conditions, and the qualitative response of brittle cracks to stress gradients and other lattice defects. Because plastic deformation involves the long-range interaction of many dislocations, plastic "ductile" fracture is best handled from either the mesocopic or macroscopic point of view.

The simplest force law useful in treating fracture problems is the piecewise-linear ("Hooke's Law") force illustrated in Figure 3.

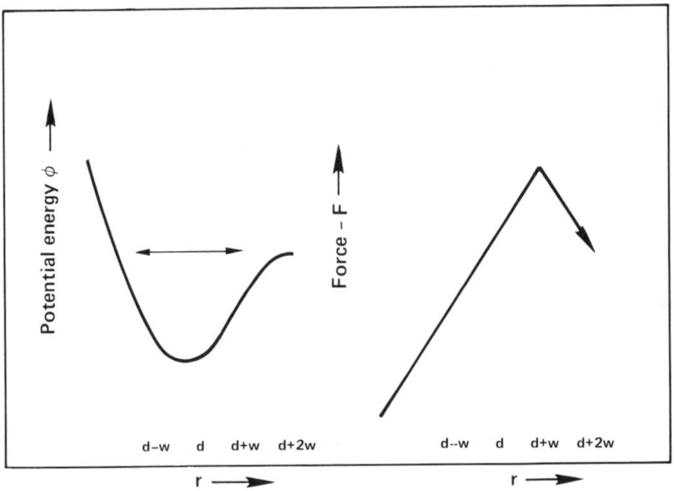

Figure 3. Potential and force used in fracture simulations.

Figure 4. Fixed-stress boundary.

The atom-atom potential has its minimum at r=d. The force resisting separation increases with r up to the mechanical instability point at r=d+w. The force then falls linearly for larger r, and vanishes for r≥d+2w.

Fracture calculations for this potential have been carried out using both fixed-stress[3] and fixed-displacement[4] boundary conditions. In the fixed-stress case 512-atom crystals, periodic in the x direction, were stressed in the y direction as shown in Figure 4.

The potential energy $\Delta\Phi$ taken up by such crystals[3] with a crack of length L is shown in Figure 5. The interparticle force constant is k, and the interparticle spacing is d. The energy data extrapolate smoothly to the elastic-theory result (marked with an x).

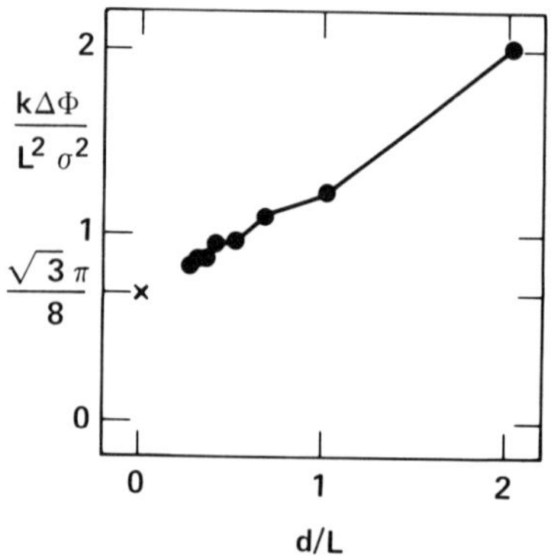

Figure 5. Energy increase in a 512-particle stressed crystal.

Fixed-displacement boundaries have been studied for the same force law. The crack shown in Figure 2 has all atoms on the top and bottom boundaries fixed, in order to produce a macrosopic vertical strain ε_{yy}. In this case too the energy extrapolates smoothly to the elastic-theory prediction as the number of atoms is increased. For the two kinds of boundary conditions shown in figures 2 and 4 the stress at which a long crack will propagate can be calculated. By considering crystals of from 700 to 10,000 atoms the limiting value of this failure stress, and thus the critical stress-intensity factor K_c was calculated with an accuracy of 1 part per thousand.[4] The results obtained by the two methods agreed nicely. The cracks require a stress exceeding the Griffith stress σ_G by about 26%. The excess energy, beyond the surface energy, ends up as thermal motion, or "heat", the signature of an irreversible process. A part of the excess energy can be used to drive fracture at stresses below the static value. The specimen shown in Figure 2 actually has a slight taper. The slope of the upper boundary is -0.016. The arrested crack passed beyond the point of arrest predicted from static analysis (indicated by arrows). A series of such calculations resulted in final arrest at stresses ranging from 2 to 12% in excess of the Griffith stress. This commonly observed feature of real cracks can occur only when the Griffith stress lies below the static failure stress.

4. ATOMISTIC MODELS FOR FRACTURE IN IRON

Most three-dimensional fracture models have dealt with the cleavage fracture of body-centered cubic iron.[2,5,6] The body-centered lattice is unstable with nearest-neighbor forces (it is a worthwhile exercise to find the shear direction demonstrating this). Therefore a potential linking second neighbors is required. The Johnson potential is most commonly used. It is shown in Figure 6.

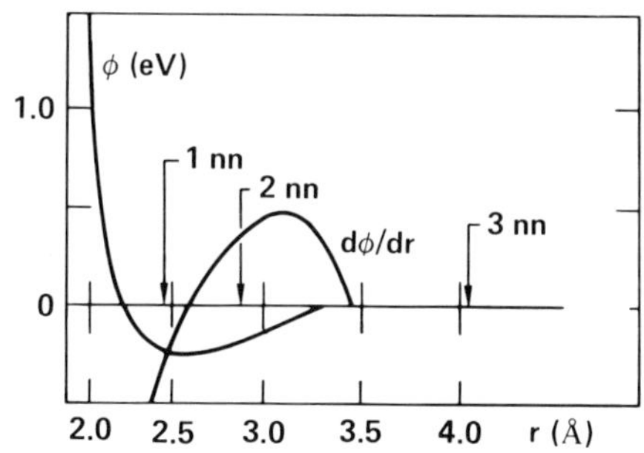

Figure 6. Johnson iron potential.

This potential underestimates the surface energy by about a factor of two, but is otherwise useful in suggesting the additonal complexities associated with three-dimensional simulations. To avoid using more than a few hundred particles, only the crack tip is described, by fixing either the displacements or the forces on the boundary particles. More sophisticated models couple the boundaries to an elastic continuum which can incorporate realistic boundary conditions. In most cases the crystals studied have been only one unit cell in thickness, and, typically, there is no mechanism for the relief of stress in the thickness direction or for the formation of jogs which might reduce the stress required for fracture.

Mullins[6] has studied the dynamics of crack tips propagating at stresses ranging from 20 to 70% above the Griffith stress. Evidently the minimum stress required for fracture, using the Johnson potential, is close to that found in two dimensions, namely between 1.2 and 1.3 times the Griffith stress. Mullins found propagation speeds ranging from 20 to 55% of the transverse sound velocity, as shown in Figure 7. These speeds lie somewhat below those found for the two-dimensional Hooke's-Law forces, and agree well with fracture data on steel.

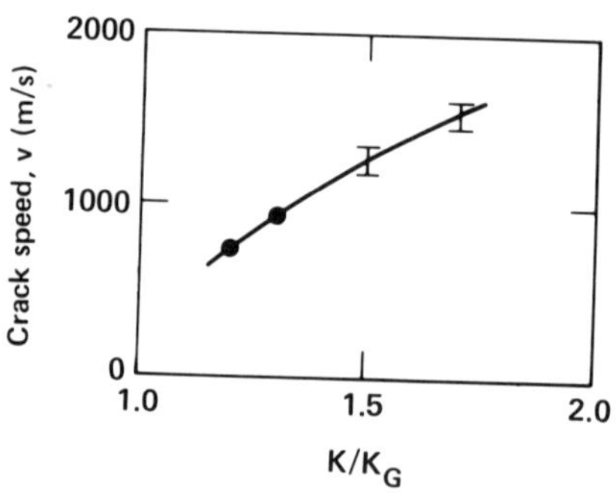

Figure 7. Crack velocity for iron.

There appear to be no computational difficulities in extending Mullins' work to include defects in the vicinity of the crack tip. The main problem lies in the uncertain nature of forces in real metals.

5. DYNAMIC STRESS INTENSITY FROM SHADOW PATTERNS

During the past ten years Kalthoff's group at Freiburg has studied dynamic fracture, in metals and plastics, using "shadow patterns". The propagation of a crack down the length of a specimen produces a moving cusp-like dimple around the crack tip. When transverse light is reflected from the specimen (as with steel) the moving dimple results in a bright "caustic" ring which can be photographed. For elastic materials, the size of the ring can be related directly to the stress intensity factor--i.e., the stress x $(2\pi r)^{1/2}$ directly in front of the crack tip. If the light is transmitted (as with transparent plastics), the dimpling acts as a lens and a similar moving ring appears behind the specimen.

By tracking the size and motion of the caustic ring Kalthoff's group has obtained considerable data indicating wide variations in stress about that which could be calculated on a static basis. Because the data indicate that fracture sometimes propagates at stresses lower than the critical static value, a quantitative understanding of the shadow pattern results has been sought by the Electric Power Research Institute.[4]

To what extent can shadow patterns be duplicated from atomistic simulations? This question has been studied by Moran.[4] By using the plane stress elastic equations to calculate displacements normal to his two-dimensional crystals, Moran could represent the surface of a three-dimensional thin fracturing strip. Next, by simulating the reflection of light from the triangular mirrors aligned according to the elastic equations, he simulated the light pattern which would be photographed in a shadow-pattern experiment. Figure 8 is a series of shadow patterns formed as the first seven bonds break. Indvidual sound waves can be seen spreading out from the moving crack tip.

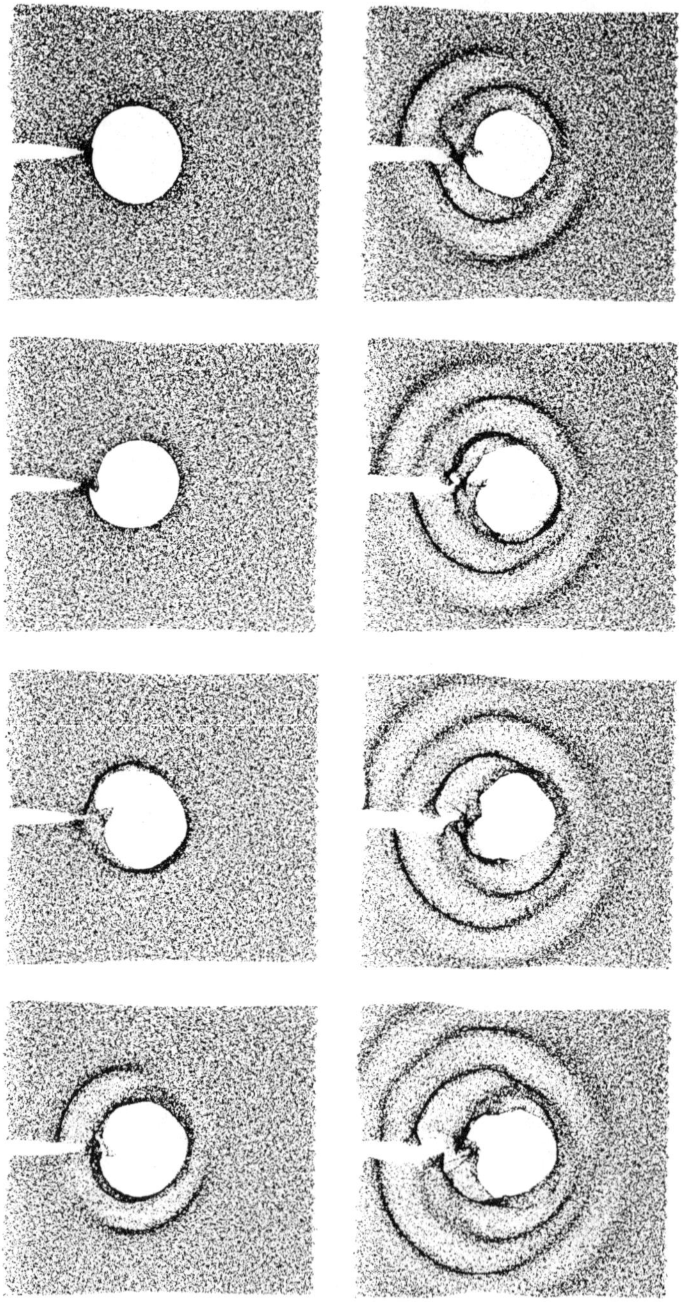

Figure 8. Shadow patterns for cracks with $(\Delta L/d) = 0,1,2,3,4,5,6,7$

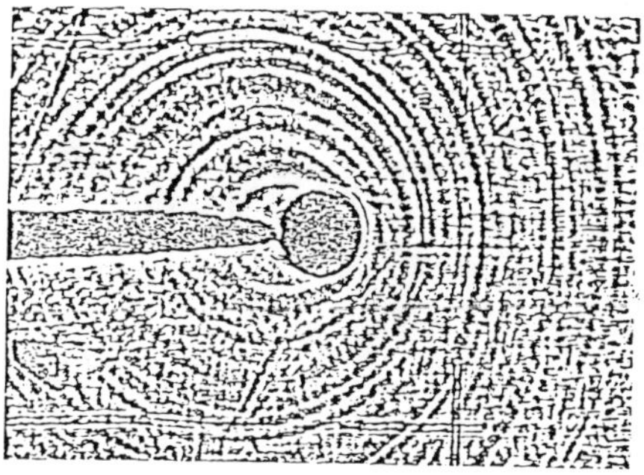

Figure 9. Experimental shadow pattern for steel.

Figure 9, taken from Kalthoff's work, is a reproduction of an experimental shadow pattern in high strength steel.

The quantitative analysis of the shadow patterns is somewhat disappointing. The <u>static</u> caustic agrees well in size with the expected value from elastic theory. The irregularity in the dynamic results (and those shown in Figure 8 were obtained with a slight viscous damping added to the equations of motion) makes quantitative comparison with experiment impossible. The steady-state caustic diameter is essentially indistinguishable (within nearly 10% fluctuations) from an estimate based on static theory.

6. DUCTILE FRACTURE

Except at low temperature, real metals normally fracture by a ductile mechanism. Voids form and coalesce as shear stress generates and moves lattice dislocations. When "plastic flow" has occurred the failed parts no longer fit together after fracture. Because real metals are made of atoms, there is no doubt that a sufficiently large atomistic model could describe ductile failure too. But how large? In ductile failure material <u>millimeters</u> away from the crack surface undergoes extensive plastic flow. A cubic millimeter of iron contains about 10^{19} atoms. Even a square milimeter exceeds current computational capacity by six orders of magnitude.

Thus the study of models following the detailed atomistic motion of dislocations is not feasible, and a continuum macroscopic approach is used instead. Rather than solving Newton's ordinary differential equations of motion,

$$ma = F \quad , \tag{2}$$

we solve instead the partial differential macroscopic equation of motion,

$$\rho a = \nabla \cdot \sigma \quad , \tag{3}$$

where σ is the stress tensor. The finite-difference method involves approximating the continuum by a discrete set of nodes (or contiguous zones) and moving these according to ordinary differential equations. If, for instance, a two-dimensional elastic continuum is divided into triangular zones (See Figure 2), and the displacement in each zone varies linearly with position, the continuum equation (3) reduces to equations (2) where the force is given by Hooke's Law.[3]

To treat fracture by continuum methods, a criterion for the breaking of nodal links or the separation of contiguous zones must be specified. Wilkins[7] has successfully used a "damage function" depending upon plastic strain and pressure as a failure criterion. Popelar and Gehlen equated the (potential) energy lost in discarding a stressed crack-tip zone to the new surface energy gained. The surface energy was a specified function of crack velocity. Other empirical approaches have been based on crack-tip radius or crack-opening displacement. Such empirical approaches can successfully correlate data for particular classes of experiments, but are not sufficiently sound to make trustworthy predictions for new geometries. A major computational difficulty is the necessity to specify in advance possible crack trajectories.

The fundamental mechanism for the propagation of ductile fracture is the growth and coalescence of voids ahead of the crack tip. The detailed numerical simulation of fracture incorporating these voids has not yet been seriously attempted.

7. OUTLOOK

Empirical short-ranged potential functions, chosen to reproduce bulk and defect properties, coupled with Mullins' boundary conditions, should help in understanding crack-tip phenomena--stress corrosion cracking, dislocation production, and the interaction of cracks with voids. The prospects for fundamental progress on ductile fracture are more remote. As a start, more emphasis on mesoscopic simulations is recommended.

8. PROBLEMS

1. Show that the shear modulus in a face-centered cubic lattice with nearest-neighbor Hooke's Law forces is not isotropic.
2. Show that the Griffith far-field stress for a triangular-lattice strip, of width 2h, with nearest-neighbor interaction energy $-\varepsilon$ is $(3\sqrt{3}\kappa\varepsilon/2h)^{1/2}$, where κ is the nearest-neighbor Hooke's-Law force constant. Hint: show first that the lattice is elastically isotropic, with both Lamé constants η and λ equal to $\sqrt{3}\kappa/4$.

REFERENCES

1. J. R. Rice, "Mathematical Analysis in the Mechanics of Fracture," in Fracture, II, Chap. 3 (Academic, 1968).

2. B. deCelis, A. S. Argon, and S. Yip, J. Appl. Phys. 54, 4864 (1983).

3. W. T. Ashurst and W. G. Hoover, Phys. Rev. B 14, 1465 (1976).

4. B. Moran, Crack Initiation and Propagation in the Two-Dimensional Triangular Lattice, Ph.D. Dissertation (Univ. of Ca. at Davis, 1983)

5. P. C. Gehlen, G. T. Hahn, and M. F. Kanninen, Scripta Met. 6, 1087 (1972).

6. M. Mullins, Scripta Met. 16, 663 (1982).

7. M. L. Wilkins, "Fracture Studies with Two- and Three-Dimensional Computer Simulation Programs", Lawrence Livermore National Laboratory Report UCRL-78376 (1977).

8. C. H. Popelar, and P. C. Gehlen, Int. J. of Fracture 15, 159 (1979).

FRACTURE STUDIES USING LINEAR ELASTIC SOLUTIONS FOR CONTINUOUSLY DISTRIBUTED DISLOCATIONS

J E Sinclair
Theoretical Physics Division
AERE Harwell, Oxfordshire
OX11 0RA
U.K.

ABSTRACT. The representation of a crack by a sheet of continuously distributed dislocations is a technique which is very well known, from the Bilby–Cottrell–Swinden model for instance. The use of similar methods with numerical rather than analytic solution proves useful to study a number of geometrically complicated situations involving the interaction with cracks of dislocations and point defects, including for instance the geometrical changes resulting from crack tip blunting. This lecture explains the dislocation method in the context of other boundary integral equation methods, and describes some applications. The models are restricted to two-dimensional models, but include arbitrary elastic anisotropy.

1. INTRODUCTION

The two major properties of a crack are that it introduces new free surface into a body, and that it acts as a stress concentrator. A great part, then, of crack modelling consists of, or involves, stress analysis. If other defects, and in particular dislocations, are treated as separate objects, then this stress analysis can usually be made in the linear-elastic approximation. On the other hand, if they are lumped in with the material's mechanical properties, then some form of anelastic model will be necessary. This lecture is concerned with the problem of linear elastic stress analysis, as it arises in typical fracture related problems.

In many such problems the body is treated as infinite, which really amounts to supposing that interactions with the external surfaces are not important. This, coupled with our restriction to linear elasticity, makes solution techniques of the 'boundary integral equation' (BIE) class very suitable. In the integrals which give these methods their name, unknown functions, usually defined over the boundary of the body, are multiplied by kernels which are in the nature of Greens functions. That is to say, they are elementary solutions throughout the domain of the body of the differential equations to be solved (the equilibrium equations of elasticity), but

G. Jacucci (ed.), Computer Simulation in Physical Metallurgy, 159–182.

they do not individually satisfy the required boundary conditions. The integral equations are a requirement that the distributed superposition of the Greens functions does satisfy the boundary conditions. After the solution is obtained, the distribution functions on the boundary have to be used to compute by integration any desired field values such as displacements or stresses. By contrast, in finite element and finite difference methods, values of the displacements or stresses at points throughout the body have to be adjusted so as to satisfy the differential equations everywhere. The boundary conditions are usually satisfied in advance by restricting the element freedoms at near-boundary points. Because of the need to dissect the whole body, the finite element and finite difference methods are not well to suited to infinite body problems.

In the following I describe work using an integral equation method in which the Greens functions (kernels) involved are the fields of dislocation lines, either in an infinite body, or in a body with a single semi-infinite crack. In this presentation, I avoid mathematical complexities such as existence theorems or solution spaces, substituting physical intuition wherever possible. On the other hand, some pitfalls which beset the over-naive will be pointed out. In principle, the method could be used in two or three dimensions, but is easy to implement only in two dimensions. Two example applications are described.

2. THE METHOD

2.1 Formulating the Integral Equations

To those involved in the field of fracture, the most familiar use of a dislocation distribution to represent a crack may be the model of Bilby Cottrell & Swinden (1963) ('BCS') for the plastic yield from a crack. Perhaps some people view such a procedure as a 'trick' of doubtful basis, while others may be misled by the often quoted similarity to the case of a pile-up of real dislocations into thinking that the dislocations in the crack representation have some individual significance. In fact, the procedure is best viewed as a convenient method for assembling solutions to the elastic equilibrium equations. Other sources, such as distributions of body force, can alternatively be used.

Such physical terms to describe the sources involved in construction of solutions do not have to be used. The traditional way to derive what is known as the 'direct' BIE method (Rizzo 1967) is in terms of Somigliana's identity, which represents the displacement field at any point interior to a bounded body in equilibrium as integrals over the displacements and tractions at the boundary — or at any other closed surface enclosing the point. This identity takes the form

$$u_i(\mathbf{x}) = \int_S \{t_j(\mathbf{x}')G_{ji}(\mathbf{x}',\mathbf{x}) - u_j(\mathbf{x}')H_{ji}(\mathbf{x}',\mathbf{x})\}\, dS \;:\; \mathbf{x} \text{ inside } S, \tag{1}$$

where $G_{ji}(\mathbf{x}',\mathbf{x})$ is the infinite-body Greens function (displacement at $\mathbf{x}'$ due to a unit force at $\mathbf{x}$), t_j is the traction acting across the boundary surface S, and H_{ji} gives the tractions corresponding to G_{ji}. In the limit that the field point $\mathbf{x}$ approaches S from the inside, (1) becomes

$$\tfrac{1}{2}u_i(\mathbf{x}) + \int_S u_j(\mathbf{x}')H_{ji}(\mathbf{x}',\mathbf{x})\, dS = \int_S t_j(\mathbf{x}')G_{ji}(\mathbf{x}',\mathbf{x})\, dS: \; \mathbf{x} \text{ on } S, \tag{2}$$

where the integral on the left-hand side is a Cauchy principal value. If the problem is one in which the tractions t_j are specified everywhere on S, the right-hand side of (2) can be evaluated, and a singular integral equation results for the boundary displacements u_i. When this is solved, (1) can be used to evaluate displacements at any interior point, or a similar integral to evaluate stresses. If boundary displacements are prescribed, the same equation can be used to solve for the unknown t_j.

Although the above is not the method which I have used in the studies to be described later, it is instructive to see how this formulation can be interpreted in terms of physical sources and response functions. This has been shown clearly by Altiero & Gavazza (1980) and others before them, but is perhaps not widely appreciated. Picture a bounded body being cut out from an infinite, unstressed medium. When this body is stressed by surface loads and/or prescribed displacements, it is distorted, so that if it were to be inserted back into the hole in the infinite body, there would be misfit (Figure 1). If this were resolved by insertion or removal of material, the result would be a continuous infinite body with a Somigliana dislocation at the cut location (equivalent to a distribution of infinitesimal dislocation loops). In addition, to hold the interior in the same state of distortion, and to leave the exterior unstressed, the surface tractions would have to be maintained, but they now appear as a sheet of body force applied to the dislocated infinite body. Therefore, the displacements or stresses at any point — in particular at a point inside the finite region — can be found by adding the fields of the Somigliana dislocation and the body force sheet. Expressions for these are known in terms of the infinite body Greens function. The displacements are

$$u_i(\mathbf{x}) = \int_S \{G_{ij}(\mathbf{x},\mathbf{x}')[t_j](\mathbf{x}') - [u_j](\mathbf{x}')H_{ji}(\mathbf{x}',\mathbf{x})\}\, dS, \tag{3}$$

where a quantity in brackets [] denotes the discontinuity across S, so that $[u_j]$ is the displacement jump introduced by the dislocation. But since the exterior region remains undisplaced, this jump is actually equal to the total displacement on the interior side of S. Thus, the

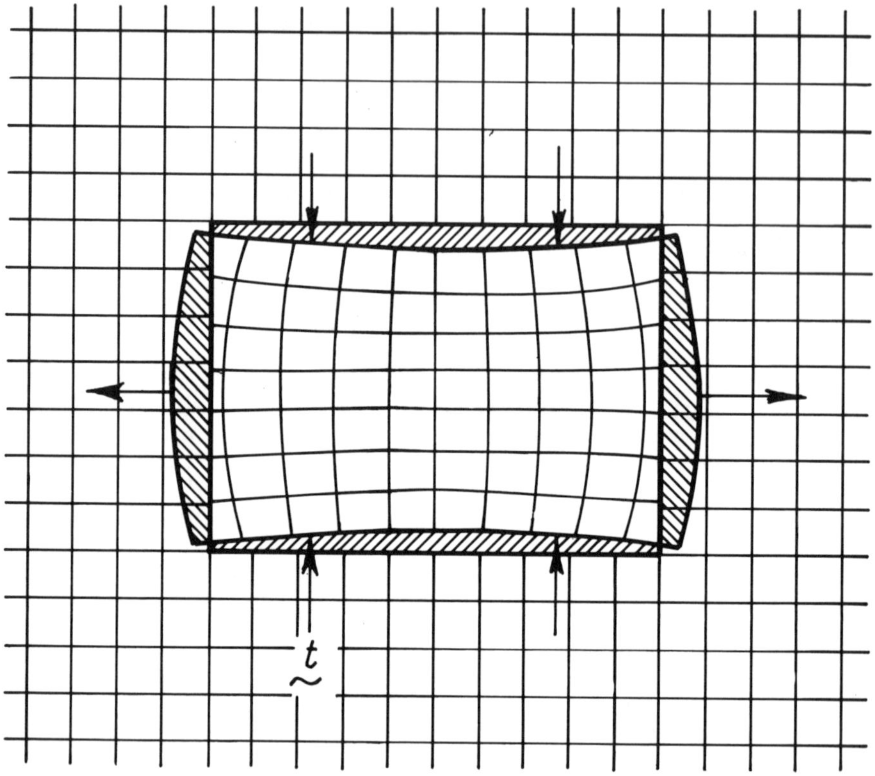

Figure 1. Embedding procedure for 'direct' BIE method.

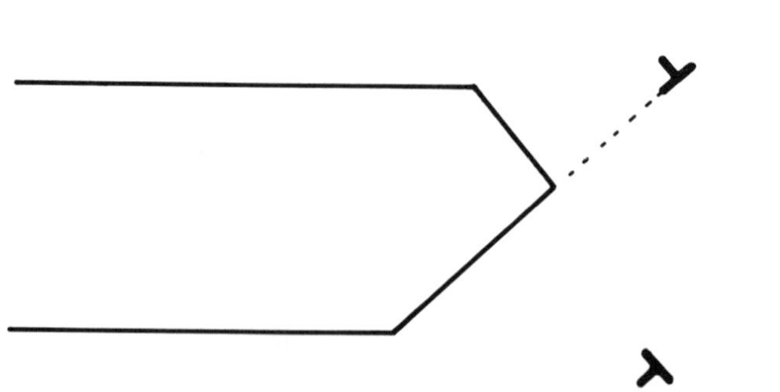

Figure 2. Blunted crack profile produced by alternating slip on two planes.

exact correspondence between (1) and (3) (taking into account the symmetries of G_{ij}) is explained.

An expression corresponding to (3) can be written down for the stresses, by differentiation and application of the elastic constants. After letting the field position **x** tend to the inside of S, and resolving the stresses to give the tractions across S, one obtains the equation

$$\int_S Q_{ij}(\mathbf{x},\mathbf{x}')[u_j](\mathbf{x}')\,dS = -\tfrac{1}{2}t_i^-(\mathbf{x}) + \int_S t_j^-(\mathbf{x}')H_{ij}(\mathbf{x},\mathbf{x}')\,dS. \tag{4}$$

Here, a superscript '−' denotes a value on the inside of S, and account has been taken of the fact that the outside values u_j^+ and t_j^+ are zero. $Q_{ij}(\mathbf{x},\mathbf{x}')$ is a response function deriving from H_{ij} in a similar way as H_{ij} does from G_{ij}. Equation (4) provides an alternative integral equation to (2), whereby the unknown part of the boundary values u_i^- and t_i^- could be solved in terms of the known part.

To take a simple example, if (4) were used to solve a traction boundary value problem, the known tractions would first be integrated with the kernel H_{ij} to form the right-hand side, leaving an integral equation for the boundary displacements u_i^-, which would be discretized in some fashion to yield a set of linear equations. Physically, one could say that the known tractions had been applied as a sheet of body force to an infinite body, then one had solved for the surface distribution of dislocations which, combined with the forces, yielded zero stresses in the exterior region, making it 'invisible' to the interior.

Such an approach has two disadvantages. One is the coding effort and computer time involved in evaluating the right-hand side integrals. The other arises in problems with sharp cracks. If the crack faces have pressure applied, the fixed tractions appearing on the right-hand side become force dipoles at the crack location in the infinite body. Thus, yet another kernel has to be worked out. Then, the variables u_i^- exist on both sides of the crack, but they do not give independent contributions to the stresses in the body, so that the equations become indeterminate.

For these reasons, I prefer a simpler approach for traction boundary value problems, in which no body force sheets are used, but only dislocations. The description in terms of embedding in an infinite body is as follows. Take the bounded body with its surface loads and resulting displacements (as yet unknown). Apply also to the hole in the infinite body equal and opposite tractions, resulting in displacements of the hole surface. Now join the two distorted parts by insertion and/or removal of material, forming a dislocated continuous infinite body, which is in equilibrium because the two equal and opposite tractions balance each other. To solve for the unknown dislocation distribution, one only needs the response function Q_{ij}. The requirement that the tractions across S (now equal on both sides) take the specified values gives the integral equation

$$\int_S Q_{ij}(\mathbf{x},\mathbf{x}')[u_j](\mathbf{x}')\, dS \;=\; t_i(\mathbf{x}). \tag{5}$$

The kernel $Q_{ij}(\mathbf{x},\mathbf{x}')$ can be described as giving the x_i–component of the tractions across S at the point $\mathbf{x}$ due to an infinitesimal dislocation loop at $\mathbf{x}'$ on S, with its plane lying in S, and Burgers vector in the x_j direction, such that the Burgers-vector loop-area product is unity. In two dimensions, where S is a cylindrical surface, the source is a dipole of dislocation lines, separated along S.

A noteworthy property of $Q_{ij}(\mathbf{x},\mathbf{x}')$ is that it is symmetric both in the indices ij and in the arguments $(\mathbf{x},\mathbf{x}')$. It thus forms a self-adjoint integral operator, and the linear equations (5) must correspond to a variational problem. In fact, it easy to show that (5) corresponds to minimisation of the strain energy of the infinite body. Since the exterior portion is just a complicated loading system for the interior, this is the potential energy of the finite body plus loading system. When the equations are discretized and solved in approximate fashion, it is thus possible to describe the obtained solution as that which minimises this potential energy subject to the constraints which the discretization scheme imposes. Other authors (Altiero & Gavazza 1980, Heise 1978) have discussed the various BIE formulations in terms of numerical efficiency and spectral character (related to conditioning), but the ability to solve in an energy-minimisation sense is a valuable property which does not appear to have been much considered by others.

Another point to note can be seen from the physical description of the embedding procedure. The insertion of the finite body into its hole could be done in many ways, related via rigid translation or rotation. The choice does not affect stresses, though it would the displacement values. Thus, in methods using equations for the tractions, with dislocation densities as variables, constraints to control rigid translation or rotation have to be applied to make the equations fully determined.

Finally, it should be mentioned that other Greens functions than those for an infinite body can equally well be used, and this may be advantageous when part of the boundary has a simple form. For instance, to treat a half-space with a cavity or crack, the analytically tractable half-space Greens function (or dislocation fields) could be used to build up a solution satisfying the additional boundary conditions, while automatically leaving the half-space surface free.

2.2 Discretizing the Integral Equations

Given a physical picture in terms of dislocation sources, it is tempting to try to represent a crack or other surface using a wall of discrete dislocations. Such an approach has been tried, by Bilby & Swinden (1965) and Vitek (1976) in two dimensions, and by Wong & Barnett (1984) in three dimensions, with apparent success. However,

Vitek (1976) admits that oscillations in his solutions had to be artificially smoothed by averaging, and my own experience is that the discrete equations are ill-conditioned, and that solutions do not converge well as the number of points is increased. The reason is that this kind of source is just 'too singular'. A single dislocation line gives stresses with a $1/r$ singularity, so the tractions certainly cannot be evaluated at the dislocation positions. Usually, mid-points are selected, but the meaning of the resulting solutions is not clear. Certainly, they are not minimum energy solutions in the sense described in the previous section, for the strain energy of a discrete dislocation is infinite.

To obtain a well-behaved set of equations, it is at least necessary to represent the dislocation distribution in a sufficiently smooth form that the strain energy is finite. Considering the two-dimensional case, the crudest possible form satisfying this requirement is that the displacement discontinuity $[u_i]$ be continuous piece-wise linear along S, which is to say that the dislocation-dipole density be piece-wise linear, or the dislocation density piece-wise constant. At the knots in such a distribution, the stresses are still singular, but only as $\ln r$, which is integrable, so that the energy is finite. Smoother forms may be required in some circumstances, but we shall discuss this form of representation to illustrate the principles of the discretization method.

In two dimensions, the boundaries are described by one or more arcs. Using s to denote an arc-length, we write the displacement discontinuity as a sum of basis functions:

$$[u_i](s) = \sum_n U_{in} f_{in}(s). \tag{6}$$

If $[u_i]$ is to be piece-wise linear in s, the f_{in} can be 'tent' functions, each having unit value at one knot and zero at all the others. The energy which is minimised when (5) is solved is

$$E = \tfrac{1}{2}\int_S \int_S [u_i](s) Q_{ij}(s,s')[u_j](s')\, ds'\, ds - \int_S [u_i](s) t_i(s) ds. \tag{7}$$

With $[u_i](s)$ formed as in (6), E becomes

$$E = \tfrac{1}{2} \sum_{i,n,j,m} U_{in} Q_{injm} U_{jm} - \sum_{i,n} U_{in} T_{in}, \tag{8}$$

where

$$Q_{injm} = \int_S \int_S f_{in}(s) Q_{ij}(s,s') f_{jm}(s')\, ds'\, ds, \tag{9}$$

$$T_{in} = \int_S f_{in}(s) t_i(s)\, ds. \tag{10}$$

Minimisation of E with respect to the knot values U_{in} is therefore achieved when

$$\sum_{j,m} Q_{injm} U_{jm} = T_{in} \text{ for all } i,\ n. \tag{11}$$

If there are N knots, and displacements and tractions can have 3 Cartesian components, Q_{injm} is a $3N \times 3N$ symmetric, positive-definite matrix. In two-dimensional isotropic or anisotropic elasticity, with the $f_{in}(s)$ linear tent functions, the integrals in (9) can be performed analytically. Expressions are given in the Appendix. Since the matrix is full, the practical limit on the degrees of freedom is smaller than in the finite element method where sparsity can be taken advantage of. On the other hand, far fewer degrees of freedom are necessary, since only the boundary needs to be discretized.

In problems where interest focusses on the behaviour near corners, crack tips or other singularities, I have found that subdivision of any interval between corners can be coarse (say 0.2 of the interval) near the middle, and progressively refined near the corners to the desired scale on which the stresses are to be examined. For instance, with a crack of length 1, if stresses are required down to 10^{-4} from one tip, then the knot spacing can be refined by halving to 10^{-4} or 10^{-5} at the tip. If stresses near the other tip are not required, only modest refinement is necessary there. Using such a procedure, excellent agreement is obtained with known analytic solutions.

3. APPLICATIONS

3.1 A Study of Crack Tip Blunting

The aim of this study was to investigate whether the geometrical changes which occur as a crack blunts by emitting dislocations play a part in determining how the blunting proceeds — whether symmetry is maintained, whether sharp corners persist, etc. Further, the question arose as to whether the stress fields as calculated with the tip shape taken into account were such as to predict conditions in which the crack might extend again by cleavage.

The question of the competition between cleavage and dislocation emission from a sharp crack tip was tackled by Rice & Thomson (1975). In the simplest form of their model, a hypothetical straight dislocation is considered, lying on a slip plane intersecting the crack front (tip), and inclined at angle θ. The two main elastic forces acting are

(i) the image force attracting the dislocation back towards the crack. This varies as r^{-1} where r is the distance of the dislocation from the tip. In fact (Asaro 1975), the radial component of the image force has a magnitude which is independent of θ and of the orientation of **b**, for arbitrary anisotropy, and the same as if the crack surface were replaced by any free plane surface passing through the origin.

(ii) the force arising from the crack loading. For suitable orientation of the dislocation, this is repulsive and varies as $r^{-\frac{1}{2}}$.

On this purely continuum-elastic basis, there is a critical distance at which the net force is zero, and within which the attractive image force dominates. When parameters are selected for a given real crystal, this critical distance may be found to be greater or less than the estimated dislocation core width. This comparison depends upon the applied load, of course. At the load given from the intrinsic surface energy by Griffith's condition, if the critical distance exceeds the core width the crack tip is judged stable against blunting by dislocation emission.

This model was extended by Sinclair & Finnis (1983) to enquire whether a crack tip which *can* emit dislocations might ever find cleavage favourable after a certain amount of blunting. The conclusion was that such a possibility did exist, but only if the dislocations were emitted on a single inclined plane, not on two symmetrically disposed planes. This calculation, like that of Rice & Thomson (1975), ignored the geometrical effects of the crack blunting, using stress solutions for a sharp crack plus dislocations. The present study sought to correct this inadequacy, and to investigate whether the supposition of asymmetry were unreasonable.

Alternating slip on two inclined planes through a crack tip can produce an angled tip as illustrated in Figure 2. Crack profiles in this class were therefore studied, calculating the stresses near the tip for various uniform stresses at infinity, and with single dislocations lying on one or other of the slip planes. The BIE method described in the previous section was used, employing infinite-body dislocation fields. Anisotropic expressions were used, but the elastic constants were chosen to be very close to isotropic. Since embedding in an infinite body was used, the surfaces relaxed had to be finite and closed. The crack (a long thin hole) was given a length some 100 times the width, and the interval subdivision was made coarsely near the left-hand end, but finely near each corner at the right-hand end.

It is valuable to make use of all the qualitative and analytic information available on any problem modelled numerically. The interaction of a straight dislocation parallel to a corner in a free surface will be much the same as if the arms of the corner extended indefinitely. Dimensional arguments show that the radial component of the image force must again be proportional to r^{-1}. In fact, Rice (1984) has recently shown that the theorem previously mentioned for a crack extends to a corner of arbitrary angle. Hence we can already say that in extending the Rice & Thomson model to a dislocation being emitted from a corner, the image attraction will be just as strong as in the sharp tip case. The driving force from the applied stresses, on the other hand, should be reduced, since the stress singularity at a corner should be weaker than at a crack tip. Williams (1952) showed that near a corner in two dimensions the stresses go like $r^{\lambda-1}$. For example, if the body occupies 270°, values $\lambda = 0.5445$, 0.9085 and $\frac{2}{3}$ are possible, with fields of form corresponding to crack modes I, II and III respectively. When the corner is flanked by a pair of close, parallel faces, it is clear that at large r, the fields must tend to the $r^{-\frac{1}{2}}$ crack-like form, so the numerical calculation is necessary to determine the strength of the $r^{\lambda-1}$ singularities.

Only a few examples from the many results generated can be given here. Figure 3 shows some stress values on the crack plane extension with two modes of remote loading on a crack formed by slip on a single plane inclined at 45° to the crack. No dislocations are present in this case. This case shows a general feature of all the tip profiles studied: that the mode I (σ_{22}) stress intensification was scarcely diminished from the sharp-tip result, whereas the mode II stresses (σ_{12}) were much smaller. The asymptotic slopes of the curves at small x agree very closely with the λ eigenvalues calculated after the fashion of Williams (1952).

For the same asymmetric profile, Figure 4 shows shear stresses on the planes at $\pm 45°$ with mode I remote loading. At small r, the shear is slightly stronger on the plane at $-45°$ than on the one at $+45°$. Thus, so far as the driving force from remote loading is concerned, dislocation emission on the lower plane would be favoured (assuming symmetric crystallography), and this would tend to produce a symmetric tip shape. Of course, dislocations already emitted tend to reduce the stress intensity at the tip from that produced by the remote loads. Calculations of the stresses produced by such dislocations in general tended also to be such as to produce symmetry. An exception was found for the larger slip plane inclinations. Figure 5 shows the shear stresses on the planes at $\pm 60°$ from a single edge dislocation at $r = 10a$, where a is the size of the tip front produced by slip only at $+60°$. In this case, the negative contribution to $\sigma_{r\theta}$ on $\theta = 60°$ is weaker than the corresponding quantity on the other plane. This contribution to the stresses thus inhibits dislocation emission on the upper plane less than it does on the lower. This is the only indication found that stress field asymmetries could ever be such as to favour the persistence of one-sided slip in a system with symmetric crystallography. Of course, for a crack unequally inclined to two slip systems, persisting asymmetry would be easy to predict.

In addition to the plain angled tip profiles, boundary configurations were also studied in which a hypothetical sharp slit emanated from a corner. Stresses were computed as a function of the slit length, so as to study the energetics of re-initiation of cleavage from a blunted tip. One way to calculate the strain energy released when an increment of cleavage occurs is to convolve the stresses existing before the increment occurs with the displacements of crack face separation existing after. Since in linear elasticity the latter always grow like $r^{\frac{1}{2}}$ from the new tip position, the only stress singularity which produces a finite energy release per unit extension is one of order $r^{-\frac{1}{2}}$. It is thus impossible on this basis to predict energetically favourable crack growth out of any notch profile other that the perfectly sharp crack. However, just as dislocation emission could be predicted if the net driving force became positive within an atomic spacing of the surface, so crack production from a

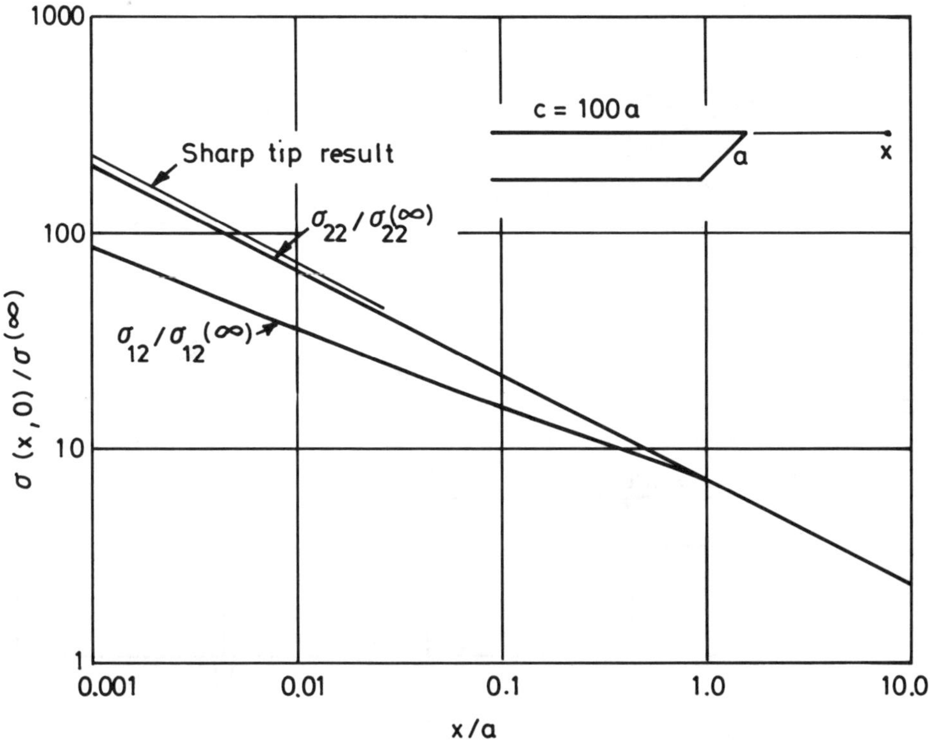

Figure 3. The asymmetric crack illustrated intensifies normal stresses σ_{22} almost as much as a sharp crack, but shear stresses σ_{12} much less.

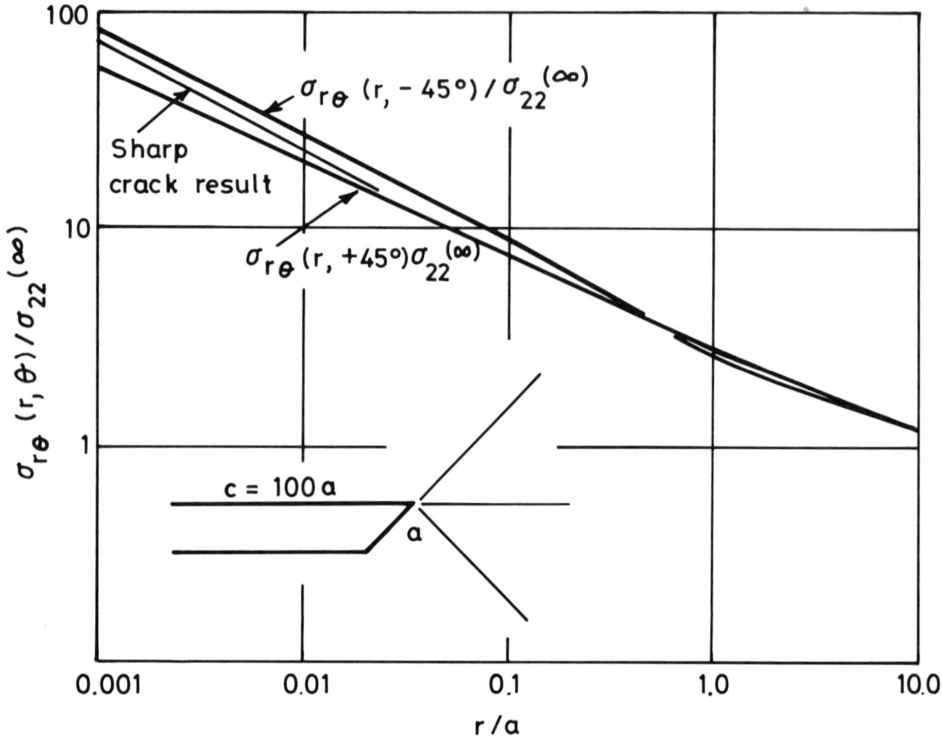

Figure 4. The shear stress from mode I loading is higher on the slip plane which has not yet slipped.

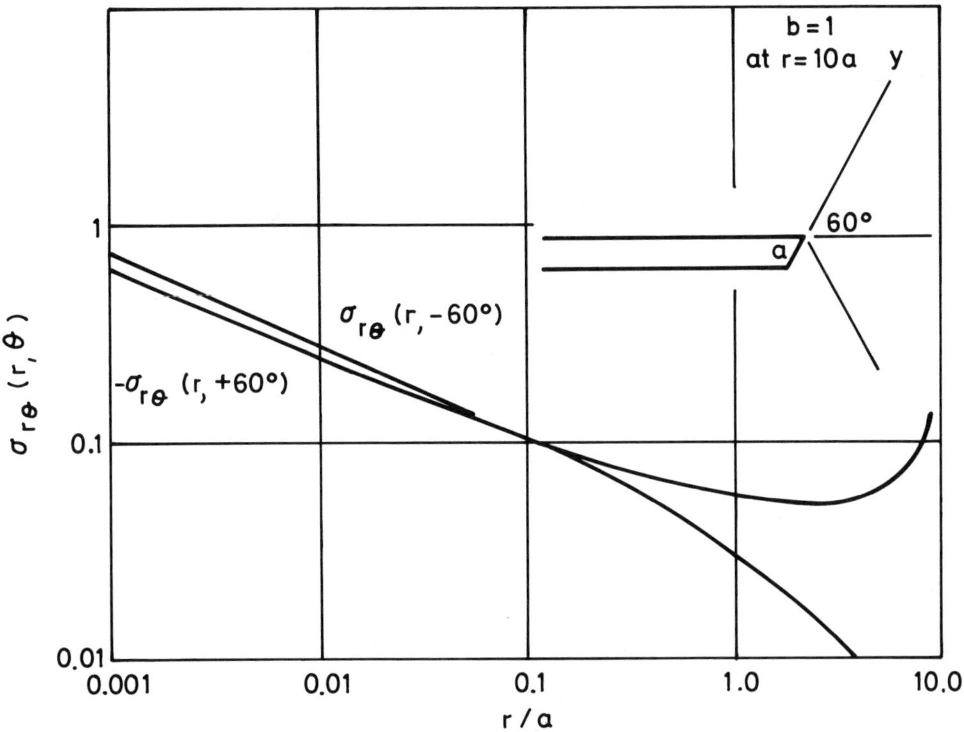

Figure 5. 'Back stress' from dislocation on plane inclined at 60° — *weaker* on the same plane than on the opposite plane at −60°.

corner or other 'weak' stress concentrator can be predicted if the stresses are sufficient to give the critical Griffith stress intensity factor on a putative sharp slit of atomic dimensions.

As a slit grows out of a loaded crack with blunted tip, the stress intensity factor at the slit tip will therefore grow from zero, and tend finally to be almost constant, when the steps left behind on the crack faces are no longer significant. Figure 6 shows computed results of this form for all three modes of loading, and an initially symmetric tip shape. Note that the mode I case shows very little effect from the initial blunting, mode II shows a large initial reduction, and mode III is intermediate.

Overall, we can conclude that the geometrical effects of crack tip blunting by dislocation emission are small, at least with mode I loading. If the sharpest feature on the tip is a corner, the shear stresses driving dislocation emission are almost unaffected, and the image forces opposing emission are completely unchanged. The tendency to cleave is also reduced by a small amount, but slightly more than the tendency to blunt further. The present studies do not address the case where the slip is such as to produce a rounded tip profile, but the same numerical techniques could be applied. However, one could anticipate that again, the effects on slip and on cleavage would be of the same order, so that the competition would not be greatly affected.

3.2 Stress Gradients near a Crack with Yield Bands

The following study is not actually an application of the BIE method as described above. It uses sheets of distributed dislocations, but instead of solving numerically for distributions with many degrees of freedom, known analytic solutions to related problems are spliced together with only simple adjustments to yield a solution of sufficient accuracy.

The problem under consideration is the diffusion of point defects in the stress field of a loaded crack. It arises both in embrittlement through stress driven solute segregation, and in the effect of irradiation on fracture. Hippsley, Rauh & Bullough (1984) have given an analysis of this diffusion using a simple model with two major approximations. One is the use of the 'pure drift' approximation in the diffusion equations. That is, random diffusion flux determined by the concentration gradient is neglected in comparison to the drift flux induced by the elastic interaction between the defects and the crack. For isotropic point stress sources, this drift is up or down the gradient of the mean normal stress σ_{ii}. The second major approximation of Hippsley et al. was to use the purely elastic crack tip stress field, neglecting the plastic yield which is certain to be present in all metals. It is this second approximation which the study below sought to remove. This, of course, was at the cost of analytic solubility.

We consider a model in which the yield is concentrated into two thin bands intersecting the crack tip (Figure 7), and take the 'small-scale yielding' case, in which the slip band length is small compared with the crack length or the distance to any other boundary.

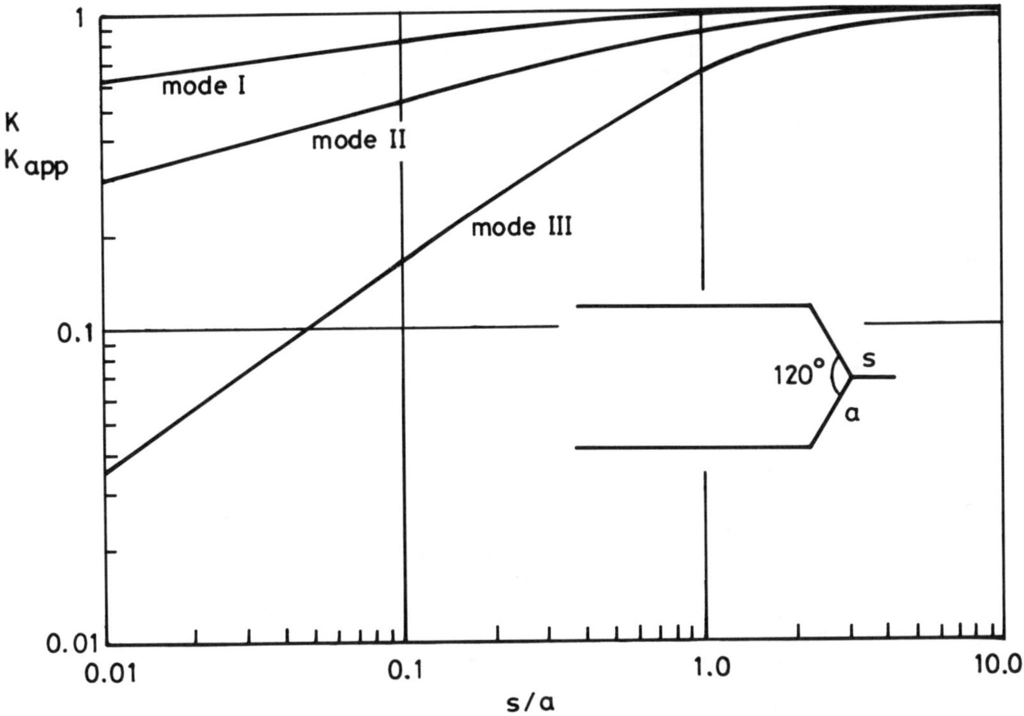

Figure 6. Stress intensity factor at tip of slit growing from blunted crack, compared to value if whole crack were sharp slit; result shown as function of slit length, *s*, for 3 modes of loading.

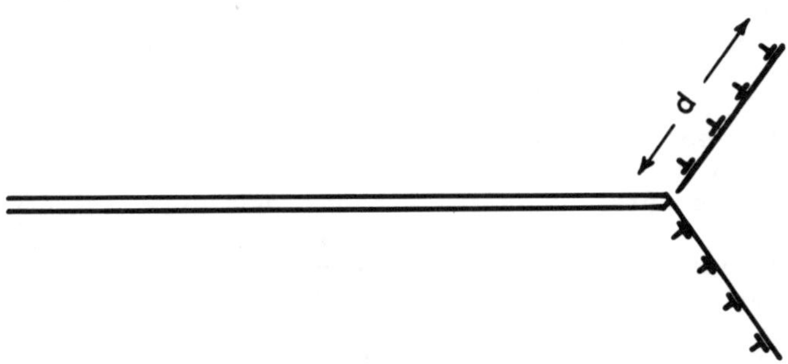

Figure 7. Crack tip with thin inclined slip bands.

The boundary conditions are then that the crack faces are free of traction, that the shear stress on each slip band is constant (it could of course vary with strain hardening), and that the stresses at distances great compared with the slip length tend to those expected around an elastic semi-infinite crack with the same remote loads. The desired field quantities are the stress gradients. These are more sensitive to discontinuities in an approximate solution than are the stresses themselves.

Rice (1974) gave a simple model for small-scale yield on two inclined planes. He considered the BCS model for yield on a band co-planar with a crack loaded in mode II, noting that it gave the slip band length and the crack-tip plastic displacement in terms of the applied load, or alternatively the stress intensity factor which would have existed in the absense of yield. Rice proposed that the same relation would hold for an inclined yield plane, if the measure of loading were taken as the intensity of the $r^{-\frac{1}{2}}$ singularity in the shear stress on the plane in the absense of yielding. The interaction between two yield planes was neglected. Numerical calculations of Sinclair (1976) confirmed that this model was extremely accurate provided the two planes were more than 60° apart and the band length less than about 0.3 times the crack half-length. However, it should be noted that this success was only for the global properties of band length and total yield. Rice's model was not intended to provide details of the stress fields.

To check what stress fields are produced by Rice's model, more than one route can be followed. The dislocation distributions of the BCS solution could be transferred — both for the free-surface crack and for the yield bands — and the stresses computed by integration using the field of a dislocation in an infinite, continuous body. Alternatively, since the field of a dislocation in a cracked body is known (Sinclair & Hirth 1975, Hirth & Wagoner 1976), the BCS expressions could be used for the dislocation distribution in the yield bands only, leaving the cracked-body response function to take care of the boundary conditions on the crack surfaces. The former procedure is not guaranteed to satisfy the desired boundary conditions anywhere, whereas the second will leave the crack faces free, but not necessarily give a constant shear stress on the yield bands.

Using the semi-infinite crack limit of both the crack-dislocation field (Sinclair & Hirth 1975) and the BCS distribution (crack length, $c \to \infty$, $K \sim \sigma\sqrt{c}$ = constant), the computed shear stress on the yield bands is shown in Figure 8 for the case of yield plane inclination at ±45°. It can be seen that the absense of stress singularity at the band ends is correctly preserved, and that the desired stress constancy holds well for the outer half of the bands, but grows progressively worse towards $r = 0$ where the two bands and the crack faces meet. Here, the $r^{-\frac{1}{2}}$ singularity is not correctly removed.

Taking the above as a first approximation, a second may be obtained, again without resort to a full-scale numerical solution, by trying to correct the singularity at the crack-tip/slip-band junction. The BCS solution gives a logarithmic singularity in the dislocation density at this point. Correspondingly, the shear stress on the common plane of the crack and yield band has a step (the imaginary part of a

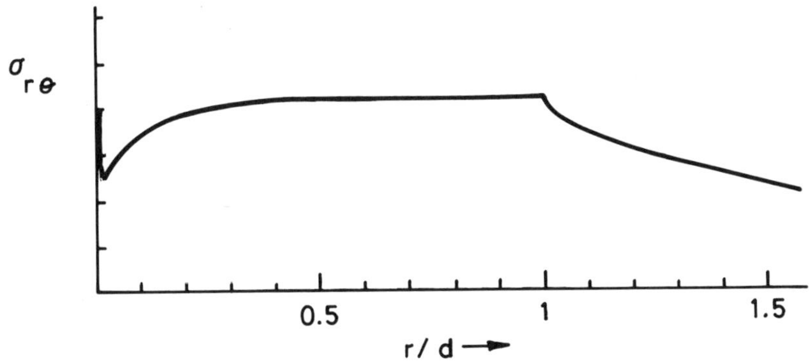

Figure 8. Shear stress $\sigma_{r\theta}$ on planes at ±45° for situation of figure 7, with dislocation distributions taken from BCS solution.

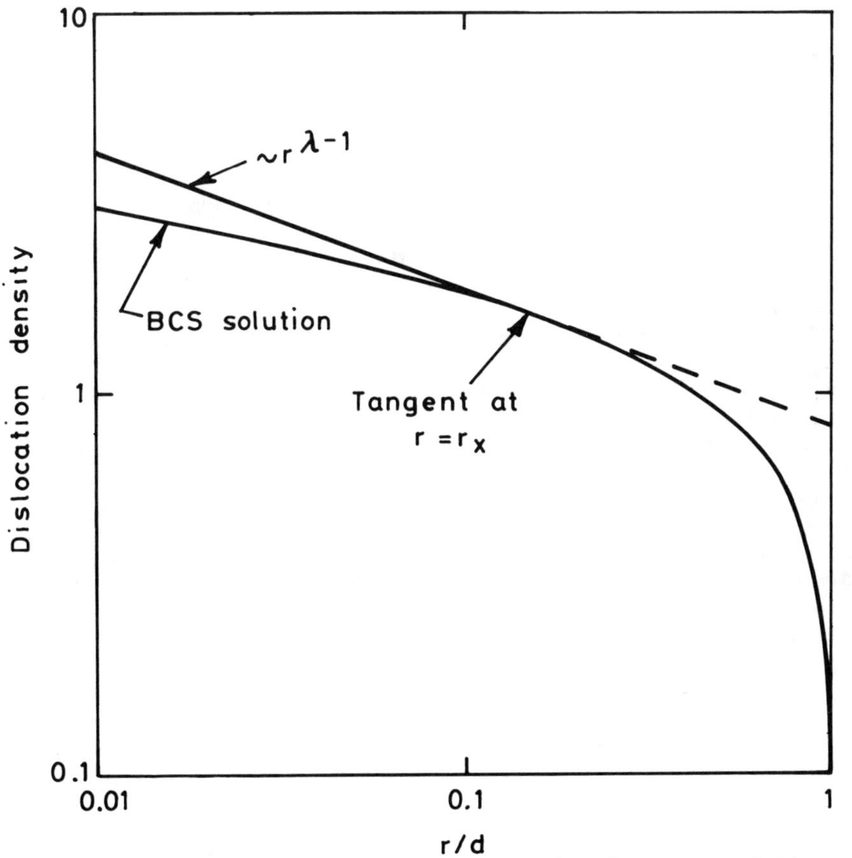

Figure 9. Splicing dislocation distribution from BCS solution with $r^{\lambda-1}$ form at point of tangency.

complex logarithm), while other stress components diverge as $\ln r$. This local form, however, is not valid in the geometry where three sectors meet. Just as Williams (1952) solved the corner problem in isotropic plane elasticity, it is possible to find local solutions to many sector problems, even in anisotropic elasticity, by postulating a solution which separates in r and θ, with the r dependence taking a power form.

In the present problem, a particular solution is provided by a uniaxial stress field parallel to the crack, with amplitude selected to give the desired constant shear on the yield planes. This leaves a homogeneous problem, since any additional terms must add zero to the stresses appearing in the boundary conditions at the crack faces and yield planes, and also satisfy certain continuity conditions (zero differences) at the yield planes. In the Appendix, a standard notation for anisotropic two-dimensional elastic solutions is introduced. Here, it is only necessary to appreciate that the 3 variables z_α are complex combinations of x_1 and x_2 (they all tend to x_1+ix_2 in the isotropic limit). If the form

$$f_\alpha(z_\alpha) = C_\alpha^{(n)} z_\alpha^\lambda \tag{12}$$

is postulated for the complex stress functions in each sector (n = 1, 2, 3), then the interface conditions result in a set of homogeneous linear equations in the coefficients $C_\alpha^{(n)}$. If displacements in 3 dimensions are allowed, there are 18 equations in 9 complex constants. These have solution only if their determinant is zero, and this condition provides a transcendental equation in λ. The smallest positive root gives the dominant singularity (stresses go like $r^{\lambda-1}$). This calculation was programmed for arbitrary anisotropy; but selecting nearly isotropic elastic constants, with Poisson's ratio 0.3, the lowest root was found to be $\lambda = 0.64702$ for the case with slip planes at $\pm 45°$. The discontinuity in tangential displacement at the slip planes varies as r^λ, which can be interpreted as a distribution of dislocations with density per unit length varying (like the stresses) as $r^{\lambda-1}$.

We now have a form of solution which should dominate at small r, and another (the BCS form) which appears to work well for larger r. It is thus worth investigating whether a simple splice of the two will yield a second approximation which significantly improves upon the first. The dislocation density in the limit of the BCS solution as crack length tends to infinity takes the form

$$f(r) = H \ln\{\sqrt{d/r} + \sqrt{d/r - 1}\}, \tag{13}$$

where d is the length of the slip band, and Hd is the total plastic displacement at $r = 0$. Plotting dislocation density vs. r on a log-log graph, Figure 9 shows the form of the BCS solution with its weak $\ln r$ singularity, and how an $r^{\lambda-1}$ form can be selected which joins with continuity of slope as well as value. It is simple enough to show that the condition for thus joining always has exactly one root. The equation to select the change-over point, r_x, cannot be solved analytically, but numerical location is very easy. In the example of 45° inclination, the solution is r_x/d = 0.1650.

Using this simple technique of splicing two analytic forms, the shear stress was computed along the slip bands for a range of slip plane angles. The results are shown in Figure 10. The stress is now finite in the limit $r \rightarrow 0$ (this only holds on the planes of course). While some 'kinks' remain between r = 0 and the splicing point, the result is certainly an improvement upon the first approximation, although it works better for some slip angles than others.

Although further improvement could be sought by splicing a third section in between the two known end forms, it would probably be better to use the full numerical treatment without any special singular elements. In principle, it is quite possible to implement the methods of section 2 using the Greens functions for a body with semi-infinite crack rather than for a continuous infinite body. However, the weighted integrals (9) cannot be performed analytically with the cracked-body form of Q_{ij} as they can in the infinite-body case. On the other hand, if the normal method is used, the crack has to be included in the boundaries to be solved, and a finite crack length must therefore be used. A further complication in the present application was that with stress gradients being required, the piece-wise linear form for $[u_i](s)$ would give singularities at the knots, making it difficult to follow defect drift paths up to the boundaries.

Finally, it may be of interest to display a sample result for the subsequent calculations of point defect drift. Figure 11 shows some computed trajectories of steepest descent in the field of σ_{ii} (these actually relate to the first approximation using only the BCS–type distributions). The markers along the paths are at selected values of drift time for trajectories which begin either at the crack tip or from the slip band, and travel in the direction of decreasing tensile stress (as would under-sized defects such as vacancies). For drift in the opposite direction (over-sized defects), the paths are identical, but the time markers would have different origins. Contours connecting points of equal drift time coincide with contours of the steady state defect concentration in the situation where the defect production rate is constant in time and space, and there is a uniform sink strength throughout the body. In the transient situation with an initial uniform concentration, and no sources or sinks other than the crack faces or tip, the drift solution is simply constant ahead of, and zero behind, a moving surface which also coincides with the above mentioned contours for the steady state case. It is of interest that although the bi-axial tension in the vicinity of the crack tip repels vacancies, their drift is directed spirally by the angular variation of the field, so that they are still captured by the crack, but on its faces behind the tip. The drift paths are fairly near to being tangential to the slip planes, so that accounting for their presence does not greatly affect the net defect fluxes around the crack tip.

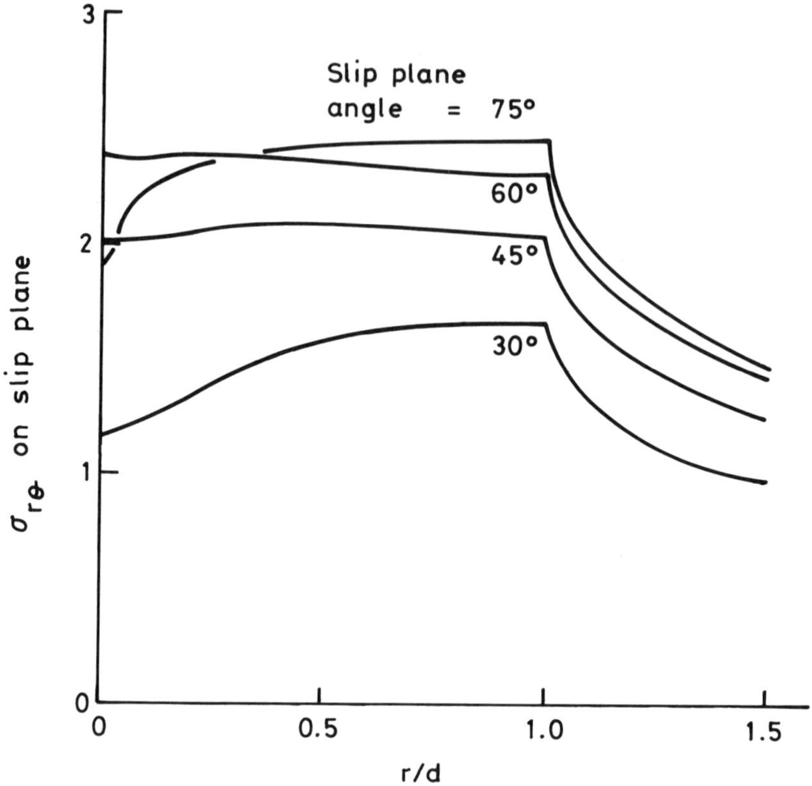

Figure 10. Shear stress $\sigma_{r\theta}$ on slip planes for situation of figure 7, with various slip plane inclinations as shown, using two-part spliced dislocation distributions.

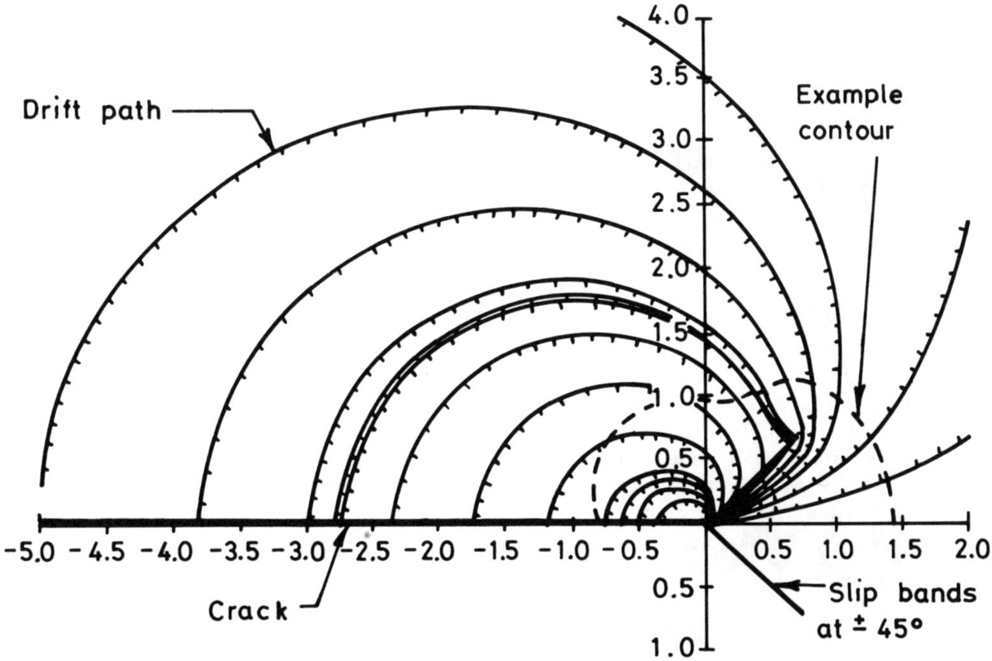

Figure 11. Selected paths of point defect drift diffusion in stress gradients produced by a loaded crack with yield bands. Contours of drift time from common source coincide with concentration contours (see text).

APPENDIX

For two-dimensional anisotropic elasticity, we use the notation of Stroh (1958, 1962). Note that minor differences of nomenclature exist between different authors using this notation. In particular, Barnett, Asaro, Lothe and their co-workers (see for instance references in Asaro 1975), take a different sign for the constants $L_{i\alpha}$.

With everything being independent of one coordinate, x_3, the six complex variables

$$z_\alpha = x_1 + p_\alpha x_2, \quad \alpha = 1,...6 \tag{A1}$$

are used, where the p_α are the roots (all complex) of a real sextic polynomial. Unless otherwise stated, all quantities with a subscript $\alpha = 4$, 5 or 6 are defined as the complex conjugates of the corresponding elements with $\alpha = 1$, 2 or 3. By convention, the first three p_α have positive imaginary parts. Any equilibrium stress field can then be written in terms of analytic functions $f_\alpha(z_\alpha)$, as follows. The stresses are given by

$$\sigma_{i2} = \partial\phi_i/\partial x_1, \qquad \sigma_{i1} = -\partial\phi_i/\partial x_2, \tag{A2}$$

with

$$\phi_i = \sum_{\alpha=1}^{6} L_{i\alpha} f_\alpha(z_\alpha) = 2\,Re \sum_{\alpha=1}^{3} L_{i\alpha} f_\alpha(z_\alpha). \tag{A3}$$

Similarly, the displacements are given by

$$u_i = 2\,Re \sum_{\alpha=1}^{3} A_{i\alpha} f_\alpha(z_\alpha). \tag{A4}$$

For definitions of $L_{i\alpha}$ and $A_{i\alpha}$, which like the p_α depend on the elastic constants, see Stroh (1958). Note that summation over α will always be indicated explicitly, whereas the implicit summation convention will be used for repeated tensor (Latin) subscripts.

The infinite-body solution for a dislocation with Burgers vector b_i lying in the x_3 direction at (x_1',x_2') takes the form

$$f_\alpha(z_\alpha) = \frac{L_{j\alpha} b_j}{\pm 4\pi i A_{k\alpha} L_{k\alpha}} \ln(z_\alpha - z_\alpha'). \tag{A5}$$

The kernel $Q_{ij}(\mathbf{x},\mathbf{x}')$, whose meaning is given following eq. (5) in the main text, can therefore be obtained from (A5) by two differentiations: one to obtain stresses from f_α (eqs. A2, A3), and the other to give the result for a dislocation dipole. Tensor resolution factors are also necessary to give the tractions across S from the stresses, and to align the dipole along S (at different points). Finally, the double integral (9) must be performed to obtain the matrix elements Q_{injm}. If the contour of S in the x_1x_2 plane consists of straight line segments joining 'knot points', then the basis functions $f_{in}(s)$ for piece-wise linear $[u_i]$ along S are in any segment varying either from 0 to 1 or from 1 to 0. Apart from summation over α and constant factors, the integrals (9) therefore take the form

$$\int_{Z_0}^{Z_1}\int_{Z_0'}^{Z_1'} \frac{z-Z_0}{Z_1-Z_0}\,\frac{z'-Z_0'}{Z_1'-Z_0'}\,\frac{dz'\,dz}{(z-z')^2}\ . \qquad (A6)$$

Here Z_0 and Z_1 represent values of some z_α at knots where the weight function takes values 0 and 1 respectively. To obtain a full Q_{injm} value, the 'descending' half of the distribution at both the source- and field-point end must be added. This gives some cancellation of terms, and the final contribution to Q_{injm} corresponding to (A6) turns out to be

$$\sum_{\alpha=1}^{6} \frac{L_{i\alpha}L_{j\alpha}}{\pm 8\pi i A_{k\alpha}L_{k\alpha}}\, F[z_\alpha(n), z_\alpha(n-1), z_\alpha'(m), z_\alpha'(m-1)], \qquad (A7)$$

where

$$\begin{aligned} F[Z_1,Z_0,Z_1',Z_0'] \;=\; & \frac{1}{(Z_1-Z_0)(Z_1'-Z_0')} \times \\ & \times\{(Z_1-Z_1')^2\ln(Z_1-Z_1') - (Z_1-Z_0')^2\ln(Z_1-Z_0') \qquad (A8)\\ & - (Z_0-Z_1')^2\ln(Z_0-Z_1') + (Z_0-Z_0')^2\ln(Z_0-Z_0')\}\cdot \end{aligned}$$

REFERENCES

Altiero, N J, & Gavazza, S D, 1980, J. Elas. **10,** 1.

Asaro, R J, 1975, J. Phys. F: Metal Phys. **5,** 2249.

Bilby, B A, & Swinden, K H, 1965, Proc. Roy. Soc. **A285,** 22.

Bilby, B A, Cottrell, A H, & Swinden, K H, 1963, Proc. Roy. Soc. **A272,** 304.

Heise, U, 1978, J. Elas. **8,** 47.

Hipplsey, C A, Rauh, H, & Bullough, R, 1984, Acta Met. in press (AERE Harwell Report R–11120).

Hirth, J P, & Wagoner, R H, 1976, Int. J. Solids Structures **12,** 117.

Rice, J R, 1974, J. Mech. Phys. Solids **22,** 17.

Rice, J R, 1984, Eshelby Memorial Symposium, University of Sheffield, 2–5 April, 1984, proceedings to be published.

Rice, J R, & Thomson, R M, 1974, Phil. Mag. **29,** 73.

Rizzo, F J, 1967, Q. appl. Math. **40,** 83.

Sinclair, J E, 1976, in *Computer Simulation for Materials Applications,* Nuclear Metallurgy vol. 20, p. 388.

Sinclair, J E, & Finnis, M W, 1983, in *Atomistics of Fracture,* ed. R M Latanision & J R Pickens, Plenum, p. 1047.

Sinclair, J E, & Hirth, J P, 1975, J. Phys. F: Metal Phys. **5,** 236.

Stroh, A N, 1958, Phil. Mag. **3,** 625.

— 1962, J. Math. & Phys. **41,** 77.

Vitek, V, 1976, J. Mech. Phys. Solids **24,** 67.

Williams, M L, 1952, J. appl. Mech. **19,** 526.

Wong, G K, & Barnett, D M, 1984, Eshelby Memorial Symposium, University of Sheffield, 2–5 April, 1984, proceedings to be published.

BASIC ASPECTS OF MICROSTRUCTURAL EVOLUTION UNDER HIGH TEMPERATURE IRRADIATION

G.MARTIN[+] et P. VALENTIN
Section de Recherches de Métallurgie Physique,
C.E.N.S. - S.R.M.P.
91191 - GIF SUR YVETTE CEDEX, FRANCE
[+] and C.E.C.M. 15, rue Georges Urbain
94400 - VITRY SUR SEINE, FRANCE

ABSTRACT. In wiew of the impossibility to propose theoretically established simple scaling laws for extrapolating microstructural evolution to unknown irradiation conditions, a full modelization of microstructural evolution at the atomistic level cannot be avoided.

We briefly review the main models available for describing : defect balance under irradiation, the nucleation of cluster of various types, the development of each component of the microstructure, synergistic effects among the latter. Attention is called on the problems which remain to be solved at each step. In particular, the swelling incubation phenomenon is just being studied from the fundamental view-point. A table of available relevant observations thereof is given.

The existence of dose-rate thresholds accross which microstructural evolution undergoes a qualitative change is stressed : such thresholds call for a detailed modelization of microstructural evolution in order to propose safe extrapolation techniques.

1. INTRODUCTION

Modelizing microstructural evolution under "high temperature" irradiation (i.e. at temperatures where point defects are mobile) is both of practical and fundamental interest.

From the fundamental view point, one has first to discover the elementary mechanisms which, at the atomistic scale, are at the origin of macroscopic changes. As far as swelling, growth, creep and embrittlement are concerned, the elementary mechanisms are understood, even if much remains to be done in the details ; moreover the degree of knowledge is much more advanced for metals than for oxydes or carbides.

The field of fundamental research, however, goes beyond these elementary aspects. Indeed, the dynamics of microstructural evolution is an interesting basic question : how ,does the microstructure evolve as a whole, once the rules of evolution of each component of the microstructure are known. The answer to this question is not trivial. Interesting synergestic effects have been studied recently.

From the practical view-point, one may wonder why macroscopic

G. Jacucci (ed.), Computer Simulation in Physical Metallurgy, 183–208.

swelling laws as a function of irradiation conditions would not suffice. Indeed if such laws can be established locally they cannot be justified theoretically and are therefore of small help when one is asked to predict macroscopic behaviours under yet unexplored irradiation conditions. The reason for this deficiency is very basic indeed : it is now well recognized that thresholds exist, for the irradiation conditions (temperature T and flux ϕ) accros which the microstructural evolution undergoes qualitative changes. As a consequence elaborating models of the macroscopic behaviour under irradiation which are based on a description of microstructural evolution at the atomistic level, is a task of real practical relevance.

In this lecture, we give the state of the art of modelizing microstructural evolution, at the atomistic scale first, then at the scale of the components of the microstructure and finaly at the macroscopic level. This paper is by no means an exhaustive review of the field but rather gives entry points in a litterature which is often redundant and where various schools sometimes ignore one another.

Reference [1] is a good introduction to the field. Every year several meetings cover this field.

2. ATOMISTIC SCALE

At this level, descriptions are simple in principle : point defect balance is written as a function of time. The defect balance may be handled by various techniques : Monte Carlo method or chemical rate theory, the latter being written to various level of sophistication. The purpose of such models is to determine the number of point defects which eliminate on various types of defect clusters and to deduce the evolution rate thereof.

2.1. Monte Carlo simulation

Monte Carlo technique has been used to simulate the evolution of cascade cores [2], the agglomeration of vacancies in quenched metals [3,4] or defect agglomeration (vacancies and interstitials) under irradiation [5,7].

The method is simple in principle : at each time, the configuration of the system is defined by the coordinates of each defect (agglomerated or not). At each time step, the configuration is modified with a probability defined a priori and reproduced by the Monte Carlo algorithm. Two versions of the method have been used. In the first version [2,3,7] the defect positions are defined on the crystalline lattice and the confirguration evolves due to defect jumps from site to site. In the second version [4-6], the medium is defined at a coarser scale by the positions of isolated defects, complexes, aggregates, the latter being defined by their size, their composition (e.g. number of trapped-solute- defects- gas atoms) and their law of capture of isolated point defects i.e. the probability for a defect at distance r to be captured by the cluster after time t.

Such methods are in principle extremely powerfull, since they take into account spatial heterogeneities, possible interference effects between various mechanisms of defect elimination, and spatial and temporal heterogeneities in the production of point defects. The price for this extreme wealth of possibilities is computer time : up to now this technique has been confined to the study of complex nucleation problems (not tractable with other techniques) such as : the role of displacement cascades or the effect of dislocation vicinity, in the nucleation of voids in the presence of low concentration insoluble gas atoms [5], the effect of dilute solute elements on void nucleation [6].

Owing to this latter technique a correlation between the swelling of pure Cu under e^-, ion, and neutron irradiation could be established [5].

2.2. Rate theory

In its simplest version, the rate theory consists in uniformizing the medium and writing balance equations for the concentrations of various species (point defects, solute atoms, transmutation products...). This is the most commonly used tool, and as such was developpped in the early days of the void swelling problem [8–10].

Typically rate equations write, e.g. for vacancy (resp. interstitial) concentrations, C_v (resp. C_i) :

$$\frac{\partial C_v}{\partial t} = G_v - RC_iC_v - \sum_\alpha k^2_{\alpha v} D_v (C_v - C^o_v) \tag{1a}$$

$$\frac{\partial C_i}{\partial t} = G_i - RC_iC_v - \sum_\alpha k^2_{\alpha i} D_i C_i \tag{1b}$$

where G_v is the vacancy production rate, R the rate constant for the interstitial-vacancy recombination reaction, D_v, D_i the diffusion coefficients for vacancies and interstitials, C^o_v the equilibrium vacancy concentration $k^2_{\alpha v}$ and $k^2_{\alpha i}$, the strength of α type sinks for respectively vacancies and interstitials.

Notice that eq. 1 describe in a deterministic manner the evolution of a vancancy concentration uniquely defined at each time. In reality, the vacancy source G_v fluctuates in time and eq.1 should be considered as governing the mean value of the vacancy concentration. Classical techniques are available to describe such fluctuations [11]. They consist either in adding a fluctuating source in the RHS of eq. 1 (Langevin type equation) or in replacing eq.1 by an evolution equation for the probability that the defect concentration has the values C at time t (Fokker-Planck type equation). A Langevin type equation was used by Krishan [12] for studying swelling : the possibility for transient swelling resulted. Gurol [13] discussed the fondations and use of such techniques. Fokker-Planck type equations are currently being used for nucleation problems [14–16], and more rarely for irradiation creep studies : indeed whenever the average dislocation velocity is controled by the time required to by-pass obstacles by climb, fluctuations in the defects arrival rate at dislocations will affect the strain rate [17].

With the exception of some works [12–17], most studies rely on equations like eq. 1. The definition of parameters in eq. 1 raise some problems which we now briefly discuss.

2.2.1. Sink strengthes : $k^2_{\alpha v}$ and $k^2_{\alpha i}$ in eq. 1 are supposed to be additive, independant of one another. They are deduced from the steady state solution of the diffusion problem around the sink of interest considered as an isolated sink. Ref. [18,19] give simple derivations of such sink strengthes. The validity of such expressions has been assessed in cases where the irradiation flux is pulsed : in most cases, the steady state values still correctly describe defect balance.

The additivity of sink strengths has been studied into details [21]. It is well known that the rate of defect capture on a population of spherical sinks increases more than linearly with the sink density [22]. As far as dislocations are concerned the stress field roughly vanishes at half the dislocation interspacing. The drift force for defects towards dislocations is therefore a function of dislocation density [19]. Sink interference effects on sink strengthes may be large [21]. The full treatment of such interference effects, including vacancy-interstitial recombination is complicated [23]. Advanced codes do include such effects ; they are however neglected in simpler models which are aimed more at revealing interesting mechanisms rather than at giving quantitative predictions.

A delicate problem deals with the difference in sink strengthes for vacancies and interstitials. Swelling implies a flux of vacancies towards cavities higher than the interstitial flux = the larger drift force of interstitials as compared to vacancies towards dislocations is taken for responsible for such an effect. The dislocation bias ϵ_d, is defined as the relative excess of dislocation sink strength for interstitials with respect to that for vacancies :

$$\epsilon_d = \frac{k^2_{di} - k^2_{dv}}{k^2_{dv}} \qquad (2)$$

Theoretical estimates of dislocation bias are still controversial. The simplest estimates compute the defect flux in the stress field of the dislocation, assuming a constant diffusion coefficient. The stress field is of course a function of the dislocation density [19]. The next approximation includes the stress dependance of the defect diffusion coefficient [24]. More recently, it was realized that the dislocation bias is a function of the solute content of the matrix. Indeed the stress field of a mixed dumbbell is smaller than that of a dumbbell : the drift force of interstitials towards the dislocation is decreased [25,26]. In some models, defect elimination at dislocations is localized at kinks : a reaction step at the dislocation core is taken into account, the rate of which is temperature dependant [27]. The dislocation bias is therefore a characteristic of the microstructure rather than of the individual dislocations. The value of ϵ_d is not well established, and ranges between 10^{-2} and 10^{-1}.

Cavities are also biased sinks, due to the image force acting on point defects in the vicinity of the free surface of the cavity [28].

Void bias must be smaller than dislocation bias for voids to grow. Void bias depends on void size. The sign of the void bias may be reversed if the cavity is coated by a stiff solute shell [29].

2.2.2. Modelizing defect production in cascades : Two modifications are brought in principle by cascades in the above description :

Defects are no more produced uniformly in space and time. At a given point of the material one rather expects to receive blasts of point defects, damped as a function of the distance to the cascade they originated from. Such effects are difficult to modelize with the rate theory but have nervertheless been studied [30].

Important effects are anticipated for nucleation of clusters and for the rate at which dislocations by-pass obstacles by climb.

A second effect of cascades is easier to modelize namely cascade core collapse into vacancy loop, which is currently observed at high temperature [31]. As a consequence, the number of mobile vacancies emitted from the cascade is decreased. In the B.E.K. model [10] such an effect is modelized by adding two equations to eq. 1 : one describes the change in the number density of vacancy loops = N_{vl} :

$$\frac{d\,N_{vl}}{dt} = KG - \frac{N_{vl}}{\tau} \qquad (3a)$$

the second one describes the evolution of the number of vacancies contained in the loop :

$$\frac{dq_v}{dt} = xKG - (\text{defect balance at the loop}) \qquad (3b)$$

in eq. 3a and 3b, G is the defect production rate, ($G = G_v$ eq. 1), K the number of collapsed cascades per defect produced, τ the life time of a loop, x the number of vacancy per collapsed cascade.

2.2.3. Solute effects. Solute are treated either as immobile or as mobile. When immobile, solute simply act as traps for point defects and therefore decrease the defect diffusion coefficient [32] :

$$D_\alpha^{eff} \simeq D_\alpha \frac{1}{1+K_\alpha C_s} \qquad (4)$$

where $\alpha = i$ or v and K_α is the association constant for the solute-defect complex and C_s is the solute concentration. When the mobility of solute is taken into account, two techniques are currently being used :

- mobile complexes are being-defined. Extra balance equations have to be added to eq. 1 The most extensive use of this technique was done by Johnson and Lam [33,34]. Six coupled equations are then used instead of eq. 1.
- the solid solution under irradiation is treated as an effective medium with four components : solvant atoms, solute atoms, vacancies and interstitials. Due to lattice site conservation, three constituants only are independant : the free and associated defect populations are supposed to be in thermal equilibrium : the defect concentrations are

therefore overall concentrations [35]. The diffusion in such a model is described by a diffusion matrix, with nine coefficients. The flux of one given species is a linear combination of the concentration gradients of the three independant species :

$$\vec{J}_\alpha = - \sum_\beta D_{\alpha\beta} \vec{\nabla} C_\beta \qquad (5)$$

with α,β = i, v. or solute. The latter diffusion coefficients may be expressed in terms of defect jump frequencies according to standard techniques [36]. Such models are currently being used for describing solute redistribution under irradiation and solute effect on the bias of various sinks [37].

3. NUCLEATION OF CLUSTERS

A large variety of clusters may nucleate under irradiation : vacancy-, interstitial-, solute- or mixed clusters (like solute decorated void or interstitial loop, precipitate associated to a void etc...). Nucleation models all rest on balance equations for the capture and emission rates of the various constituants of the cluster. As an example, a cluster made of v vacancies and s solute evolves in time according to :

$$\begin{aligned} \frac{dv}{dt} &= k_v^2\,(s,v)\; D_v C_v - k_i^2\,(s,v)\; D_i C_i - e_v\,(s,v) \\ \frac{ds}{dt} &= k_s^2\,(s,v)\; D_s C_s - e_s\,(s,v) \end{aligned} \qquad (6)$$

$k_\alpha^2(s,v)$ is the sink strength for the element α (α = i,v,s) of a cluster made of s solute atoms and v vacancies ; C_α is the concentration and D_α the diffusion coefficient of specie α in the matrix ; $e_\alpha(s,v)$ is the emission frequency of specie α from a cluster (s,v). When e_v and e_s are known, a master equation may be built for the concentration of clusters ρ (s,v) [38] ; the latter equation is then solved numerically :

$$\begin{aligned} \frac{d}{dt}\,\rho\,(s,v) = &- [e_s(s,v) + e_v(s,v) + a_s(s,v) + a_v(s,v) + a_i(s,v)]\rho\,(s,v) \\ &+ a_s\,(s-1,v)\rho\,(s-1,v) + a_v\,(s,v-1)\rho\,(s,v-1) \\ &+ e_s\,(s+1,v)\rho\,(s+1,v) + e_v\,(s,v+1)\rho\,(s,v+1) \\ &+ a_i\,(s,v+1)\rho\,(s,v+1) \end{aligned} \qquad (7)$$

In eq. 7 a_α (s,v) is the absorption frequency of an α monomer on the cluster (s,v). Numerous studies based on this technique have been published [39-41].

Another technique, more suggestive, helps visualizing the possible nucleation mechanisms. It was developped by Russell and coworkers [42]. Indeed eq. (6) define in the plane (s * v) the velocity field of the point which represents the cluster (s,v). This space is divided by the so called nodal lines (lines along which dv/dt = 0 or ds/dt =0 into domains inside which the cluster evolves in a unique direction. (e.g. growth of s, decay of v, or growth of s and v etc...). Figure 1 is an example of the a priori possible configuration of the nucleation pathes for a cavity in the presence of inert gas atoms (x = number of gas atoms, n = number of vacancies) [42]. In (a) no barrier opposes cavity growth ; in (b) the accumulation of gas atoms in the clusters will drive the nucleation process ; (c) is a case of homogeneous nucleation. Based on this formalism, Russel proposed expressions for the stationary value of the nucleation flux, for the incubation time, and for the asymptotic void number density. Russell also uses this technique for quantitative estimates in cases of practical interest [43-45].

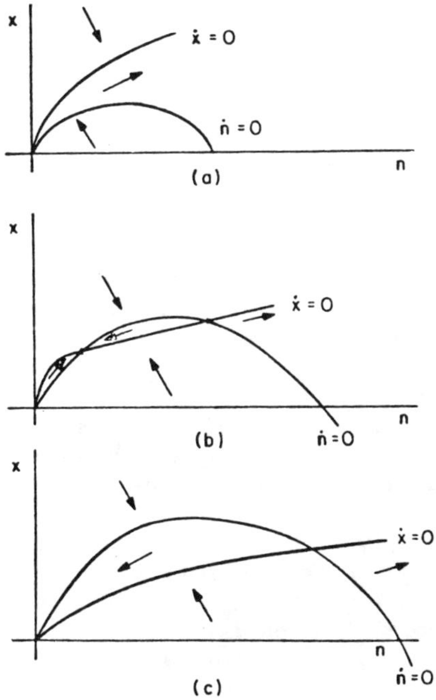

FIG. 1 - Void nucleation in the presence of gas atoms : [42]
x : number of gas atoms
n : number of vacancies
Along the nodal line, one of the components of the velocity of the representative point of the cluster goes to zero.

One important difficulty in this type of model is to evaluate the emission rates e_α (s,v). The latter coefficients are deduced from the absorption frequencies, according to the detailled balance argument : under equilibrium conditions : e.g.

$$\bar{\rho}\,(s,v)\; k_s^2\,(s,v)\; D_s \bar{C}_s = \bar{\rho}\,(s+1,v)\; e_s\,(s+1,v) \qquad (8)$$

where the quantities marked ($\bar{.}$) have their equilibrium value. It is easily shown that :

$$e_s\,(s+1,v) = k_s^2\,(s,v)\; D_s \bar{C}_s \exp \frac{\delta \Delta F(s,v)}{\delta s} \times \frac{1}{kT} \qquad (9)$$

where ΔF (s,v) is the free energy change on forming a cluster of the type (s,v).

At the time being no general formula is available for ΔF (s,v). For single constituant clusters (e.g. voids, solute clusters...) the capillarity approximation may be used :

$$\Delta F\ (n) = n\ \Delta\mu + n^{2/3}\Sigma \qquad (10)$$

where n is the number of particles in the cluster, $\Delta\mu$ the chemical potential difference of the particle between the two phases (dilute and dense) and Σ the specific surface free energy. This latter model is based on a macroscopic representation of the cluster. It was recently shown that several corrections had to be added to the RHS of eq. 10 [46]: in a crystal, the cluster free energy contains together with the bulk and surface terms, contributions of the edges and of the vortices : noticeable corrections result which affect the nucleation rate [47].

For gas containing clusters, various equations of state have been used for the gas such as, perfect gas approximation, van der Waals type equations or liquid [48].

For more complex clusters, cluster free energy models do not exist. In practice empiricaly estimated binding energies are used.

4. ALLOYS

The question of phase stability under irradiation is addressed in another lecture. Points relevant to this chapter are briefly summarized here. Under irradiation, the alloy structure is affected because of three different processes :

- point defects are being stored in the lattice
- whenever point defects are mobile, a dynamic equilibrium is established between the chemical mixing forced by atomic collisions and the radiation enhanced diffusion back to lower energy configurations.
- whenever the defect fluxes are strongly coupled to solute fluxes, solutes atoms accumulate at point defect sinks (or migrate away therefrom) or in regions of preferential vacancy-interstitial recombination. A non uniform alloy composition results.

The stored energy is now taken to be responsible for radiation induced amorphization.

The dynamical equilibrium between forced disorder and enhanced recovery results in enhanced solubility, lowering of the degree of long range order in ordered alloys, and in destabilization of compounds with high ordering energies. This problem has been recently reformulated [50]. In simple cases, it is shown that such effects may be treated as an increase in effective temperature under irradiation : the configuration to be stabilized is the configuration the alloy would developp outside irradiation at a temperature T' larger than the irradiation temperature T.

Fig. 2. is an example of the relationship between T' and T for a given irradiation flux, and with parameters values typical of HVEM irradiation of a refractory alloy [58]. At high temperature, T' ~ T the alloy configuration is essentially unaffected. At lower temperature, T' increases abruptly which means that more and more disordered configurations are being developped. According to Fig. (2b), configurations may be reached where an amorphous phase would coexist with an ordered phase.

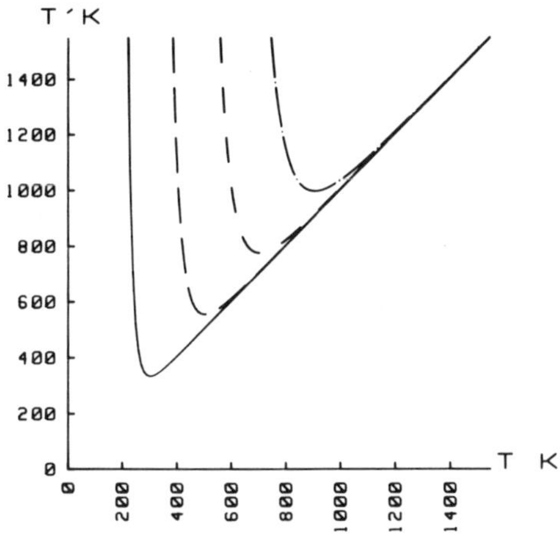

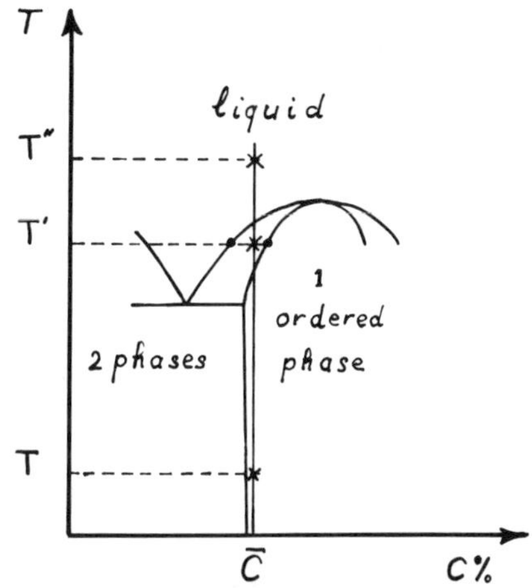

FIG. 2 – Ballistic effects on phase stability [50] :

a) under irradiation at temperature T, the alloy reaches the configuration it would have at equilibrium outside irradiation at temperature T'. The curves correspond to decreasing vacancy migration energies from the right to the left.
b) depending on irradiation temperature and flux, the effective temperature may be T'' or T'. At T'' one expects an amorphous alloy structure, while at T' a two-phase equilibrium is expected with the coexistence of an amorphous and an ordered phase with different compositions.

As far as solute flux – point defect flux coupling is concerned, it is now recognized to be at the origin of irradiation induced precipitation in undersaturated alloys.

As shown on Fig.3, the space of the control parameters of the state of an alloy under irradiation (temperature * irradiation flux) is divided into domains. The microstructure of the alloy undergoes a qualitative change when crossing the borders of these domains : e.g. inbetween the boundaries drawn on Fig. 3, the solid-solution is made of two phases, while it is single phase outside these boundaries [35].

The very existence of such boundaries casts doubts on dose equivalencies between various irradiation conditions. Indeed in pure metals, the swelling peak is known to be shifted by about 50 K to higher temperatures when the dose-rate is increasesd by a factor of 10. As shown on Fig.3, such a rule may not apply to alloys : e. g. at 600°C with a flux of 10^{-5} dpa s^{-1} the alloy is a single phase one, while at 500°C 10^{-7} dpa s^{-1} the alloy is in a two phases state.

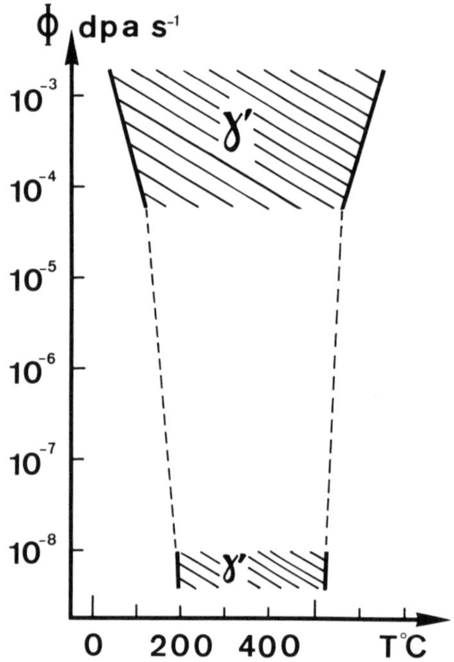

FIG. 3 - 1 MeV electron irradiation conditions (irradiation flux G and temperature T) for which a Ni- 4 at% Si alloy is in a two phase state (between the borders) or in a single phase state (outside the borders) [35].

Let us stress also that the location of the phase stability boundaries as shown on Fig. 3, is sensitive to the defect production mode : under ion irradiations (cascades) the two phases domain is narrower than under electron irradiations (Frenkel pairs), the shift of the low temperature boundary being larger than that of the high temperature one. Finally, the incubation dose for precipitation is weakly flux dependant (some 10^{-2} dpa for 10^{-4} dpa s^{-1} versus 10^{-3} dpa for 5 10^{-5} dpa s^{-1}).

A more complete account of the above phenomena is given in ref. [51].

5. SCALE OF THE MICROSTRUCTURE

Void and loop growth have been described extensively. Models for the growth of precipitates are more rare, and very few models exist for the development of the dislocation network. We briefly review the field.

5.1. Loop and void growth

Measuring interstitial loop growth-rate under irradiation is a good tool for studying point defect elementary properties and production mechanism [52]. Such studies produced displacement threshold energy values [53], interstitial migration energy values (as deduced from the thickness of the denuded zone at the surface of a thin foil in the HVEM [54], vacancy migration energy and dislocation bias (as deduced from the growth rate of loops at temperatures where both vacancies and interstitials are mobile [55]).

As far as void growth is concerned, such studies are typical swelling studies (cf. Ch. 6). Let us call attention to the models of precipitate void association [56], (where the precipitate matrix interface acts as a collector for point defects which then diffuse to the associated void), and to the models of coated voids (where a stiff coating may bias the void against interstitial capture [57]).

5.2. Precipitate coarsening

The following effects have been studied quantitatively :

- irradiation enhanced solute diffusion will accelerate coarsening [58]
- the competition between the above effect and radiation induced resolution of peripheral solute atoms [59].
- the irreversible deposition of solute atoms at the periphery of precipitates as a consequence of the recombination of a mixed-dumbbell and a vacancy trapped at the precipitate [60].
- vacancy elimination at a precipitate-matrix incoherent interface favoring the release of the stress developped by precipitate growth [61].

It is difficult to assess such models experimentaly due to the simultaneous evolution of other elements of the microstructure. In particular, in simple alloys, one often observes a renucleation of precipitates due to various mechanisms of solute resolution [62]. Let us quote a recent experiment in CuBe where the saturation of precipitate size at large dose is attributed to solute ejection in the matrix, by irradiation [63].

5.3. Dislocation network

Few experimental studies are available, concerning the development of the dislocation network under irradiation [64-66]. As a first approximation, the dislocation density seems to evolve towards a stationnary value at large dose, whatever the starting configuration (Fig.4).

Apart from M. Finnis et al. model to be discussed into details in a next lecture, phenomenological models have been proposed [67-68]. The dislocation density evolves due to the competition between irradiation induced densification and irradiation enhanced annealing of the network:

$$\frac{d\rho}{dt} = A\rho^{m} - B\rho^{n} \qquad (11)$$

The condition $m < n$ insures that the steady state solution $\rho = (A/B)^{n-m}$ is a stable one, while $\rho = 0$ is not.

At the atomistic scale, the driving force for dislocation motion is the preferential absorption of interstitials at dislocations which results in an effective climb force [69]. The exponents m and n may achieve several values depending on the controling mechanism :

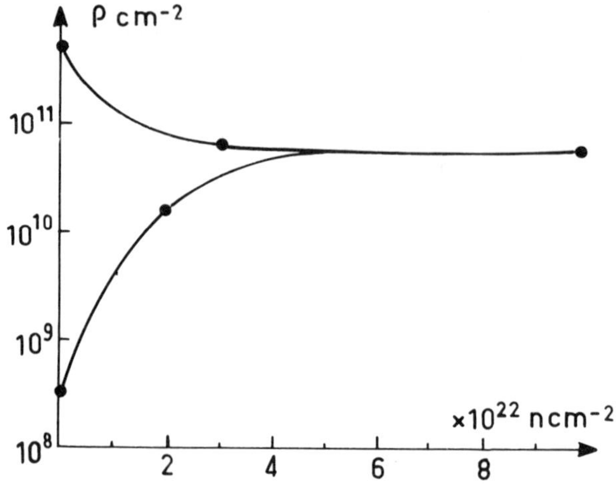

FIG. 4 – Stationnary dislocation density as predicted by eq. 11 [68]

– m = 1 if the increase in dislocation density is due to dislocation arc bending under the effect of the effective climb force just discussed [68]

– m = 1/2 if the densification of the dislocation network results from dislocation cell division by nucleation and growth of loops inside the cells [67]

– n = 2 if network annealing is controled by the thermal emission of point defects at jogs [68]

– n = 3/2 if it occurs by dislocation dipoles annihilation [67].

Available data may be fitted with m = 1 and n ≃ 1,2 [68]. Eq. 11 however is certainly too naïve. It has been shown [68] that the dislocation density under irradiation is not uniquely defined. In a dislocation network which had been well stabilized by HVEM irradiation, a short duration irradiation at a lower temperature promoted dislocation loop nucleation in the dislocation cells. Further irradiation at the nominal temperature promoted loop growth and subsequent division of the cells i.e. an increase in the dislocation density. A simple model for this phenomenon revealed that for each irradiation flux, at a given temperature, a range of stable steady state dislocation densities is accessible. Hysteresis effects on changing the irradiation flux are to be expected [68].

5.4. Synergisms in microstructural evolution

Many models, are available for describing the coupled evolution of the components of the microstructure. Most models, however, are numerical and do not give clear view of the mechanisms by which such evolutions are intercorrelated.

In order to overcome this difficulty, we proposed a general technique which we have applied to a simple case [70] : the microstructure is represented by sink densities and sink sizes (e.g. cavity and loop radii). Rate equations are used to describe the time evolution of the vacancy and interstitial concentrations on the one hand and of the radii (dR_c/dt, dR_l/dt) on the other hand. Due to the rapid relaxation of point defect concentrations to their steady state value in a fixed microstructure, it is justified to make a quasi steady state assumption for the defect concentration in the presence of the slowly varying microstructure ($dC_i/dt = 0$). Such an approximation allows for the elimination of the defect concentrations from the rate equations describing the radii. As a consequence the rate of change of the radii are expressed as a function of the values of the radii only. A map of a priori possible evolutions of the microstructure can thus be drawn : Fig. 5 exhibits the configuration space of a microstructure which is defined by the curvatures of the cavities and loops. The space is divided into four domains which differ by the direction of microstructural evolution. Studying such maps revealed unexpected properties of a priori possible microstructural evolutions.

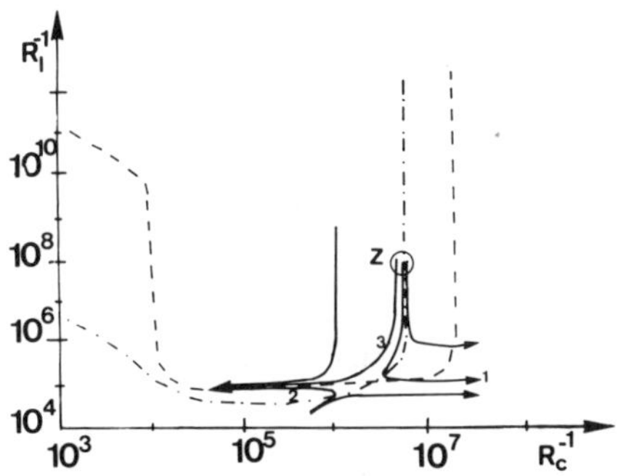

FIG. 5 – Trajectories of the representative point of a microstructure in the space of the void and loop curvature (R_c^{-1}, R_l^{-1}). The doted lines represent the nodal lines.

Let us quote [70] :

- the impossibility for permanent swelling to occur below a dose-rate threshold in certain microstructures
- the possibility for a transient swelling (trajectory number 1 on Fig.5) or a transient sintering (trajectory number 2)
- the possibility for certain microstructures to developp along very similar pathes at low doses and then diverge at larger doses (e.g. microstructures starting from region Z in Fig. 5)

This is a possible origin for the large scatter of swelling in swelling resistant materials.

Other effects may be studied using this technique, in particular the effect of dose rate or temperature changes in the course of irradiation : a shift of the domains would result in Fig. 5.

6. DESCRIBING SWELLING AT THE MACROSCOPIC LEVEL

Recent studies [91,91-bis] of the swelling of cladding steels in Rapsodie and Phenix reactors have revealed that "swelling resistant" steels are indeed steels with a large incubation dose for swelling. Indeed the amount of swelling which is achieved at large dose is dominated by the transient behaviour at lower dose (fig. 6). As shown by table I given in the appendix, the incubation phenomenon for swelling is very usual indeed : it is neither a pecularity of industrial alloys nor

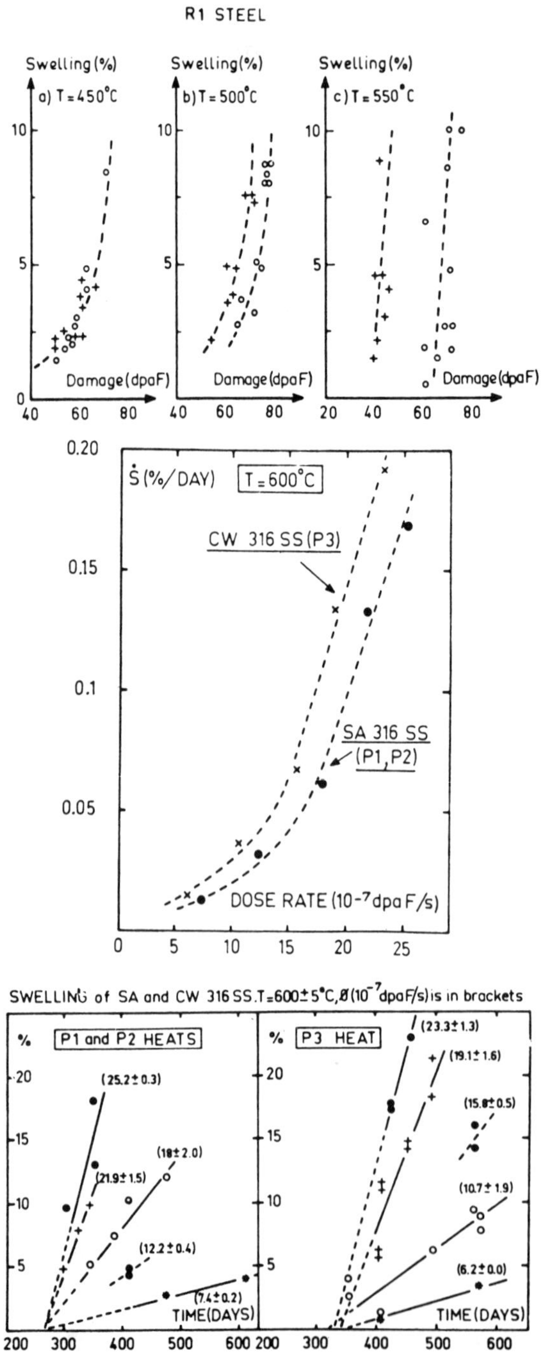

FIG. 6 - Swelling of fuel claddings :

6a Rapsodie [91]
6b Phenix [91bis]
6c Phenix [91bis] = the time axis is labelled in days, and not in dpa.

a pecularity of neutron irradiations. In simple alloys, the transient period is longer the lower the purity or the less aged is alloy. Three types of mechanisms may contribute to the incubation dose :

a) developping a dislocation network which allows for swelling : the dislocation density must be high enough, but the pinning points must be distant enough or weak enough for continuous dislocation climb to be possible

b) gas atom production or detrapping for the nucleation of voids to be abundant

c) microchemical evolution, i.e. the solute redistribution between the matrix, the precipitates or segregation shells around point defects sinks.

Mechanisms (b) and (c) have been modelized into a quantitative manner [71 - 74]. The effect of helium generation and detrapping from dislocations on the growth of preexisting void nuclei has been studied into details [71 - 72]. A large incubation dose is reproduced, together with the effects of the neutron energy spectrum on the latter (i.e. helium production to dpa ratio), provided that initial microstructures typical of those obtained under the various irradiation conditions be chosen.

In a highly quantitative manner, Bullough et al. [72] could extrapolate the swelling of 316 steels under neutron irradiation from 39 dpa to 62.5 dpa, once the parameters of the model had been calibrated from swelling studies under electron and ion irradiations.

The effect of microchemical evolution on swelling incubation has been recently modelized [73 - 74]. Solute in the matrix affects the driving force for swelling in various ways :

a) slowing down interstitials increases the fraction of recombined Frenkel pairs (Ch. 2.2.3.)

b) forming mixed dumbbells the strain field of which is smaller than that of pure dumbbells decreases the drift flux of interstitials towards dislocations [75 - 76]

c) segregating at point defect sinks, solute may create a diffusion barrier (or diffusion short circuit [77] for point defects arriving at the sink, or alter the bias of the sink [57].

A simple rate theory model of items (a) and (b) has been proposed [74] ; solute drag by the flux of point defects towards sinks, and solute agglomeration at the sinks result in a decay of the solute content of the matrix and therefore increase the swelling rate. Fig. 7 exhibits typical swelling dose curves as deduced from the model. One notices an incubation period followed by a high swelling - rate regime. For a given dose, the differences in swelling essentially result from differences in the incubation dose. Finally dose rate effects in the incubation dose are complex in the details but may be seen as resulting from a shift of the swelling peak towards higher temperatures. The latter model qualitatively reproduces many trends observed under technological conditions.

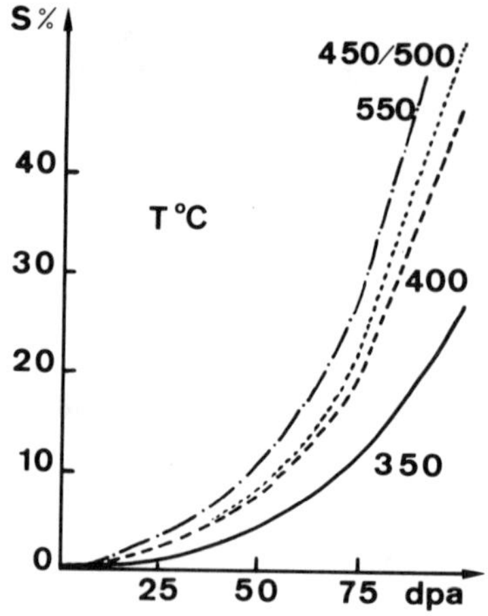

FIG. 7 – Typical variation of swelling with dose at various temperatues [74].

7. CONCLUSION

Let us stress the following points :

1) good mathematical tools are available for describing microstructural evolution. Improvements remain to be brought to Monte Carlo codes for dealing with large doses, and to the chemical rate theory for taking fluctuations into account ;

2) as far as nucleation is concerned, the main problem remains the choice of a good model for cluster – monomer binding energies ;

3) modelizing the evolution of the dislocation network under irradiation is just improving ;

4) very good models for the growth of a given microstructure are available and allow for quantitative extrapolation of the swelling from medium to large doses ;

5) fundamental studies of the initial transient swelling are just beginning ;

6) when extrapolating known microstructural evolution to different irradiation conditions, one must take into account the possible existence of dose rate or temperature thresholds accross which microstructural evolution may follow a different route.

ACKNOWLEDGEMENTS

The interest of Dr. Y Adda in this work and many usefull discussions with our colleagues are gratefully acknowledged.

TABLE I

Métal	Irradiation conditions	Observations	Références
Al	n 55°C, 10^{14} ncm^{-2} s^{-1} E > 0.1MeV	Short incubation dose in Al doped with 650 weight ppm impurities; Not in high purity Al	78
Mo, Mo + Re	1MeV e^-, 630°C at 1130°C 4 10-4 dpa s^{-1}	Pure and alloyed Mo up to 2.6 at.% Re. Adding Re increases the dislocation density and the incubation dose	79
Mo, Mo + Zr	14MeV Cu^+ 100 to 900°C 5 to 8 dpa s^{-1}	Zero incubation dose in pure Mo at 900°C; 2 dpa incubation dose in Mo + 9.1 at.% Zr (two phases alloy)	80
Cu, Cu + Fe	1MeV e^- 300°C 9 10-4 dpa s^{-1}	Zero incubation dose in pure Cu. 7 dpa incubation dose in Cu-1.5 weight % Fe solution annealed incubation dose decreases on aging the alloy	81
Fe, Mo, Nb	MeV e^-, 10^{-3} dpa s^{-1} 300 to 450°C	Some dpa incubation dose in Fe and Mo	82
Fe	n, 1MeV e^-, 2.8 to 5MeV Ni^+, 350 to 450°C	Incubation dose decreases in going from irradiation by ions, to electrons, to neutrons	83
Fe 15Cr 25Ni	n, 1MeV	"ADIP" programm. The incubation dose decreases on going from ion, to electron, to neutron irradiation	84
Fe 15Cr 20Ni 85Fe 15Cr High purity 316 Commercial 316	Ni^+	Under Ni^+ irradiation, the incubation dose is extremely high. 316 commercial steel has a higher incubation dose than high purity steel	84bis

316 steels	1MeV e^-, 500°C 10^{-3} dpa s^{-1}	Incubation dose is about 10 dpa i.e. roughly one fifth that for neutron irradiations	85
316 AISI	n. 400 to 620°C 10^{23} n cm^{-2}	Whether solution annealed or 20% cold worked, such steels exhibit a variation of incubation dose with Si content	86
316 and 304 steels	n. 440 to 650°C 1023 n cm^{-2}	Cold work increases the incubation doseat 440°C with respect to solution annealed alloys. 316 steels exhibit larger incubation dose.	87
304 steel	n.	In the range 465-530°C the incubation dose is constant while the swelling rate increases	88
316 steels	1MeV e^- 1.8 10^{-3} dpa s^{-1}	Some dpa incubation. Incubation dose is sensitive to the content in minor elements	89
316 steels	n. DFR	In the solution treated state, temperature effect on swelling results from the increase in incubation dose with temperature	90
316 steels	n. Phenix and Rapsodie	Solution treated fuel cladding : below 500°C, the incubation dose changes with temperature, not with dose rate; swelling inhibitors increase the duration of the incubation period. Above 500°C, incubation dose increase with dose rate, swelling inhibitors are less effective.	91
316 steels	n. Phenix	Incubation dose is proportionnal to irradiation flux and is weakly temperature dependant. It is larger for cold worked materials; swelling rate increases with the flux.	91bis

316 steels	n. 510 to 610°C	The incubation dose of solution annealed steels range from 15 to 30 dpa.	92
316 steels	n. 370 to 650°C	Certain heats exhibit a highly temperature sensitive incubation dose.	93
316 steels	Ni^+ and He^+ dual irradiation 625°C $3\,10^{-3}$ dpa s^{-1} 5,15,50 appm He/dpa	5 and 15 appm He/dpa result in about 15 dpa incubation dose while 50 appm He/dpa results in a lower incubation dose.	94
316 steels 316 + Ti 316 + Si	1MeV e^- 1.3 dpa s^{-1} 300 to 700°C	20% cold worked steels. Adding solute decreases the swelling rate without affecting the incubation dose (≃ 2 dpa). Aging increases the incubation dose.	95
316 steels	N^+ 1.1.MeV 450 to 650°C	Long incubation period during which the void number density seems to be stable.	96
316 steels	C^+ 200keV Ni^+ 5MeV e^- 1MeV	10 to 20 dpa incubation dose; increased by cold working.	97
Modified 316 Nimonic Incalloy	Ni^+ 5 10^{-3} dpa s^{-1} e^- 1MeV	Very large incubation dose (≃ 100 dpa) for Incalloys at the swelling peak, under Ni^+ ion irradiation. Incubation dose is shorter inder e^- irradiation.	98
Austenitic Nimonic Incalloy	C^{2+} 20MeV, Ni^{6+}46.5MeV, Cr^{2+}52MeV: 1 to 3.10^{-2} dpa s^{-1} e^-1MeV: 8 10^{-3} dpa s^{-1} 450 to 900°C	Incubation dose is ≃ 10 dpa in 316 steels under 1MeV e^- irradiation; very large incubation dose in PE 16 under Ni^+ ion irradiations.	99

Austenitic steels	Ni^+	Cold work enhances the incubation dose (40 dpa vs 20 dpa for solution annealed steels).	100
304,316,321 type steels	neutrons	Incubation dose and swelling rate depend on the type of steel, on the degree of cold work and on temperature.	101
A 286 Commercial Fe 25Ni 15Cr	Ni^+ 3 to 7 10^{-3} dpa s^{-1} 450 to 650°C	10 to 20 dpa incubation dose depending on the heat treatment, and the preinjection.	102
Fe Ni Cr precipitate Strenglhened	Ni^+ 5MeV 5 10^{-3} dpa s^{-1} 500 to 700°C	145 dpa incubation dose. Aging shortens the incubation period.	103
Commercial Fe Ni Cr	HVEM 4 10^{-3} dpa s^{-1} 450 to 650°C	Large variety of alloys, some of which with large incubation. For PE 16, incubation dose under e^- irradiation (10 dpa) is much shorter than under ion irradiation.	104
PE 16	Neutron irradiations followed by 4MeV Ni^+ ion irradiations with 7 10^{-3} dpa s^{-1}	Neutron irradiations at 454, 538 and 593°C inducing swelling ranging from 0.28 to 1.48% are followed by Ni^+ ion irradiations at temperature ranging from 640 to 700°C: swelling restarts only after a very large incubation dose (50 to 100 dpa). At the higher temperatures, very large swelling heterogeneities are observed (20% of the observed areas have a short incubation and a large swelling rate; the reverse is true for the 80% remaining areas).	105
304 steels	HVEM	N doped steels. The incubation dose is independent on the N content.	106

REFERENCES

[1] For an introduction : cf "Radiation damage in metals" by PETERSON and HARKNESS eds, American Society for Metals, Metals Park, Ohio, USA, 1976

[2] DORAN (D.G.), Rad. Eff., 1970,**2**, 119

[3] a) BEELER (R.J.) and JOHSON (R.A.), Phys. Rev.1967,**156**, 677
b) BEELER (R.J.), in : "Radiation Effects Computer Experiments", Defects in Solids, vol.**13**, North Holland, Amsterdam, 1983

[4] LEVY (V.), LANORE (J), Phil. Mag., 1973,**28**, 389

[5] GLOWINSKI (L.D.), LANORE (J.M.), FICHE (C.), ADDA (Y), J. Nucl. Mat., 1976,**61**, 41

[6] MAURICE (F.), DOAN (N.V.), Rapport C.E.A. R 5101, 1981

[7] MUROGA (T.), ISHINO (S.), in : "Point defects and defects interactions in Metals", TAKAMURA, DOYAMA and KIRITANI eds, University of Tokyo Press, 1982, p. 895 ; idem, J. Nucl. Mater., 1972, **43**, 59

[9] WIEDERSICH (H.), Rad. Eff., 1972, **12**, 111

[10] BULLOUGH (R.), EYRE (B.L), KRISHAN (K.), Proc. Roy. Soc., 1975, **A346**, 81

[11] For a presentation cf. HAKEN (H.), "Advanced Synergistics", Springer, Berlin, 1983, p. 282

[12] KRISHAN (K.), Rad. Eff., 1980, **45**, 169 ; Phil. Mag. 1981, **A43**, 71

[13] GUROL (H.), J. Appl. Phys. 1979, **50**,2705 and 1980, **51**, 15 ; J. Nucl. Mater. 1980, **90**, 133

[14] FICHE (C.), LOTT (M.), Note Interne C.E.A DEDR 32/73 SPE

[15] WOLFER (W.G.), MANSUR (L.K.) SPRAGUE (J.A.), in : "Radiation Effects in Breeder Reactor Structural Materials", BLEIBERG and BENNETT eds, AIME, NEW-YORK, 1977, p. 841

[16] KITAJIMA (K.), cf. Ref. 7, p. 847

[17] GUROL (H.), WOLFER (W.G.), J. Nucl. Mater., 1983, **117**, 244

[18] NICHOLS (F.A.), J. Nucl. Mater., 1978, **75**, 32

[19] HEALD (P.T.) and SPEIGHT (M.V.), Acta Met., 1975, **23**, 1389

[20] RAUH (H.), BULLOUGH (R.) and WOOD (M.H.), Proc. Roy. Soc. (London) 1983, **A388**, 311

[21] BRAILSFORD (A.D.) and BULLOUGH (R.), Phil. Trans. Roy. Soc. (London) 1981, **A302**, 87

[22] FLYNN (C.P.), Point Deffects and Diffusion, Clarendon, Oxford, 1972, 487

[23] RAUH (H.), WOOD (M.H.), BULLOUGH (R.), Phil. Mag., 1981, **44**, 1255

[24] a) DEDERICHS (P.H.) and SCHROEDER (K.), Phys. Rev., 1978, **B17**, 2524
b) WOO (C.H.) and SAVINO (E.J.), J. Nucl. Mater., 1983, **116**, 17

[25] WERTMAN (J.) and PARKIN (D.M.), J. Nucl. Mater., 1981, **99**, 66

[26] VALENTIN (P.) and MARTIN (G.), Phil. Mag., in the press

[27] BULLOUGH (R.) and QUIGLEY (T.M), cf. Ref.[7], p. 839

[28] WOLFER (W.G.) and ASHKIN (M.), J. Appl. Phys., 1975, **46**, 547 and 4108

[29] WOLFER (W.G.) and MANSUR (L.K.), Phys. Stat. Sol., 1976, **a37**, 211

[30] a) MANUSUR (L.K), COGHLAN (W.A.) and BRAILSFORD (A.D.), J. Nucl. Mater. 1979, **85** and **86**, 591
b) MARWICK (A.D.), J. Nucl. Mater., 1983, **116**, 40

[31] For a review : MERKLE (K.L.), Radiation Damage in Metals, Ref.[1], p. 58

[32] MANSUR (L.K.), Acta Met., 1980, **29**, 375

[33] JOHNSON (R.A.) and LAM (N.Q.), Phys. Rev., 1976, **B13**, 4364

[34] WIEDERSICH (H.) and LAM (N.Q.), in : "Phase Transformations during Irradiation", NOLFI eds, Appl. Sci., London, 1983, p. 1

[35] MARTIN (G.), CAUVIN (R.) and BARBU (A.), Ref.[34], p. 47

[36] ALLNATT (A.R), BARBU (A.), FRANKLIN (A.D.), LIDIARD (A.B.), Acta Met., 1983, **31**, 1307

[37] WOLFER (W.G.), J. Nucl. Mater., 1983, **114**, 292

[38] For a review : WIEDERSICH (H.) and KATZ (J.L.), Adv. Col. Interf. Sci. 1979, **10**, 33

[39] HALL (B.O.), J. Nucl. Mater., 1980, **91**, 63

[40] GHONIEM (N.M.), SHARAFAT (S.), J. Nucl. Mater., 1980, **92**, 121

[41] WOLFER (W.G.), SI-AHMED (A.), Phil. Mag., 1982, **46**, 723

[42] RUSSEL (K.C.), Acta Met., 1978, **26**, 1615

[43] RUSSEL (K.C.), Acta Met., 1971, 19, 753 : 1972, **20**, 899

[44] MAYDET (S.I.) RUSSEL (K.C.), J. Nucl. Mater., 1979, **82** 271

[45] PARKER (C.A.), RUSSEL (K.C.), J. Nucl. Mater., 1983, **119**, 82

[46] PERINI (A.), JACUCCI (G.), MARTIN (G.), Phys. Rev., 1984, **B29**, 2689

[47] PERINI (A.), JACUCCI (G.), MARTIN (G.), Surf. Sci., 1984, **144**, 53

[48] WOLFER (W.G.), GLASGOW (B.B.), WEHNER (M.F.), TRINKAUS (H.), J. Nucl. Mater., 1984, in the press

[49] LIMOGES (Y.), BARBU (A.), Phys. Rev., 1984, **B30**, 2212

[50] MARTIN (G.), Phys. Rev., 1984, **B30**, 1424

[51] MARTIN (G.), BARBU (A.),Acta Scripta Met. : "Early Stages of Alloy Decomposition", 1984, in the press

[52] For a review : REGNIER (P.), J. Microsc. Spectrosc. Electron., 1981, **6**, 345

[53] e.g. BOURRET (A.), Phys. Stat. Sol., 1971, **4**, 813

[54] HOUSSEAU (N.), PELLISSIER (J.), J. Nucl. Mater., 1983, **119**, 317

[55] STATHOPOULOS (A.Y.), MURPHY (S.M.), WOOD (M.H.), BULLOUGH (R.) ENGLISH (C.A.), J. Nucl. Mater., 1982, **110**, 301

[56] MANSUR (L.K.), Phil. Mag., 1981, **44**, 867

[57] WOLFER (W.G.), MANSUR (L.K.), J. Nucl. Mater., 1980, **91**, 265

[58] POTTER (D.I.),WIEDERSICH (H.), J. Nucl. Mater., 1979, **83**, 208

[59] a) SKLAD (P.S.), MITCHELL (T.E.), Acta Met., 1975, **23**, 1287
b) FROST (H.J.), RUSSELL (K.C.), Acta Met., 1981, **30**, 953

[60] URBAN (K.), MARTIN (G.), Acta Met., 1982, **30**, 1218

[61] MAYDET (S.I.), RUSSELL (K.C.), J. Nucl. Mater., 1977, **64**, 101

[62] GELLES (D.S.), in : "Solid – Solid Phase Transformations", ARRONSON (H.I.) et col. eds, AIME, New York, 1982, p. 293

[63] KOCH (R.), WAHI (R.P.), WOLLENBERGER (H.), J. Nucl. Mater., 1981, **103** and **104**, 1211

[64] AZAM (N.), LE NAOUR (L.), DELAPLACE (J.), J. Nucl. Mater., 1973/74 **49**, 197

[65] HUDSON (J.A.), J. Nucl. Mater., 1976, **60**, 89

[66] BRAGER (H.R.), GARNER (F.A.), GILBERT (E.R.), FLINN (J.E.), WOLFER (W.G.), cf ref.[15], p. 727

[67] GARNER (F.A.), and WOLFER (W.G.), in : "Effects of Radiation on Materials", BRAGER and PERRIN eds, ASTM–STP 782, Philadelphia (U.S.A.), 1982, p. 1073

[68] VALENTIN (P.) and MARTIN (G.), Note technique interne C.E.A. SRMP 5698, 1980

[69] WIEDERSICH (H.), HERSCHBACH (K.), Scripta Met., 1972, **6**, 453

[70] VALENTIN (P.) and MARTIN (G.), Phil. Mag., 1982, **A46**, 971

[71] STOLLER (R.E.), ODETTE (G.R.), J. Nucl. Mater., 1981, **103–104**, 1361

[72] BULLOUGH (R.), MURPHY (S.M.), QUIGLEY (T.M.), WOOD (M.H.), J. Nucl. Mater., 1983, **117**, 78

[73] VALENTIN (P.) and MARTIN (G.), in : "Dimensional Stability and Mechanical Behaviour of Irradiated Metals and Alloys", British. Nucl. Soc. London, 1983, vol 2, p. 17

[74] VALENTIN (P.) and MARTIN (G.), Phil. Mag., in the press

[75] WEERTMAN (J.) and PARKIN (D.M.), J. Nucl. Mater., 1981,**99**, 66

[76] VALENTIN (P.) and MARTIN (G.), Phil. Mag. in the press

[77] BOQUET (J.L.), Phil. Mag., 1983, **A47**, 547

[78] RISBET (A.), BREBEC (G.), LANORE (J.M.), LEVY (V.), J. Nucl. Mater. 1975, **56**, 348

[79] IGATA (I.), KOHYAMA (A.), NOMURA (S.), J. Nucl. Mater., 1981, **103** and **104**, 1157

[80] LIOU (K.Y.), WILKES (P.), KULCINSKI (G.L.) and BILLEN (J.H.), J. Nucl. Mater., 1979, **85** and **86**, 735

[81] TAKEYAMA (T.), OHNUKI (S.) and TAKAHASHI (H.), J. Nucl. Mater., 1980,**89**, 253

[82] KITAJIMA (K.), FUTAGAMI (K.) and KURAMOTO (E.), J. Nucl. Mater., 1979,**85** and **86**, 725

[83] KURAMOTO (E.), YOSHIDA (N.), TSUKUDA (N.), KITAJIMA (K.K), PACKAN (N.H.), LEWIS (M.B.) and MANSUR (L.K.), 2nd Topical Meeting Fusion React. Mat., Seattle 1981, p. 1091

[84] GARNER (F.A.), POWELL (R.W.), DIAMOND (S.), LAURITZEN (T.), ROWCLIFFE (A.F.), SPRAGUE (J.A.), KEEFER (D.), cf. Ref.[15], p. 543

[84bis] JOHNSTON (W.G.), LAURITZEN (T.), ROSOLOWSKI (J.H.), TURKALO (A.M.), Ref. [1], p. 227

[85] LEVY (V.), J. Micros. Spectr Electr., 1981, **6**, 415

[86] BATES (J.F.), POWELL (R.W.) and GILBERT (E.R.), in : "Effects of Radiation on Materials", DRAMER et al. eds, ASTM-STP 725, Philadelphia (U.S.A.), 1981, p. 713

[87] WEINER (R.A.), BOLTAX (A.), cf. Ref. [86], p. 484

[88] BRAGER (H.R.), GARNER (F.A.), GILBERT (E.R.), FLINN (J.E.), WOLFER (W.G.), in : "Radiation Effects in Breeder Reactor Structural Materials", BLEIBERG et al. eds, AIME, New-York, 1977, p. 727

[89] LAIDLER (J.J.), GARNER (F.A.), THOMAS (L.E.), Ref. [1], p. 194

[90] BRAMMAN (J.I.), BROWN (C.), WATKIN (J.S.), COWTHORNE (C.), FULTON (E.J.), BARTON (P.J.), LITTLE (E.A.), Ref. [88], p. 479

[91] SERAN (J.L.), DUPOUY (J.M.), in : "Effects of Radiation on Materials", BRAGER and al. eds, ASTM-STP 782, Philadelphia, (U.S.A.), 1982, p. 5

[91bis] SERAN (J.L.), DUPOUY (J.M.), cf. Ref. [73] vol. 1, p. 15

[92] LE NAOUR (L.), VOUILLON (N.), LEVY (V.), Ref. [91], p. 310

[93] YANG (W.J.S.), GARNER (F.A.), Ref. [91], p. 186

[94] AYRAULT (G.), HOFF (H.A.), NOLFI (F.V.), TURNER (A.P.L.), J. Nucl. Mater., 1981, **103** and **104**, 1035

[95] IGATA (N.), KOHNO (Y.), SAITO (M.), TSUNAKAWA (H.), J. Nucl. Mater., 1981, **103** and **104**, 1047

[96] SHIRAISHI (K.), ARUGA (T.), KATANO (Y.), J. Nucl. Mater., 1981, **103** and **104**, 1053

[97] TERASAWA (M.), SHIMADA (M.), KAKUMA (T.), YUKITOSHI (T.), SHIRAISHI (K.) and UEMATSU (K.), cf. Ref. [88], p. 687

[98] BLEIBERG (M.L.), BAJAJ (R.), BELL (W.L.) and THOMAS (L.E.), cf. Ref. [88], p. 667

[99] MAKIN (M.J.), HUDSON (J.A.), MAZEY (D.J.), NELSON (R.S.), WALTERS (G.P.) and WILLIAMS (T.M.), Ref. [88], p. 645

[100] EHRLICH (K.), GROSZ (R.), and SCHNEIDER (W.), Ref. [88],p. 529

[101] APPLEBY (W.K.), BLOMM (E.E.), FLINN (J.E.) and GARNER (F.A.), Ref. [88], p. 509

[102] CHICKERING (R.W.), BAJAJ (R.), LALLY (J.S.), Ref. [88], p. 405

[103] BAJAJ (R.), DIAMOND (S.) and BLEIBERG (M.L.), Ref. [91], p. 856

[104] THOMAS (L.E.) and GELLES (D.S.), J. Nucl. Mater., 1982, **108** and **109**, 490

[105] BAJAJ (R.), DIAMOND (S.), CHICKERING (R.W.) and BLEIBERG (M.L.), cf. Ref. [86], p. 541

[106] IGATA (N.), KOHNO (Y.), TANABE (N.), FUJIHIRA (T.) TSUNAKAWA (H.), Ref. [91], p. 906

[107] GARNER (F.A.), J. Nucl. Mater., 1983, **117**, 177

NUMERICAL SOLUTION OF THE MASTER EQUATION FOR CLUSTER POPULATION IN THE CLASSICAL NUCLEATION THEORY.

A. PERINI
Istituto per la Ricerca Scientifica e Tecnologica
38050 Povo
Italy

ABSTRACT. We review the classical description of the decomposition processes in a low supersaturated alloy (classical nucleation theory). The rate equation for the cluster population is derived. First we write the steady state solution of this rate equation and discuss its meaning in the nucleation theory and in the interpretation of the real nucleation processes. The crucial dependence of the steady state nucleation rate on the choice of the model for the cluster free energy is stressed. The time evolution of cluster distribution obtained solving numerically the rate equation is compared with that resulting from a simulation experiment of decomposition processes in the Ising model.

1. INTRODUCTION

The classical nucleation theory is commonly used to describe the decomposition processes which characterize the metastable states of systems like undercooled liquids or supersaturated alloy. This phenomenological approach appeared, about 40 years ago /1/, in the literature which concerns the description of the initial stage of vapour condensation but it can be adapted to describe the kinetics of the beginning of any first phase transition. In this contribution we will consider the decomposition in two phases with different solute concentration which appears when an initially homogeneous binary alloy is quenched under, but not too far from the coexistence line. The density fluctuations which characterize this metastable state are described as non interacting clusters of the nucleating phase, hence the time evolution of the system can be analysed in terms of time evolution of the cluster population. Let us say a few words about the definition of cluster. Of course it depends on the model assumed to describe the system : if continuous potentials are considered we may require that each atom of the cluster lies within a

G. Jacucci (ed.), Computer Simulation in Physical Metallurgy, 209–223.

distance r_c of another atom of the same cluster(r_c may be related to the range of interaction); on a lattice we may require that each atom of the cluster is nearest neighbour of another atom of the same cluster. The rate equation for cluster population will be derived in sec. 2. The steady state solution of this equation with particular boundary conditions, is discussed in sec. 3 while its time dependent solution is numerically computed as specified in sec. 4. The comparison of the time evolution of cluster population obtained from the numerical solution of the rate equation with that resulting from a direct computer simulation of the decomposition processes, is discussed.

2. RATE EQUATION FOR CLUSTER POPULATION

In its classical form the nucleation theory rests mainly on two assumptions :

a) the clusters are characterized only by the number of contained atoms. Parameters like shape, surface/volume ratio, kinetic energy... do not individuate different classes of clusters hence the system evolution will be represented as cluster size pattern.

b) clusters grow and shrink due to absorption and emission of single solute atoms.

From a) and b) the rate equation which gives the time evolution of the concentration of N-particles cluster, ϱ_N is :

$$\frac{\partial}{\partial t}\varrho_N(t)=-(b_N(t)+f_N(t)\varrho_N(t))+\varrho_{N-1}(t)f_{N-1}(t)+\varrho_{N+1}(t)b_{N+1}(t) \qquad N\geq 2$$

$$\frac{\partial}{\partial t}\varrho_1(t)=\sum_2(b_N(t)-f_N(t))+b_2\ \varrho_2-2f_1\varrho_1 \tag{1}$$

$b_N(t)$ and $f_N(t)$ represent the probability that the N-particles cluster shrinks to size N-1 or grows to size N+1 respectively (monomer emission and absorption coefficient). Usually eq. (1) is rewritten in terms of the nucleation rate $J_N(t)$, defined by :

$$J_N(t) = \varrho_N(t)f_N(t)-\varrho_{N+1}(t)b_{N+1}(t) \tag{2}$$

J_N represents the net number of N+1 clusters formed from N-clusters, per unit volume and per unit time. Substituting J_N in eq.(1) we obtain the equivalent equation :

$$\frac{\partial}{\partial t}\varrho_N(t) = J_{N-1}(t)-J_N(t)\ .$$

$$\frac{\partial}{\partial t}\varrho_1(t) = -\sum_{N=1} J_N - J_1 \tag{3}$$

The typical problem one finds using the nucleation theory is the definition of the coefficients b_N and f_N (see eq.(1) and eq.(2)). Furthermore even with the simplest choice for these coefficients, numerical methods are needed to solve eq.(3). In section 2. we discuss the steady state solution of eq.(3) without any other particular assumption on b_N and f_N besides the requirement that they will be time-independent in order to assume the steady state.
In sec.4 we present the results of the numerical solution of eq.(1) which involves particular models for b_N and f_N.
Concluding this section let us mention a more general approach to the description of nucleation processes due to Binder and Stauffer /2/. Here we recall the main ideas of this model without going into deeper details. The hypothesis a) and b) are removed :
- a cluster is defined by some "relevant coordinates", (for example cluster volume, surface area...), and some "irrelevant coordinates";
- reactions involving a couple of generic clusters are considered. Memory effects (respect to the relevant coordinates) on the probability of these reactions are excluded. Binder and Stauffer's rate equation for cluster population contains a first term analogous to the r.h.s. of eq. (3) and a second term which takes into account processes of coagulation of two generic clusters. Neglecting the coagulation term and using only one relevant parameter (the cluster size) they obtain the results of the classical eq.(3).

3. STEADY STATE SOLUTION

In this section we will consider the steady state solution of eq.(3), that is the solution which satisfy the condition :

$$\frac{\partial}{\partial t} \varrho_N = 0 \qquad J_N = const. \tag{4}$$

for each value of N, and its crucial dependence on the cluster free energy. Some considerations about the validity of the steady state concept in the nucleation theory and in the nucleation experiments are appropriate.
The decomposition phenomena are typical non-steady state, time-dependent processes : the system evolves from states with a given concentration of solute atoms toward final states where most part of the solute clusters coagulate in macroscopic domains of the new phase, the original phase resulting depleted in its solute atoms concentration. In other words the supersaturation decreases during the experiments and the conditions defining our problem are no more constant. This is not the case, for example, of voids nucleation under irradiation, where the

radiation supplies a continuum source of point defects.
The following considerations could justify the use of the steady state solution which allows one to avoid the problems involved in the direct solution of eq.3.
A typical nucleation experiment can be divided in 3 periods :
- a short period (for example in a vapour condensation process it can last about ten μsec.) when the cluster population grows rapidly up to a not too large size. The duration of this "incubation" period depends on the observation techniques.
- a longer period (for the vapour condensation from ten to hundred msec) characterized by a quasi steady-state production of clusters of a supercritical size.
- a last period when supercritical clusters coalesce. It is during this "coarsening growth" regime that the supersaturation falls rapidly.
The second period is successfully described by a steady state where the concentration of clusters of each size does not change with time. The nucleation theory describes this situation solving eq.(3) provided an artificial steady state is defined where a "demon" continuously reevaporates the clusters much larger than the critical size ($\varrho_G=0, G>N^*$) and the solute concentration has nearly the thermal equilibrium value. The solution is insensitive to changes in the limit size G as verified a posteriori :
Let us derive the classical expression for the steady state nucleation rate J_S.
From the steady state conditions (eq.4)), we obtain :

$$J_1=J_2=\dots=J_S \tag{5}$$

The absorption and emission coefficients will be time independent moreover they must satisfy the detailed balance condition at thermal equilibrium (when $J_S=0$):

$$\bar{\varrho}_N \cdot \bar{f}_N=\bar{\varrho}_{N+1}\,\bar{b}_{N+1} \tag{6}$$

$\bar{\varrho}_N$ and $\bar{\varrho}_{N+1}$ refer to equilibrium concentrations.
Using eq.(6) in eq.(3) the steady state nucleation is written as follows :

$$J_S=\bar{\varrho}_N\bar{f}_N\left(\frac{\varrho_N f_N}{\bar{\varrho}_N\bar{f}_N}-\frac{\varrho_{N+1}b_{N+1}}{\bar{\varrho}_N\bar{b}_N}\right)=\bar{\varrho}_N\,\bar{f}_N\left(\frac{\varrho_N}{\bar{\varrho}_N}-\frac{\varrho_{N+1}}{\bar{\varrho}_{N+1}}\right) \tag{7}$$

On summing all sizes up to G and using the boundary conditions :

$$\varrho_G=0,\ \varrho_1/\bar{\varrho}_1=1,\ \text{we obtain:}\ \sum_{N=1}^{G-1} J_S/\bar{\varrho}_N\bar{f}_N = 1 \tag{8}$$

so that we obtain the typical expression for J_S

$$J_S = \left(\sum_{N=1}^{G-1} 1/\bar{\varrho}_N \bar{f}_N \right)^{-1} \tag{9}$$

The crucial dependence of J_S on the cluster free energy is hidden in the $\bar{\varrho}_N$ terms of the sum.
Assuming the classical result /3/ :

$$\bar{\varrho}_N = exp(- \Delta F(N)) \tag{10}$$

where $\Delta F(N)$ is the free energy of formation of an N-cluster, we will procede by first discussing the dependence of $\Delta F(N)$ on N and then by rewriting eq.(9) in terms of this quantity.

3.1. Dependence of $\Delta F(N)$ on the cluster size N.

The free energy of formation of an N-cluster is defined in terms of cluster free energy F_N and chemical potential of the parent phase μ^I :

$$\Delta F(N) = F_N - \mu^I \cdot N \tag{11}$$

Phenomenological models are proposed for F_N (as discussed elsewhere /4/, the crudest one individuates 2 contributions in F_N : a bulk term proportional to the chemical potential of the solute rich phase (μ^{II}) and a surface term related to the surface free energy ($\sum$) :

$$\Delta F(N) = (\mu^{II} - \mu^I) + \sum N^{2/3} \tag{12}$$

We can clearly see from fig. 1 that :
if $\mu^{II} > \mu^I$ (undersaturated system), $\Delta F(N)$ grows rapidly with N implying that the probability of observing an N-cluster falls rapidly to zero.
If $\mu^{II} < \mu^I$ (supersaturated system), $\Delta F(N)$ grows with N up to a critical size N^* and then decreases. The critical size is said to be in unstable equilibrium respect to growth and shrinkage, a supercritical cluster can grow "spontaneously". Let us stress that the maximum in $\Delta F(N)$, (free energy barrier), is due to the surface term which represents the work of formation of the interface between the two phases.
In fig. 2 we report some examples of the behaviour of $\Delta F(N)$ respect to N depending on the model assumed for F_N /4/.
The general expression for $\Delta F(N)$ can be written as :

$$\Delta F(N) = AN + BN^{2/3} + CN^{1/3} + D \cdot lg(N/C^{coex}) + E \tag{13}$$

Putting the suitable coefficients equal to zero we consider the crudest model. The curves plotted refer to the coefficient values of table I which were obtained fitting the Monte Carlo estimation of F_N as cited in /4/.
Depending on the model choosen for F_N a change in the free energy bar-

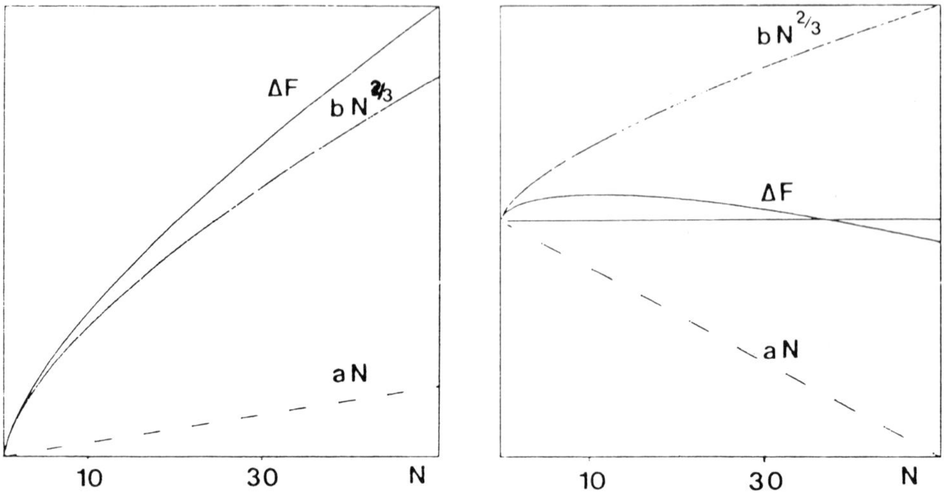

Fig. 1 : Free energy of formation of a cluster containing N particles when the basic "capillarity model", eq.12, is assumed :
a) undersaturated system $\mu^{II} - \mu^{I} = 1.$ $\Sigma = 1.996$
b) supersaturated system $\mu^{II} - \mu^{I} = -.6$ $\Sigma = 1.996$

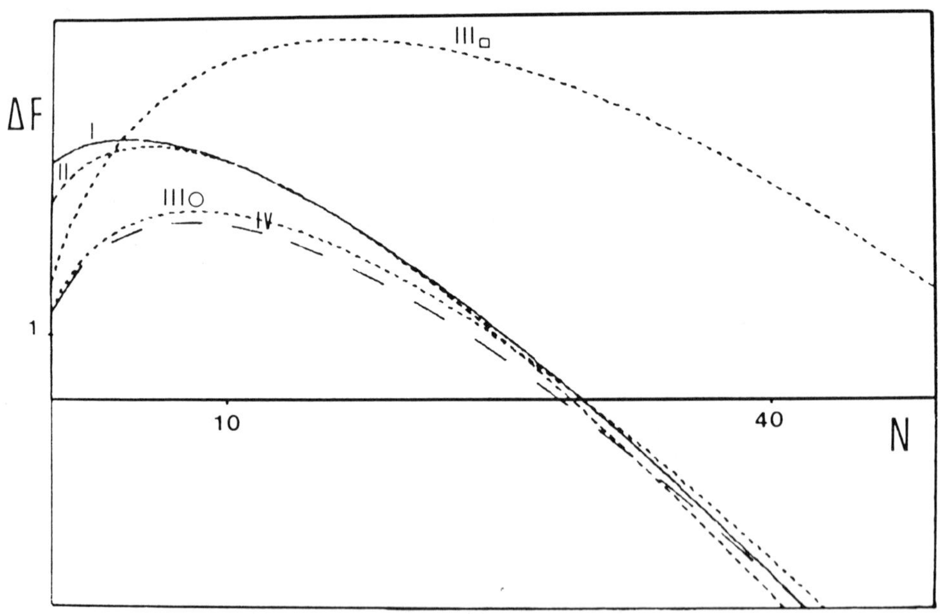

Fig. 2 : $\Delta F(N)$ as resulting from the different capillarity models obtained on changing the coefficient values in eq.(13). The coefficient values, for each model are reported in TABLE I.

rier and in the critical size is obtained.
We will see how these small variations in ΔF will reflect drastically on J_s.

3.2. Dependence of J_s on $\Delta F(N)$.

Let us study the dependence on N of the term $\frac{1}{\bar{\varrho}_N \bar{f}_N}$ of the sum in eq. (9) :

$\bar{\varrho}_N$ has a pronounced minimum at N^* due to its exponential dependence on N.

$\bar{f}_N$ does not depend strongly on N (in section 4 we will consider $f_N \propto N^{1/3}$ these considerations allow us to write :

$$1/\bar{\varrho}_{N^*}\,\bar{f}_{N^*} \gg 1/\bar{\varrho}_N\,\bar{f}_N \tag{14}$$

and the well known expression for J_s :

$$J_s = Z \cdot \bar{f}_{N^*} \exp(-\beta\,\Delta F(N^*)) \tag{15}$$

Z is the so-called Zeldovich factor :

$$Z = \left(\sum_N \bar{\varrho}_{N^*}\,\bar{f}_{N^*}/\bar{\varrho}_N \bar{f}_N\right)^{-1} \tag{16}$$

It can be seen that $\frac{1}{Z}$ is related to the width of $\Delta F(N)$ at a distance of the order of kT from the maximum.
Eq.(15) for J_s contains an equilibrium term : the concentration of critical clusters and a kinetic term : the probability that a critical cluster absorbs a monomer. The Zeldovich factor takes into account the fact that a critical cluster can also shrink (unstable equilibrium). Its presence results in decreasing the critical cluster concentration respect to the equilibrium one. The most evident fact is the exponential dependence of J_s on $\Delta F(N^*)$ which implies that big variations of J_s correspond to small changes of $\Delta F(N^*)$. Table II reports the values of J_s corresponding to the different model assumed for F_N.

4. TIME EVOLUTION OF CLUSTER POPULATION

In this section we will present the results obtained solving numerically eq.(1) and the comparison with the results obtained by Kalos et al. /5/ from a direct simulation of the decomposition of a supersaturated binary alloy (modelled as an Ising lattice) /8/.

4.1. A numerical solution of the rate equation

As already said the crucial problem on solving eq.(1) is the definition

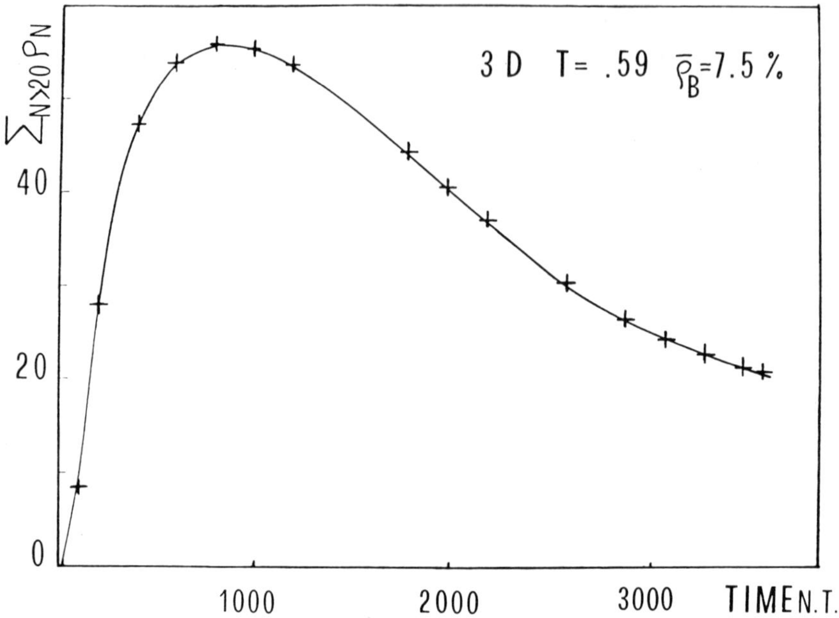

Fig. 3 : Number of cluster containing more than N'B atoms on a lattice of M sites, as a function of time. T/T_C=.59
3-D N'=20 M=*125000* $\bar{\rho}_B$= *7.5%*

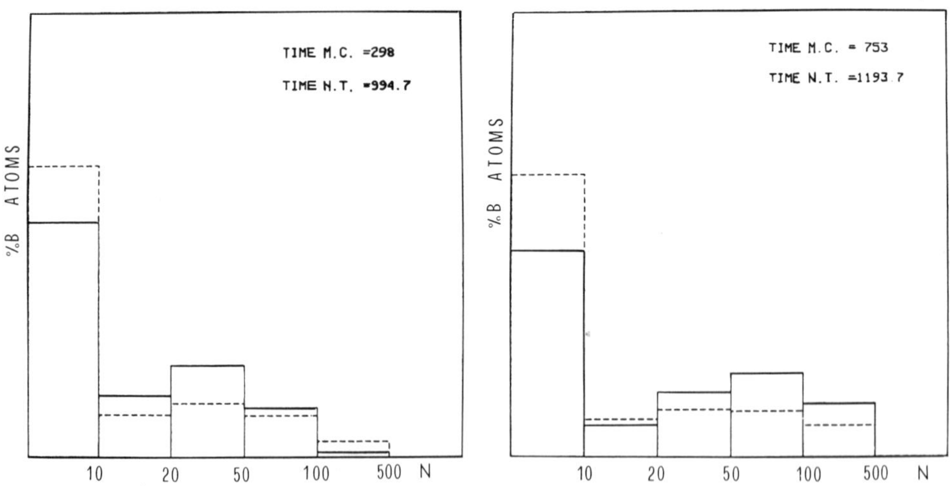

Fig. 4 : Histograms showing % of B atoms belonging to clusters of N atoms as a function of the cluster size.
—— from the kinetic Ising model /5/;----- from eq. 1.
T/T_c= *.59*, $\bar{\rho}_B$=*7.5%;* time is expressed in M.C. units.

of the emission ($b_N(t)$)and absorption ($f_N(t)$) probability. The classical theory suggests, for $f_N(t)$, the use of the steady state solution of the diffusion problem of monomers from a spherical region, (with radius R) at monomer concentration $\varrho_1(t)$, toward a spherical sink of radius r_N (the cluster radius) where the monomer concentration is equal to zero /6/ (note that r_N is defined by $4/3\pi\ r_N^3 = Na^3$, a^3 atomic volume) The assumption made, assuming clusters of spherical shape, is in accord with the hypothesis a)about the relevant characteristics of the cluster. The absorption probability writes :

$$f_n(t) = g_N D.\varrho_1(t).\alpha \tag{17}$$

α is the sticking coefficient which is supposed to be equal to unity which means that all the monomers which collide with the cluster are absorbed.

D is the monomer diffusion coefficient.

g_N is a geometrical factor :

$$g_N = 4\pi/a^2 \ .\ r_N/(1-\frac{r_N}{R}) \qquad in\ 3\text{-}D \tag{18}$$

The emission coefficient is usually considered to be a property of the cluster itself that is it does not depend on the state of the surrounding medium and a simple choice is to compute b_N considering the system in an equilibrium state (on the coexistence line) where the detailed balance conditions implies :

$$b_N = \overline{\varrho}_{N-1}/\overline{\varrho}_N \ .\ \overline{f}_{N-1} = g_N.D.\alpha \ .\ exp(\beta\,(F_N - F_{N-1} - F_1) \tag{19}$$

(use is made of eq.(10) and eq.(17)).

With f_N and b_N defined by eq.(17) and (19) respectively and measuring the time in units of D^{-1} ,eq.(1)writes in terms of the unique parameter $F_N - F_{N-1} - F_1$. Using accurate M.C. values of the free energy differences /7/ we solved the non-linear eq.(1) numerically for reduced temperature $T/T_C = .59$ and several values of solute concentration $\overline{\varrho}_B$ ($\overline{\varrho}_B = 5\%$, $\overline{\varrho}_B = 7.5\%$, $\overline{\varrho}_B = 10\%$).

The starting configuration of the system corresponds to : $\varrho_1(0) = \overline{\varrho}_B$, $\varrho_N = 0,\ N \geq 2$. The histograms of the percentage of the solute atoms belonging to the cluster vs. the cluster size, at a given instant (fig.4) or the time evolution of the cluster concentration (fig.5) are examples of the form in which the results can be presented. The time evolution of the "large" cluster concentration is depicted in fig. 3 for the 3-D case at $\overline{\varrho}_B$ =7.5%. Two regimes are clearly distinguisheable : a first nucleation stage followed by a coarsening period.

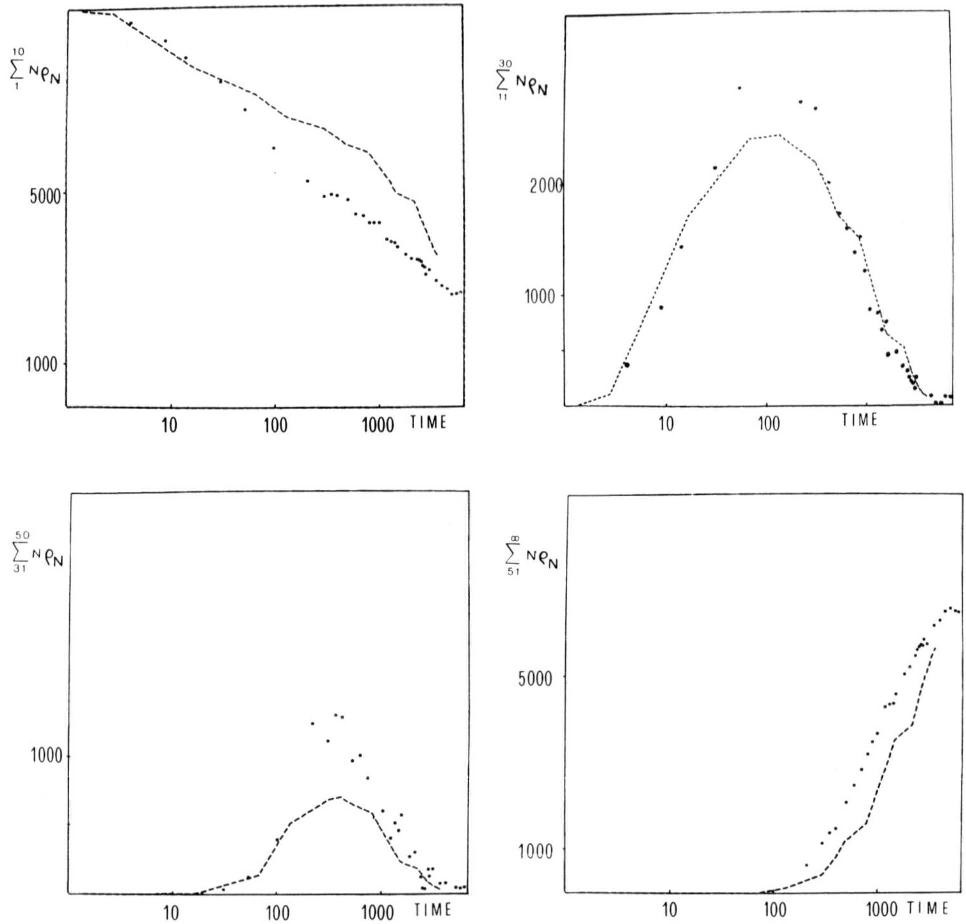

Fig. 5 : Time evolution of the number of particles in clusters of sizes S_1 through S_S : $\sum_{S_1}^{S_S} N\cdot\rho_N$, for the S.C. Ising model at $T/T_C=.59$ and $\bar{\rho}_B=7.5\%$.

The dotted line corresponds to the numerical solution if its time scale is multiplied by a factor $\delta(t)$ which varies from 1/7, for $t < 20$ M.C. time units, and tends to 1.

4.2. Simulation experiment

A direct simulation of the time evolution of a model of a binary alloy following quenching, was performed by Kalos et al./5/ using Monte Carlo simulation. The model system is an Ising lattice the sites of which can be occupied by A or B atoms. Nearest neighbours interactions are considered. Starting from a random configuration (corresponding to an "infinite" temperature) and with a given solute concentration $\bar{\rho}_B$, the system is quenched to a temperature corresponding to a phase point under the coexistence line. After this rapid cooling the system is still in a homogeneous state which now results thermodinamically unstable. Processes of phase segregation take place. The microscopic dynamics is simulated via exchange between A and B atoms on nearest neighbour sites. The probability per unit time of such an exchange is assumed to be

$$Wij = {}^{1}/\tau \ .exp(-\beta \Delta E) \ / \ (1+exp(-\beta \ \Delta E)) \tag{20}$$

where $\Delta_E = E_j - E_i$ is the change in energy resulting from the exchange; τ sets the time unit.
The time evolution of the cluster population is computed.

4.3. Comparison between the numerical solution and the direct simulation.

The nucleation process described by the nucleation theory can be tested against the simulation experiment by comparing the cluster concentration at a given instant or the time evolution of the cluster population resulting from the two approaches.
The correspondence between the time scale of the simulation experiment and the time scale of the numerical solution is derived in appendix A.
In fig. 4 we report the histograms of the percentage of solute atoms vs. the cluster size, which show the best overlap. (Observe that for a rigorous quantitative comparison we need to know the indetermination on the simulation data).
The principal points coming out from this kind of analysis are the following :
- the nucleation process described by the classical nucleation theory (eq.(1)) is slower than the one observed in the M.C. experiment.The delay becomes negligible on longer times when the system is characterized by a "slow" growth of the clusters due to coagulation of large clusters, by evaporation of monomers from one cluster and by their condensation on other clusters. The delay is more pronounced at higher densities.

An analogous results is found by Penrose et al. /9/. They observe a good agreement between numerical solution and simulation provided the time scale of the numerical solution is increased by an appropriate factor which depends on the solute concentration. The authors interpret this discrepancy assuming an effective value for the monomer diffusion coefficient depending on the solute concentration (D increases with the solute concentration).

- the concentration of cluster containing less than 10 B - atoms is constantly higher in the numerical solution than in the simulation.
- the maximum value of the concentration of clusters containing more than 10 solute atoms is smaller in the numerical solution than in the simulation.

In order to individuate possible reasons for these discrepancies let us analyse how we have treated the 3 basic models of the nucleation theory:

- solute emission from the cluster : we have used the classical approach which consists of computing b_N in a reference equilibrium state where the detailed balance on each cluster can be applied. Using the accurate M.C. values for the free energy differences /7/ this step has been solved to the highest precision available today.
- growth rate model for each cluster : expression (17) is valid in the limit of very high dilution of the "cluster gas" while in the cases studied the cluster density may be high enough for higher order terms to become significant.
- diffusion model for the solute in the alloy : only monomer diffusion was considered while the contributions of dimer, trimer, ... diffusion and "reactions" involving more than 2 small clusters may be relevant in the kinetics of the early stages of decomposition, when the concentration of these small clusters is high.

This assumption is supposed to be the principal cause of the observed delay in the nucleation kinetics and the increase of the delay in systems at higher solute concentrations supports this idea. Also the difference in the cluster concentration values (higher for small clusters and consequently lower for clusters containing more than 10 solute atoms) can be attributed to this assumption.

5. CONCLUSION

The results of the numerical solution of the classical rate equation (eq.(1)) are compared with the results of a simulation experiment of the decomposition processes in a supersaturated alloy. The Ising model for the binary alloy is assumed. Several solute densities (5%,7.5%,10%) at T/T_C=.59 are considered.

Good agreement is found for the shapes of the cluster distributions but the first stage of decomposition within the classical theory is slower than in the M.C. experiments. This delay is more evident for the cases of higher solute concentration as might be expected since the basic assumptions of the classical theory are valid at very high dilution. Further analysis on the classical solute diffusion model which includes monomer diffusion only are necessary in order to assess the dependence of this discrepancy on this assumption and to improve the agreement with the simulation results.

APPENDIX A

We show how to derive the correspondence between the time scale of the simulation experiment and the time scale of the numerical solution. In the numerical solution the time unit is expressed in term of D^{-1}, more precisely we assume as frequency unit :

$$\frac{4\pi D}{N_{site}\,a^2} = 1 \tag{21}$$

In order to derive the frequency unit in the M.C. experiment in a comparable form let us suppose that the monomer diffusion coefficient is the same and let us consider the usual relation with the average quadratic displacement :

$$D = \langle r^2 \rangle / 6t \tag{22}$$

in terms of the mean number of elementary jumps (jumps of length equal to the lattice parameter a) the average quadratic displacement and the corresponding time interval can be written as :

$$\langle r^2 \rangle = \langle x^2 \rangle + \langle y^2 \rangle + \langle z^2 \rangle = N/3\ a^2 + N/3\ a^2 + N/3\ a^2 = Na^2 \tag{23}$$

and

$$t = N/\,\Gamma_1 \tag{24}$$

Γ_1 is the jump frequency of the monomer for any of the 6 possible directions and it is derived from eq.(20) remembering that configurations which differ by a monomer jump have the same energy $(\Delta E=0)$

$$\Gamma_1 = 6Wij\,(\Delta E=0) = 3/\tau \tag{25}$$

The frequency unit for the M.C. experiment writes :

$$2D/a^2 = 1 \tag{26}$$

From eq.(21) and eq.(26) the ratio of the two time scales are given by :

$$t_{M.C.}/t_{N.T.} = N_{Site}/2\pi \qquad (27)$$

REFERENCES

/1/ A review of the classical nucleation theory can be found in :
A.C. Zettlemoyer ed. Nucleation Marcel Dekker,New York (1969)
F.F. Abraham Homogeneous Nucleation Theory H.Eyring, D.Henderson eds. New York and London Academic (1974)

/2/ K. Binder and D. Stauffer, Adv.Phys. 25,343, (1976)

/3/ J. Frenkel J.Chem.Phys., 7, 200 (1939); ibidem 7, 538 (1939)
W. Band, J.Chem.Phys. 7, 324, (1939); ibidem 7, 927 (1939)

/4/ See contribution on "Monte Carlo studies of interface free energy in nucleation theory"

/5/ K. Binder, M.H.Kalos, J.L. Lebowitz and J.Marro, Adv.Coll.Int.Sci., 10 173-214 (1979)
M. Rao, M.H. Kalos, J.L. Lebowitz and J. Marro, Phys.Rev.B13,4328, (1976)

/6/ C.P. Flynn, Point Defectsand Diffusion, p.467, Clarendon Press, Oxford (1972)

/7/ G. Jacucci, A. Perini and G. Martin, J.Phys. A: Math Gen 16, 369 (1983)
A. Perini, G. Jacucci and G. Martin, Phys. Rev.B 1, March (1984)

/8/ A. Perini, G. Jacucci and G. Martin, Surf. Sci. 144, 53 (1984)

/9/ O. Penrose, J. Lebowitz, J. Marro, M. Kalos and J. Tobochnik, J.Stat. Phy., 34, 399 (1984).

Table I

MODEL	A	B	C	D	E
I	-.6	1.9964	-1.4001	-.0739	3.4416
II	-.6	1.7060	0	-.0739	1.7588
$III_{\bigcirc}$	-.6	1.8758	0	0	0
$III_{\square}$	-.6	2.3273	0	0	0
IV	-.6	1.8758	0	-.0739	0

Table I : values of the coefficients of eq.(13), corresponding to different approximations of the capillarity model. (T/T_c=.59 C_{coex} =.985).

Table II

MODEL	N^*	$\Delta F(N^*)$	$\bar{f}N^*$	Zeld	J_s	J_s/J_s^I	$e^{\Delta F} / e^{\Delta F^*}$
I	6.1–6.5	6.04	1.85	.078	.000341	1	1
II	6.3–6.6	5.50	1.86	.084	.000641	1.9	1.7
$III_{\bigcirc}$	9.1–9.2	4.08	2.09	.072	.00254	7.4	7.1
$III_{\square}$	17.3	7.80	2.59	.052	.000056	.16	.17
IV	8.5–8.9	3.84	2.05	.073	.00323	8.6	9.0

Table II : values for the critical size N^*, the free energy barrier $\Delta F(N^*)$, the absorption coeff. of a critical cluster $\bar{f}N^*$, the Zeldovich factor Z, and the steady state nucleation rate (eq.15-16) as resulting by assuming the $\Delta F(N)$ models plotted in Fig. 2.

RADIATION DAMAGE AND CREEP USING RATE THEORY

M. W. Finnis
Theoretical Physics Division
A.E.R.E. Harwell
Oxfordshire OX11 ORA
United Kingdom

ABSTRACT. The general principles of the effective-medium rate theory of radiation damage are described. The derivation of sink strengths is discussed and illustrated for the simplest case of a spherical cavity. A model of creep under irradiation is presented in which the dislocations climb and their density evolves according to rate equations. Creation of dislocations is assumed to be due to Bardeen–Herring sources and recovery is modelled by annihilation of straight dislocation lines, as in Nabarro's theory of thermal creep. The resulting rate equations exhibit terms which can be interpreted as creep, swelling and creep-swelling coupling, with both irradiation and thermal driving forces. An example is presented in which observed dislocation densities and secondary thermal creep rates in a steel are explained with the theory.

1. INTRODUCTION – PHENOMENOLOGY AND MODELS OF CREEP

The aim of this lecture is to explain one of the roles of computer simulation in the study of creep. I emphasize the word one, because the lecture covers only a part of the field which has been investigated in some detail at Harwell, which is the rate theory of the evolution of voids and dislocations in the presence of irradiation. The scope of computer simulation is actually much wider than this, for broadly speaking, theoretical studies of creep can proceed in two ways.

The first is to develop phenomenological models, which in their simplest form look like:

$$d\varepsilon/dt = A_0 \nu_D (\mu\Omega/kT)(\sigma/\mu)^n \exp(-Q_c/kT) \ . \tag{1}$$

where μ is the shear modulus, ν_D the atomic frequency, Ω the atomic volume and A_0 and n are dimensionless constants[1]. As discussed in reference (1), this equation can be fitted to a wide range of experimental data. However, it is not satisfactory for all aspects of the high-temperature deformation of metals. For example, it does not take into account the effect of prior deformation history on subsequent creep behaviour and would not describe the effect of a stress reversal. It does not recognise the tensorial nature of stresses and strains, and therefore while it may be fitted to a tensile test it contains no prescription for creep rate under a different loading. These considerations have led to more elaborate phenomenological models,of which that of Hart[2–3] has been very succesful. The problem of deformation history is dealt with by including an additional anelastic strain and several related parameters while the symmetry requirements are imposed by working in terms of scalar invariants of the tensor stresses and strains. I do not propose to discuss this model further; it is reviewed in more detail in ref (4). The point is that it is practical to use these equations in a computational scheme for solving two and three dimensional boundary value problems. Both the techniques of boundary and finite element methods have been applied[4],at the heart of which the problem has become that of solving a system of N ordinary differential equations of the form:

G. Jacucci (ed.), Computer Simulation in Physical Metallurgy, 225–244.

$$dy_i/dt = F_i(y_1 ... y_N, t) \ . \tag{2}$$

These are usually solved by standard numerical methods, such as those predictor-corrector methods for stiff systems described in refs.(5)–(8).

The second way of proceeding with a theoretical study of creep is to attempt to develop microscopic models for the processes which take place. This has a twofold attraction over the purely phenomenological approach. Firstly, it meets the purely scientific objective, which is not simply to fit experimental data but to understand the physical processes of creep. In a practical sense it is only this kind of theory which could give any understanding of how creep properties are modified by different microstructure, and how they might therefore be controlled. Secondly, we are especially interested in the effect of reactor neutron damage on creep rates, but for economic reasons most data is obtained from ion beam irradiations; we are more likely to develop a realistic model for extrapolating from the high damage rates to low damage rates in a reactor core if the atomic processes are understood and properly included in the model. The microscopic approach is even more important when void swelling is present as well as creep, since their effects are not simply additive, as we shall see.

Most of this lecture is about how we set up a particular microscopic model of creep suitable for computation. We have developed a theory in which an increment of strain appears as a result of dislocation climb by the emission or absorption of a point defect. Consider for example a self interstitial atom diffusing around a lattice in a ferritic steel. The creation of the interstitial as a result of some incident particle of irradiation has caused a temporary strain in the lattice, but this only becomes directional when the interstitial is absorbed at a *sink* of a particular orientation. If the sink is an edge dislocation, the lattice receives an increment of longitudinal strain in the direction of its Burgers vector. Together with the interstitial, a vacancy must have been created, the fate of which has to be considered too. For example, if it diffuses to the surface the volume of the specimen is reduced by Ω; but creation and absorption of the interstitial increased the volume by Ω, so the net result is deviatoric (volume conserving) strain. On the other hand if the vacancy disappears into a cavity, its creation and annihilation has produced no net volume change; but there is now still a net volume change or swelling Ω contributed by the interstitial absorption. It is easy to see that the relaxation volumes of the point defects themselves do not enter the argument. This is the microscopic picture of dimensional changes, elaborated by Wiedersich[(9)], which is made quantitative in what follows. The quantitative theory is a rate theory, which simply means that it contains microstructural parameters such as void size and dislocation density which obey rate equations of the form of equations (2) and which enter expressions for creep rate and swelling rate. Although it is far from a complete theory of dimensional changes for all materials and conditions, its importance has been demonstrated in many cases.

As the preceeding discussion suggests, the occurrence of creep or swelling by the mechanism of point defect absorption depends on unequal absorption of vacancies and interstitials by particular sinks. The theory of elastic interactions between inclusions and the stress fields of cavities and dislocations suggests that the sinks to which interstitials prefer to migrate are dislocations. The distortion of the host lattice which surrounds them both has the effect of favouring interstitial motion towards the dislocation core compared to other neutral sinks such as cavities. The result is a *bias* of dislocations for interstitials compared to vacancies upon which is based the rate theory of void swelling[(10–12)]. For deviatoric dimensional changes, which we refer to as growth (in the absence of stress) or creep (in the presence of an applied stress), it is proposed that two effects contribute within this model. The first is simply anisotropy in the distribution of orientations of dislocations, which may have its origin in the fabrication process, in the nucleation of dislocation loops[13–15], or during the evolution of the microstructure. The second is a dependence of the bias of dislocations on their orientation to the applied stress, an effect which gives rise to so-called SIPA creep (Stress Induced Preferential Absorption), which has been discussed by several authors[16–19].

In order to construct a model which can describe power-law creep, which means $n \geqslant 1$ in equation (1), as well as the linear stress dependence observed in irradiation creep, it is necessary to develop equations for the evolution of the dislocation network, for the power law arises when the density of mobile dislocations is determined self-consistently by balancing dislocation generation and annihilation under the applied stress. An attempt at this[20] will be described here, but there are still difficulties as I will indicate and it is an aspect which is currently under development. This leads to an important general point : the framework of rate theory is sufficiently general, as I hope to convey, that new physical effects can be incorporated in a natural way if it is thought that they might be important. The rate theory is not a finished model but a framework for developing and testing ideas. Thus although the computations have become more complicated over the past ten years as equations have been modified or added, their basic structure and the underlying philosophy is unchanged.

2.THE RATE THEORY

The objective of the rate theory is to calculate the time and spacial average rates of net absorption of point defects by sinks such as edge dislocations, dislocation loops, cavities, precipitates, grain boundaries and surfaces. A thorough review of the methodology which I summarise here is given in reference (21). The starting point is the pair of continuity equations for vacancies and self interstitials :

$$\frac{\partial c_v}{\partial t} + div\ j_v = K - \alpha c_i c_v, \tag{3}$$

$$\frac{\partial c_i}{\partial t} + div\ j_i = K - \alpha c_i c_v. \tag{4}$$

c_v and c_i are the atomic concentrations of vacancies and interstitials. At this stage these quantities are locally varying. Their fluxes j_v and j_i are determined by the usual relations of irreversible thermodynamics[22] :

$$j_v = -D_v \nabla c_v - \frac{D_v c_v}{kT} \nabla \phi_v, \tag{5}$$

$$j_i = -D_i \nabla c_i - \frac{D_i c_i}{kT} \nabla \phi_i. \tag{6}$$

The potentials ϕ_v and ϕ_i represent the effect of stresses in the matrix; in practice only the stress field of disloca tions and their interaction with interstitials is considered, since the interstitial misfit is much greater than that of the vacancy. No account is taken of the effect of temperature gradients or possible local variation of the diffusion coefficients. The quantity K is the defect production rate in displacements per atom per second, and α is known as the intrinsic recombination coefficient. Boundary expression for the sink strength[23]. The first step is to obtain the average point defect concentration in the continuum, which for vacancies follows from equation (7) when we set the time derivative to zero (steady-state) :

$$c_v = (K + K_v)/Dk_v^2. \tag{9}$$

Equation (9) introduces the total vacancy sink strength :

$$k_v^2 = \sum_L k_{Lv}^2, \tag{10}$$

and the total thermal emission rate :

$$K_v = \sum_L K_{Lv}. \tag{11}$$

The second stage is to consider a single spherical cavity embedded within the lossy continuum. Strictly speaking, the spherical cavity should be surrounded by some sink-free region. However, it is not clear what the 'best' radius is for the sink free region and the result is not very sensitive to this choice[24] so we shall assume that the complete lossy medium extends right in to the surface of the cavity. The diffusion equation to be solved is :

$$(D_v/r^2)\frac{d}{dr}r^2\frac{dc}{dr} - D_v k_v^2 c + K + K_v = 0. \tag{12}$$

The boundary conditions are :

$c = c_{Cv}$ at the cavity radius $r = r_C$

and

$c = c_v$ as $r \to \infty$.

The solution is easily shown to be :

$$c = c_v - (c_v - c_{Cv})(r_C/r)exp(-k_v(r-r_C)). \tag{13}$$

Next we obtain the flux of vacancies into the cavity :

$$\begin{aligned} 4\pi r_C^2 j_C &= 4\pi r_C^2 D_v (dc/dr)_{r_C} \\ &= 4\pi r_C D_v (1+k_v r_C)(c_v - c_{Cv}). \end{aligned} \tag{14}$$

Compare this with the flux of vacancies into cavities in the rate theory equation (7) ; we identify the sink strength by equating coefficients of c_v :

$$k_{Cv}^2 = 4\pi r_C(1+k_v r_C)C_C, \tag{15}$$

where we have multiplied the flux to a single cavity by the volume concentration of cavities C_C. The thermal emission rate from cavities K_{Cv} is identified as the remaining term in the vacancy balance which arises because of the boundary concentration c_{Cv}, and by comparing (14) with (7) we find that it is given by :

conditions have to be specified over surfaces surrounding each of the sinks. In the case of interstitials, which are not generated thermally, their concentration should vanish at a sink surface unless there is some limitation to the rate of atomic jumps into the surface, in which case a mixed boundary condition applies. Since vacancies on the other hand may be generated thermally at sinks, there is an appropriate concentration boundary condition which is the equilibrium vacancy concentration modified by any local stress which can assist the absorption or emission process. No possibility is envisaged that vacancy absorption may be rate limited.

The solution of equations (3)–(6) within a matrix containing randomly distributed sinks is an impossible task. Even if the sinks are arranged periodically it is still a daunting prospect for cases with more than one type of sink. This is why procedures for 'smearing out' the sinks and defining a 'lossy effective medium' were developed. If the averaging is performed with care it is possible to preserve most of the interesting physics of the original problem, including the interactions between sinks of the same and different types. For details of the story at this point, the reader may consult reference(21). The conclusion is that under a remarkably unrestrictive set of assumptions, equations (3)–(6) can be replaced by two equations for c_v and c_i which are now interpreted as spacial averages of the quantities originally defined. These rate equations are :

$$\frac{dc_v}{dt} + \sum_L (D_v k_{Lv}^2 c_v - K_{Lv}) = K - \alpha c_i c_v, \tag{7}$$

$$\frac{dc_i}{dt} + \sum_L D_i k_{Li}^2 c_i = K - \alpha c_i c_v. \tag{8}$$

The summation is over all the types of sink which are labelled L, and k_{Lv}^2, k_{Li}^2 are their sink strengths. Calculation of the sink strengths will be discussed in section 3. The term K_{Lv} is included to account for the fact that sinks can emit vacancies as well absorb them, and it is referred to as the thermal emisson rate of vacancies.

3. SINK STRENGTHS

When the sink strengths in equations (7) and (8) are known, the concentrations of point defects in the steady state are easily evaluated. In fact, it is generally assumed that the timescale for adjustment of c_v and c_i is much faster than the timescale of variation of any of the sink strengths or other terms in these rate equations. They then become simple algebraic equations for the point defect concentrations.

The most satisfactory procedure for calculating the sink strengths is a self-consistent effective medium method, which works as follows. A particular sink is surrounded by a region of sink-free matrix, which is embedded within a lossy continuum described by equations (7) and (8). The flows of point defects to the sink are calculated by solving the diffusion problem, and when multiplied by the concentration of similar sinks these flows define their sink strength. Self-consistency is achieved by making the sink strength so deduced equal to the value assumed in the lossy continuum. The following example shows how this is done in practice.

3.1 The spherical cavity

This is the simplest case which illustrates the main principles of embedding and self-consistency, so I will present its derivation in full. We make the simplification of ignoring intrinsic recombination, which introduces a negligible error into the final

$$K_{Cv} = D_v k_{Cv}^2 c_{Cv} . \tag{16}$$

The self-consistency step is taken when k_v is evaluated from equation (10). k_v^2 contains k_{Cv}^2 together with the sink strengths of all the other sinks in the system, which themselves should depend on all the sink strengths.

The above derivation begs a number of questions which have been addressed in the literature, besides the proper extent of the sink-free region to be used to which I have referred. For example, the importance of the spacial distribution of the sinks [25], the importance of time-dependence and transients [26] and the importance of the stress field about a cavity [27]. These refinements introduce a great deal of complexity but fortunately under most conditions they have been shown to introduce relatively minor corrections to the cavity sink strength.

3.2 The edge dislocation

The main problem in deriving a sink strength for dislocations along the same lines as for the spherical cavity is how to deal with the long ranged stress field. It is essential to include its effect, even if this necessitates some further approximations, because it determines the relative uptake of vacancies and interstitials and the relative uptake of interstitials on dislocations of different orientation with respect to an applied stress. The interaction of the misfit strain of interstitials with the strain field around dislocations drives an excess of interstitials to the dislocations, causing swelling. The orientation dependence of the interaction with respect to an applied stress causes an important contribution to creep, by the SIPA mechanism. Let us jump straight to the result for the interstitial sink strength of edge dislocations of density ρ_K :

$$k_{Ki}^2 = Z_{Ki} \rho_K . \tag{17}$$

The suffix K indicates the orientation of the dislocation with respect to an applied tensile stress σ.

ρ_1 = density with Burgers vector and line direction normal to σ,
ρ_2 = density with Burgers vector normal to σ and line direction parallel to σ,
ρ_3 = density with Burgers vector parallel to σ and line direction normal to σ.

Equation (17) is simply a definition of the dimensionless parameters Z_{Ki} , known as bias parameters, which contain all the effects of internal and applied stress. Recent calculations with an embedding model[28] have also included the fact that absorption of interstitials depends on the presence of jogs; assuming that interstitials cannot instantaneously diffuse along a dislocation core, the absorption rate is limited by the concentration of thermally activated jogs. The resulting bias parameters are therefore temperature dependent, and quite complicated formulae have been obtained for them. Comparison with experimental data[29] has indicated that this is an essential complication, otherwise swelling rates would be overestimated. To include the effect of stress, the interaction field must be modified by terms obtained with elasticity theory[30]. All the calculations are made of necessity for infinitely long, straight dislocations (or for small dislocation loops, which I do not discuss), and the results must therefore be taken as approximate. I will not describe the formulae here, except to say that they are readily handled by subroutine modules in the computer codes which have been written to solve the equations of rate theory[31].

For a better understanding of the models it will be convenient to split the bias parameters into three terms :

$$Z_{Ki} = Z_{Kv} + \Delta Z + \delta Z_K . \tag{18}$$

Z_{Kv} represents the bias parameter of network dislocations for vacancies, for which it is assumed in practice that there is no dependence on the applied stress, and hence the subscript K is only retained for formal completeness. ΔZ is the bias of dislocations for interstitials in the absence of any deviatoric stress. It may be modified by an applied hydrostatic component of stress. An applied deviatoric component of stress induces the additional bias δK_K. The biases are found to be small compared to unity, which justifies this perturbative expression.

By analogy with equation (17) the sink strength for vacancy absorption is given by :

$$k_{Kv}^2 = Z_{Kv} \rho_K , \tag{19}$$

and by analogy with equation (16) the thermal emission rate from dislocations is given by

$$K_{Kv} = D_v k_{Kv}^2 c_{Kv} . \tag{20}$$

4. EQUATIONS FOR SWELLING CREEP AND GROWTH

4.1 Swelling and creep rates

I am now going to describe the model of dimensional changes due to interstitial absorption at edge dislocations. The only sinks considered so far are voids and edge dislocations, and dimensional changes due to dislocation glide are not included. It should be noted that these are not limitations of the rate theory, which has previously been applied to other sink types. However, the structure of the theory even with these restricted assumptions is worth exploring before further complications are introduced.

The swelling rate in this model is equal to the net rate of absorption of interstitials by dislocations. To calculate both swelling and creep rates we need the net rates of absorption of interstitials by dislocations of each orientation, which we denote by dq_K/dt, measured in units of number per second per lattice site of the medium. These are net of vacancy absorption, which for this purpose is regarded as negative interstitial emission. The swelling rate is given by :

$$dS/dt = dq_1/dt + dq_2/dt + dq_3/dt, \tag{21}$$

and the deviatoric creep rate in the direction of the applied stress is given by :

$$d\varepsilon_3/dt = \tfrac{2}{3} dq_3/dt - \tfrac{1}{3}(dq_1/dt + dq_2/dt). \tag{22}$$

It should be noted that the deviatoric creep rates in directions normal to the applied stress are not simply given by permuting the suffices of equation (22) as incorrectly stated in reference (20). This is because these suffices refer to particular combinations of Burgers vector and line direction, defined in section 3.2, and are not Cartesian vector indices. It is left as an exercise to show that if the deformation is symmetric about the tensile axis 3 then :

$$d\varepsilon_1/dt = d\varepsilon_2/dt = -\tfrac{1}{3} dq_3/dt + \tfrac{1}{6}(dq_1/dt + dq_2/dt). \tag{23}$$

$$g_K = 2w_K/(1 + w_K^2), \tag{30}$$

where the w_K satisfy rate equations :

$$dw_K/dt = 2(1 - |w_K|)^2/(\lambda_K \rho_K b^2). \tag{31}$$

4.2 The generation of dislocations

In this and the following sections I describe a physical model which leads to rate equations for the dislocation density. It must be emphasised that theoretical understanding of the evolution of the dislocation density is at present still primitive and incomplete, and much remains to be explained. Our present model, which we regard as a working hypothesis, is that the dislocation network is climbing as individual line segments bow out from between pinning points. They act as Bardeen–Herring sources, as in Nabarro's theory[32]. We assume throughout that it is valid to resolve the dislocation population distribution along Cartesian axes, otherwise we would have to work with integrals over solid angle of distribution functions. Even so, for a general orthorhombic symmetry, e.g. under triaxial stress for an isotropic medium, we need more than the three K indices introduced so far, and we cannot avoid some rather complex equations. Let us denote the density of dislocations with Burgers vector in direction i and line direction j as ρ_{ij}. Figure (1) illustrates this notation, which includes screw dislocations, which have $i=j$.

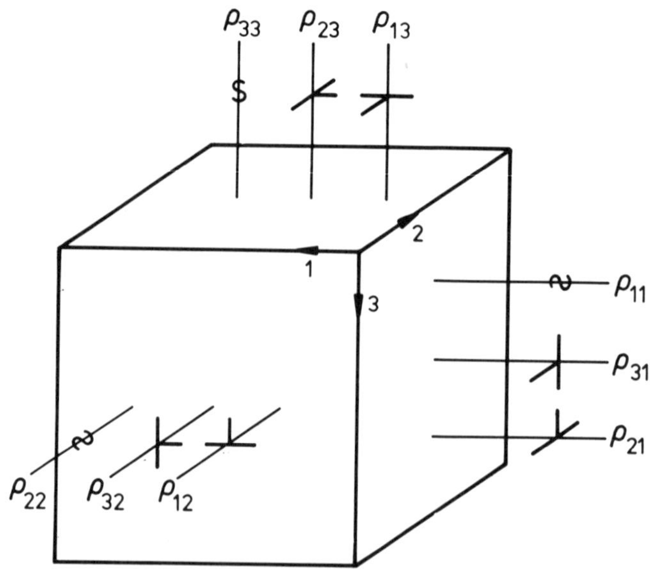

Figure 1 : Notation for dislocation density

For the case of uniaxial stress along axis 3, the dislocation densities introduced previously are given by :

The net interstitial capture rates (less vacancy emission rates) are given by the rate theory expression :

$$dq_K/dt = D_i k_{Ki}^2 c_i - D_v k_{Kv}^2 (c_v - c_{Kv}), \tag{24}$$

in which the terms on the right hand side are defined in section 3. The trickiest problem is to write down a physically sensible and useful formula for the concentration of vacancies at the surface of the dislocation, c_{Kv}, where by 'surface' I mean the rate theory sense of the surface of a cylinder enclosing the dislocation core. This determines the vacancy emission rate. It is approximately equal to the equilibrium vacancy concentration c_v^e, where :

$$c_v^e = exp(-H_v^f/kT), \tag{25}$$

and H_v^f is the enthalpy of formation of a vacancy. However, there are two significant modifications which must be taken into account.

Firstly, the applied stress does work when a vacancy is emitted from an edge dislocation oriented with its Burgers vector normal to the applied stress. This modifies the concentration at such a dislocation by a factor $exp(\sigma\Omega/kT)$, in which if we ignore vacancy relaxation Ω is the atomic volume. Consequently there is a concentration gradient between such dislocations and those which are oriented with their Burgers vector perpendicular to the applied stress, which is the driving force for thermal creep of the Nabarro type[32].

Secondly, if the dislocations are bowing by interstitial absorption they have an effective line tension which favours the state in which a vacancy is absorbed and the dislocation thereby less bowed compared to the state in which the vacancy is sitting outside the dislocation core. This situation would modify the concentration next to the dislocation by the Boltzmann factor $exp(-\lambda_K^{-1}\mu\Omega/kT)$, in which λ_K^{-1} is a dimensionless measure of the curvature of the line segment. We have taken the radius of curvature to be given by the line segment length between pinning points $\lambda_K b$ where b is the Burgers vector. This is really a lower bound to the radius of curvature which is not suitable for describing transient behaviour as dislocation bowing develops. I will refer to this difficulty again later. We have so far only circumvented it in the following ad-hoc fashion. Combining the effects of stress and line tension we write :

$$c_{Kv} = c_v^e exp(-(g_K/\lambda_K)\mu\Omega/kT) \quad \text{for } K = 1,2 \tag{26}$$

and

$$c_{3v} = c_v^e exp((\sigma/\mu - g_K/\lambda_K)\mu\Omega/kT), \tag{27}$$

where the factor g_K which measures the completeness of bowing must satisfy :

$$-1 \leq g_K \leq +1. \tag{28}$$

From the characteristic velocity of climb (see section 4.2) :

$$v_K = (dq_K/dt)/(\rho_K b), \tag{29}$$

we see that dislocations should bow out in a characteristic time $\lambda_K \rho_K b^2/(2|dq_K/dt|)$. We require that the g_K tend to ± 1 on this timescale, which we ensure by writing them in the form :

$$\rho_1 = \rho_{21} + \rho_{12}, \tag{32}$$

$$\rho_2 = \rho_{23} + \rho_{13}, \tag{33}$$

$$\rho_3 = \rho_{31} + \rho_{32}. \tag{34}$$

The pinning of a particular type of dislocation is assumed to be due to intersections with any of the three types of dislocations running parallel to its Burgers vector, i.e. normal to its climb plane. We can thus obtain characteristic pin spacings or segment lengths $\lambda_{ij}b$ for each type of dislocation density, e.g. :

$$\lambda_{12}b = 1/\sqrt{(\rho_{11} + \rho_{21} + \rho_{31})}. \tag{35}$$

The rate of edge dislocation generation is then estimated as follows :

No. of interstitials per unit volume per second captured by type jk dislocations (including net vacancy emission)

$$= dq_{jk}/dt \div \Omega. \tag{36}$$

Hence no. of interstitials per unit length per second captured by type jk dislocations (including net vacancy emission)

$$= dq_{jk}/dt \div (\rho_{jk}\Omega). \tag{37}$$

Hence velocity of climb of segment

$$v_{jk} = dq_{jk}/dt \div (\rho_{jk}b). \tag{38}$$

We assume for simplicity that each source operates $|v_{jk}|/\lambda_{jk}b)$ times per second and produces new length $\pi\lambda_{jk}b$ in each operation. This new length has the same Burgers vector as the source but only half of it is on average of the same line direction and half is oriented with perpendicular line direction. Thus the rate of increase of ρ_{ij} from these type jk sources is given by :

$$d\rho_{jk}/dt = \tfrac{1}{2}\pi b\lambda_{jk} \times \textit{frequency of operation} \times \textit{density of sources}$$

$$= \tfrac{1}{2}\pi|dq_{jk}/dt|/(b^2\lambda_{jk}). \tag{39}$$

4.3 Recovery

We adopt here the simplest model for annihilation of dislocations, in which they are infinitely long, parallel and randomly distributed in the medium. Type jk dislocations are assumed to climb with velocity v_{jk} until they encounter the strain field of an opposite dislocation which is sufficiently close for the pair to annihilate by slip. For this to happen the two dislocations must be within a distance $\zeta_{jk}b$ on their mutual glide plane. The Nabarro assumption was that $\zeta_{jk}b$ is the mean dislocation spacing for dislocations of line direction k. Further apart than this,the fluctuations of the internal stress field due to all the other dislocations would intervene and the two would not automatically glide together. We have slightly modified this assumption, in case the dislocation spacing is significantly greater than l, which is the distance within which the two would glide together in the absence of interference from other stress fields. The length l would be determined by the balance between the attractive elastic force and the lattice frictional force. We regard it as an adjustable parameter. The dislocation spacing sets an upper limit to the glide-together range of dislocations. Taking account of the orientation dependence of the dislocation spacing and using the three index notation appropriate to cylindrical symmetry, these considerations were combined in simple analytic formulae :

$$\zeta_1 b = \zeta_3 b = (1/l^2 + \tfrac{1}{2}\rho_1 + \tfrac{1}{2}\rho_3)^{-\frac{1}{2}}, \tag{40}$$

$$\zeta_2 b = (1/l^2 + \rho_2)^{-\frac{1}{2}}. \tag{41}$$

The annihilation rate within this model, illustrated in figure 2, is easily shown to be :

$$d\rho_{jk}/dt = -\rho_{jk}^2 |v_{jk}| \zeta_{jk} b, \tag{42}$$

or in terms of the net interstitial absorption rate (using eqn. 29) :

$$d\rho_{jk}/dt = -\zeta_{jk}\rho_{jk} |dq_{jk}/dt| . \tag{43}$$

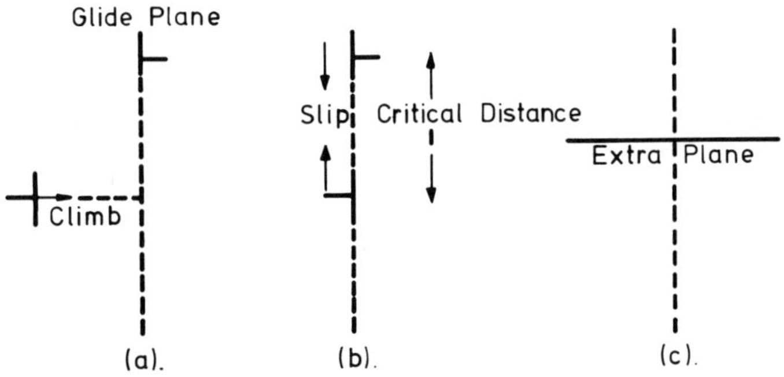

Figure 2 : Annihilation of a pair of dislocations

At this point we have to deal with an inherent shortcoming of the effective medium theory, which is its failure to deal properly with internal stresses. We want the model to describe thermal recovery but this only happens because of internal vacancy fluxes between dislocations which are driven by local variations in the internal stress. For example, in a medium in which dislocations were the only sources and sinks of vacancies, the effective medium average value of dq_{jk}/dt would be zero, which according to equation (43) gives us no recovery. The effect we are looking for is thrown away by spacial averaging. The solution is to treat dq_{jk}/dt in (43) as a local, spacially varying quantity. Details of the procedure are given in reference (20). We end up with an expression which combines random motion of dislocations with directed motion under the influence of internal stresses :

$$d\rho_{jk}/dt = -\zeta_{jk}\rho_{jk} \surd[(dq_{jk}/dt)^2 + b^2\rho_{jk}^3(D_v c_v^e \mu\Omega/kT)^2] . \tag{44}$$

These two components are represented by the first and second terms in the bracket respectively.

4.4 Summary

We have now assembled all the ingredients for a rate theory of irradiation and thermal creep. The equations for dislocation generation and recovery derived above must be combined into rate equations for the dislocation density, which in the case of uniaxial stress and cylindrical symmetry about the stress axis take the form :

$$d\rho_1/dt = \pi(|dq_1/dt| + |dq_2/dt|)/(2\lambda_1 b^2) - (\rho_1\zeta_1/4)\sqrt{[(dq_1/dt)^2 + \tfrac{1}{2}b^2\rho_1^3(D_v c_v^e \mu\Omega/kT)^2]}\,. \tag{45}$$

$$d\rho_2/dt = \pi(|dq_1/dt| + |dq_2/dt|)/(2\lambda_2 b^2) - (\rho_2\zeta_2/4)\sqrt{[(dq_2/dt)^2 + \tfrac{1}{2}b^2\rho_2^3(D_v c_v^e \mu\Omega/kT)^2]}\,. \tag{46}$$

$$d\rho_3/dt = \pi(|dq_3/dt|)/(\lambda_3 b^2) - (\rho_3\zeta_3/4)\sqrt{[(dq_3/dt)^2 + \tfrac{1}{2}b^2\rho_3^3(D_v c_v^e \mu\Omega/kT)^2]}\,. \tag{47}$$

These equations must be solved together with the rate equations for q_K described in section 4.1, and the results give swelling and creep rates according to equations (21)–(23). This has been carried out by the numerical methods described in reference (5) and in the case of irradiation creep leads to an unexpected new growth effect which I will discuss in section 6. In the absence of irradiation, the equations simplify but still require the use of a computer to perform numerical integration. This case will be described further in the following section. The point to be made here is that with the computational model as we have described it, variations of irradiation dose rate from zero upwards can be studied on the computer, as can the effect of temperature and other parameters which are less well known, such as the spacing of pins or the glide distance. Improvements to the model and the incorporation of new physical effects can be carried out within the existing general framework of rate equations.

5. AN EXAMPLE – APPLICATION TO THERMAL CREEP

In the absence of irradiation or cavities the rate equations become simpler. This case has been studied in detail[33] and compared with experimental results. I will summarise the main features here.

The absence of cavities means that the net flow of vacancies or interstitials (regarding one as the negative of the other) to the dislocation network vanishes. The equation (21) for swelling becomes

$$0 = dq_1/dt + dq_2/dt + dq_3/dt, \tag{48}$$

and the deviatoric creep rate in the direction of the applied stress is given from equation (22) by :

$$d\varepsilon_3/dt = dq_3/dt = -(dq_1/dt + dq_2/dt). \tag{49}$$

Since there is no irradiation, there are no interstitials, so equation (24) becomes (inserting equation (19) for the sink strength) :

$$dq_K/dt = D_v Z_{Kv}\rho_K(c_{Kv} - c_v). \tag{50}$$

Equations (48)–(50) are necessary steps on the way to deriving analytic approximations and understanding the behaviour of the numerical solutions, but the latter are obtained with just the same computer code which contains the full equations. This is done by specifying zero irradiation flux and void density; that is $K = 0$ and $C_C = 0$.

Calculated results for creep of M316 stainless steel at 700⁰C are shown in figure 3.

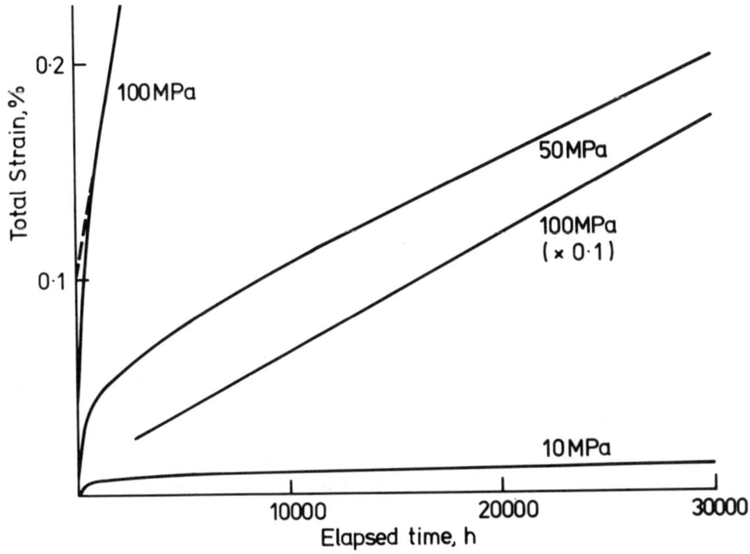

Figure 3 : Theoretical constant load creep curves for CW M316 tested uniaxially at 700^{0}C.

They were based on the input data listed in table 1.

Table 1

Input data for integration of rate equations

$b \quad = 2.57 \times 10^{-9}\ m.$

$\Omega \quad = 0.1206 \times 10^{-28}\ m^3.$

$c_v^e \quad = exp(-0.2564 \times 10^{-18} Joules/kT).$

$D_v \quad = 0.6 \times 10^{-4}\ exp(-0.2083 \times 10^{-18} Joules/kT)\ m^2 s^{-1}.$

$\mu \quad = 0.58 \times 10^{11}\ Jm^{-3}.$

$\rho \quad = 2.5 \times 10^{14}\ m^{-2}.$

$l \quad = 1000b$

The initial dislocation density ρ was assumed to be isotropic.

After a transient period, approximately constant dislocation densities and creep rates are achieved. These were plotted in figures 4 and 5 respectively.

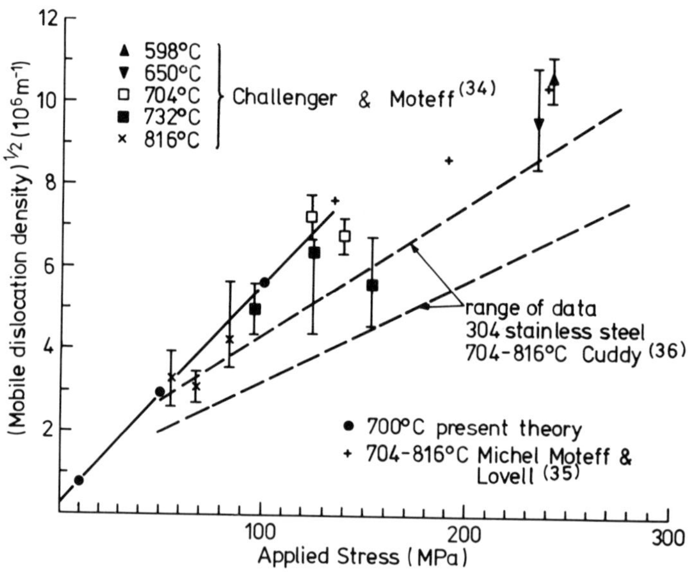

Figure 4 : Dislocation density as a function of applied tensile stress.

The steady-state dislocation densities agree well with experimental values, also shown in figure 4. An approximate σ^2 dependence is apparent.
The experimental data[34–36] have been adjusted to ensure that the dislocation counts quoted are converted to densities in the manner recommended by Hirsch *et al* [37].

Experimental creep rates for 20% cold worked M316 reported by Standring, Wilson and Clayden[38] reach a similar steady-state value but it is of relatively short duration. As the load is maintained the creep rates accelerate. The minimum creep rates are compared with the theoretical steady-state values in figure 5. There is satisfactory agreement with experiment for the 40–90 MPa range of stresses, over which an approximate σ^3 dependence of the creep rate is observed. For higher stresses and longer times beyond the so-called secondary or steady-state creep regime the model is clearly inadequate. Dislocation glide mechanisms must certainly be considered for these situations. Also at lower stresses the experimental results show a linear stress dependence and a higher creep rate than the present model. This can be explained by including the grain boundaries as sources and sinks of vacancies and as a vacancy diffusion path (Coble creep[39]). The straightforward Coble mechanism predicts creep rates which are much too high, but it has been invoked to explain the

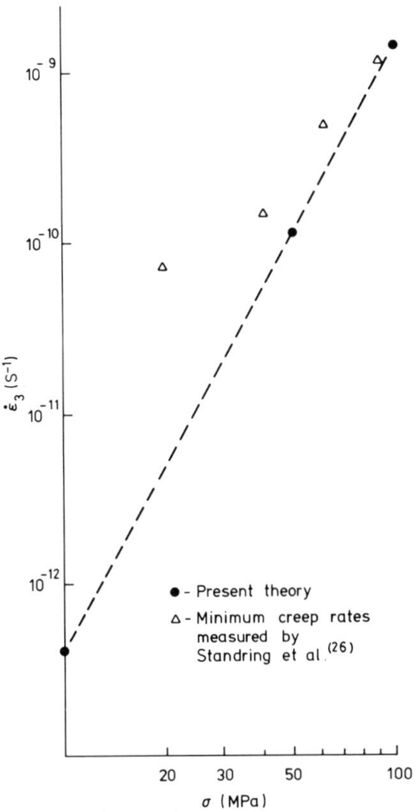

Figure 5 : Steady state thermal creep at 700°C.

present data by assuming that carbide particles on the grain boundary inhibit its ability to act as an efficient source and sink[40].

Some comments on the sensitivity of the calculations to the parameters are pertinent. The glide distance l was varied by a factor of 10, down to $100b$, which produced variations in minimum creep rate within the range observed experimentally, so it is a significant but not a dominant parameter. The initial dislocation density was only significant at the lowest stress value of 10 MPa; at higher stress the dislocation density reached the same steady-state value whatever its initial value. Finally the approximate power law behaviour of dislocation densities and creep rates can be derived analytically in the steady-state limit, as discussed in reference (33). The theory in this limit is physically the same as Nabarro's[32], although the mathematical details are different.

6. THE EFFECT OF IRRADIATION

The preceeding equations of rate theory can help us to understand many of the radiation damage effects. An area of special current interest is the coupling of creep and swelling. Although calculations and comparisons with experiment are not yet available, some possibilities can be foreseen by examining the structure of the equations. For example there is experimental evidence that swelling enhances creep. It has been suggested that the SIPA mechanism predicts the opposite tendency[(41)], but this criticism does not apply to the present form of the SIPA model. I shall show here how the description of anisotropic dislocation densities is of particular importance in this regard, since it introduces a term in the creep rate which is directly proportional to the swelling.

We start with equation (24) for the net interstitial absorption rate and substitute for c_v and c_i. These point defect concentrations are given by equation (9) and its interstitial analogue, where recombination is ignored, or at least subsumed into an effective point defect production rate K. This gives :

$$dq_K/dt = \left(\frac{k_{Ki}^2}{k_i^2} - \frac{k_{Kv}^2}{k_v^2}\right)K + k_{Kv}^2\left(D_v c_{Kv} - \frac{K_v}{k_v^2}\right). \tag{51}$$

The next step is to substitute for the sink strengths in terms of the bias parameters, using the relations in sections 3.1 and 3.2. If terms of order $(\Delta Z + \delta Z_K)^2$ are neglected we obtain after some algebra :

$$\begin{aligned} dq_K/dt = {} & (K/k_v^2)\rho_K\delta Z_K - (K/k_v^4)\rho_K Z_{Kv}\sum_K \rho_K \delta Z_K \\ & + (Kk_C^2/k_v^4)\rho_K \Delta Z \\ & + (Z_{Kv}k_C^2/k_v^2)D_v(c_{Kv} - c_{Cv})\rho_K \\ & + (Z_{Kv}/k_v^2)D_v[(k_v^2 - k_C^2)c_{Kv} - \sum_K k_{Kv}^2 c_{Kv}]\rho_K. \end{aligned} \tag{52}$$

We have dropped the distinction between the sink strengths of cavities for interstitials and vacancies, denoting both by k_C^2. We shall also drop the K suffix on Z_{Kv}.

Finally we can combine the dq_K/dt terms using equations (21) and (22) for the swelling and creep rates. The fractions f_K will be introduced, such that $\rho_K = f_K\rho$. To see the physical significance of the rather lengthy expressions which result it is helpful to split them into a number of terms, thus :

$$dS/dt = dS^{PA}/dt + dS^N/dt + dS^T/dt, \tag{53}$$

where :

$$\begin{aligned} dS^{PA}/dt = {} & (Kk_C^2/k_v^4)\rho\{[\tfrac{2}{3}f_1 - \tfrac{1}{3}(f_2+f_3)]\delta Z_1 \\ & + [\tfrac{2}{3}f_2 - \tfrac{1}{3}(f_1+f_3)]\delta Z_2 + [\tfrac{2}{3}f_3 - \tfrac{1}{3}(f_1+f_2)]\delta Z_3\}, \end{aligned} \tag{54}$$

$$dS^N/dt = (Kk_C^2/k_v^4)\rho\Delta Z, \tag{55}$$

$$dS^T/dt = (Z_v k_C^2/k_v^2)D_v\rho(\sum_K f_K c_{Kv} - c_{Cv}). \tag{56}$$

Similarly for the creep rate :

$$d\varepsilon_3/dt = d\varepsilon_3^{PA}/dt + d\varepsilon_3^{N}/dt + d\varepsilon_3^{CS}/dt + d\varepsilon_3^{T}/dt, \tag{57}$$

where :

$$\begin{aligned} d\varepsilon_3^{PA}/dt = {} & (K/k_v^2)\rho(\tfrac{2}{3}f_3\delta Z_3 - \tfrac{1}{3}f_1\delta Z_1 - \tfrac{1}{3}f_2\delta Z_2) \\ & - (K/k_v^4)\rho^2 Z_v\,(\tfrac{2}{3}f_3 - \tfrac{1}{3}f_1 - \tfrac{1}{3}f_2)\{[\tfrac{2}{3}f_1 - \tfrac{1}{3}(f_2+f_3)]\delta Z_1 \\ & + [\tfrac{2}{3}f_2 - \tfrac{1}{3}(f_1+f_3)]\delta Z_2 + [\tfrac{2}{3}f_3 - \tfrac{1}{3}(f_1+f_2)]\delta Z_3\}, \end{aligned} \tag{58}$$

$$d\varepsilon_3^{N}/dt = (Kk_C^2/k_v^4)\rho\Delta Z(\tfrac{2}{3}f_3 - \tfrac{1}{3}f_1 - \tfrac{1}{3}f_2), \tag{59}$$

$$\begin{aligned} d\varepsilon_3^{CS}/dt = {} & (Z_v k_C^2/k_v^2)D_v\rho[\tfrac{2}{3}f_3(c_{3v} - c_{Cv}) \\ & - \tfrac{1}{3}f_1(c_{1v} - c_{Cv}) - \tfrac{1}{3}f_2(c_{2v} - c_{Cv})], \end{aligned} \tag{60}$$

$$\begin{aligned} d\varepsilon_3^{T}/dt = {} & (Z_v/k_v^2)D_v\rho[\sum_K k_{Kv}^2(\tfrac{2}{3}f_3c_{3v} - \tfrac{1}{3}f_1c_{1v} - \tfrac{1}{3}f_2c_{2v}) \\ & - \sum_K k_{Kv}^2 c_{Kv}(\tfrac{2}{3}f_3 - \tfrac{1}{3}f_1 - \tfrac{1}{3}f_2)]. \end{aligned} \tag{61}$$

Consider the physical significance of these expressions. The swelling term dS^{PA}/dt is due to preferential absorption, δZ_K, and it only contributes when the dislocation distribution is anisotropic, that is when

$$f_K \neq \tfrac{1}{3}. \tag{62}$$

The swelling term dS^N/dt is the normal swelling result, proportional to the damage rate, the cavity sink strength and the bias ΔZ. In general this bias depends on the hydrostatic component of stress p , since the dislocation – interstitial interaction depends on p[30]; dependence on the deviatoric component of stress is by definition confined to δZ_K. The final swelling term dS^T/dt can be described as 'thermal swelling', since it is proportional to the difference in emission rates between dislocations and cavities.

The creep terms have the following meaning. The first, $d\varepsilon_3^{PA}/dt$, is of the familiar SIPA type, depending on the differences between the induced biases δZ_K, but not on the anisotropy of the dislocation density. The second, $d\varepsilon_3^{N}/dt$, is proportional to the anisotropy and to dS^N/dt; this represents direct coupling between creep and swelling. It is also the form expected due to preferential nucleation of dislocation loops aligned with the stress axis, so-called N–creep. The third, $d\varepsilon_3^{CS}/dt$, represents a thermal creep-swelling coupling. The fourth, $d\varepsilon_3^{T}/dt$, represents pure thermal creep, in which vacancies diffuse between dislocations of different orientations.

The relative importance of these terms will depend on the temperature, irradiation history and degree of anisotropy of the dislocation structure.

7. CURRENT THEORETICAL RESEARCH

I conclude this brief exposition of the application of rate theory to creep by mentioning some areas of current theoretical interest, which have yet to be developed sufficiently to include in computational models.

I have already referred to the importance of a realistic description of the evolution of the dislocation density, particularly its anisotropy. The model described in section 4 is not gererally adequate in this respect, since it does not deal with the processes occuring during the evolution of sub-grain boundaries and their role in recovery. Nabarro *et al.* [42] have considered some of the climb processes in such a sub-grain boundary which enhance the recovery rate, and one ought to incorporate these processes into new rate equations.

I have not mentioned the role of dislocation glide in dislocation generation, nor its contribution to creep strain. Although at higher stresses it is probably most important, systematic rate equations have not yet been formulated to include glide processes.

The dislocation sink strength is a long standing topic for theoretical research, which may in future have to draw more upon the discipline of atomistic simulation to learn about the strength of the point defect – dislocation interaction, because several aspects are beyond the scope of elasticity theory. The rate limitation of point defect absorption and emission at the dislocation core has already been mentioned. Another effect is anisotropy of the diffusion coefficient induced by stress. Woo[43] has estimated that this may cause preferential absorption and creep up to thirty times the magnitude of the SIPA contribution. Theoretical improvements of this kind can be included in simulation by defining appropriate equations for the bias parameters; the form of the creep and swelling equations discussed in the previous section is not modified.

It is not surprising that the dislocation density develops anisotropy under the influence of an applied stress according to the rate equations described above. An unexpected effect, however, when we integrated the creep equations including irradiation, was that anisotropy developed even without an applied stress! This growth phenomenon would previously only have been expected if the system was asymmetric to start with, either by having a preferred crystallographic axis, as in hexagonal crystals such as zirconium, or by dislocation structure or texture introduced during the fabrication or previous stress history. We should not have been too surprised, however, since it is now well known in other branches of physics that symmetric equations commonly have asymmetric solutions if the equations are non-linear[44], a phenomenon studied under the name of bifurcation theory. This appears to be the case with our rate equations. The concept is not completely new to metallurgy. Equations for dislocation density in the context of a model of yielding have been studied by Ananthakrishna and Valsakumar [45–46] and shown to have analogous symmetry-breaking solutions, in their case of a periodic type, which were invoked to explain the phenomenon of repeated yield drop. In our case, in order to study the symmetry-breaking solutions properly it may be necessary to retain the full set of components of dislocation density ρ_{JK}, as required in any case for a generalisation of the model to triaxial stress states.

Finally, there have recently been applications of bifurcation theory to the problem of the origin and characteristic spacing of inhomogeneities in microstructure, such as the famous void lattice [47]. Studies have also been made of instabilities in the rate equations of cavity growth[48] and of instabilities in the equations of chemical segregation under irradiation[49] as described in the lectures of G.Martin at this Course. These developments lead one to expect that this aspect of the mathematics will be of increasing importance in computer simulations.

ACKNOWLEDGEMENTS

I am grateful for discussions with my colleagues Drs. R.Bullough, J.R.Matthews and M.H.Wood who collaborated in the development of the theory described here, and to M.Windsor who coded the equations with the FACSIMILE package.

REFERENCES

(1) Takeuchi S and Argon A S 1976, *J. Mater. Sci.* **11** 1542.

(2) Hart E W 1976, *J. Eng. Mater. and Technology* **98** 193.

(3) Hart E W, Li C Y, Yamada H and Wire G L 1976, *Constitutive Equations in Plasticity,* edited by A.S.Argon, MIT Press, Cambridge MA, 149.

(4) Mukherjee S 1982, *Boundary Element Methods in Creep and Fracture,* Applied Science Publishers, Barking, England.

(5) Curtis A R 1980 The FACSIMILE numerical integrator for stiff initial valuc problems. *Proc. conf. on Computational techniques for ordinary differential equations, Manchester, 18–20 Dec 1978,* I.Gladwell and D.K.Sayers (eds.), Academic Press, London, 47.

(6) Gear C W 1969 The automatic integration of stiff ordinary differential equations. *Information Processing 68,* A.J.H.Morrell (ed.), North Holland, Amsterdam, 187.

(7) Gear C W 1971 The automatic integration of ordinary differential equations. *Communications of the ACM,* **14,** 176.

(8) Gear C W 1971 Algorithm 407 DIFSUB for solution of ordinary differential equations [D2], *Communications of the ACM,* **14,** 185.

(9) Wiedersich H 1970 *Proc. 2nd Int. Conf. on the Strength of Metals and Alloys* , American Society for Metals, 142.

(10) Harkness S D and Li Che–Yu 1971 *Metall. Trans.* **2,** 1457.

(11) Brailsford A D and Bullough R 1972 *J. Nucl. Mater.* **44,** 121.

(12) Heald P T and Speight M V 1975 *Acta Metall.* **23,** 1389.

(13) Hesketh R V 1962 *Phil. Mag.* **7,** 1417.

(14) Lewthwaite G W 1973 *J. Nucl. Mater.* **46** 324.

(15) Brailsford A.D. and Bullough R 1973 *Phil. Mag.* **27** 49.

(16) Wolfer W G and Ashkin M 1973 *Scripta Metall.* **7** 1175.

(17) Heald P T and Speight M V 1974 *Phil. Mag.* **29** 1075.

(18) Heald P T and Speight M V 1974 *Phil. Mag.* **30** 869.

(19) Bullough R and Willis J R 1975 *Phil. Mag.* **31** 855.

(20) Bullough R Finnis M W and Wood M H 1981 *J. Nucl. Mater.* **103/104** 1263.

(21) Brailsford A D and Bullough R 1981 *Phil. Trans. Roy. Soc.* **302** 87.

(22) De Groot S R and Mazur P 1962 *Non-equilibrium Thermodynamics* North Holland, Amsterdam.

(23) Rauh H Wood M H and Bullough R 1981 *Phil. Mag.* **A44** 1255.

(24) Brailsford A D Bullough R and Hayns M R 1976 *J. Nucl. Mater.* **60** 246.

(25) Talbot D R S and Willis J R 1980 *Proc. R. Soc. Lond. A* **370** 351.

(26) Rauh H Bullough R and Wood M H 1983 *Proc. R. Soc. Lond. A* **388** 311.

(27) Bullough R Murphy S M Quigley T M and Wood M H 1983 *J. Nucl. Mater.* **117** 78.

(28) Bullough R and Quigley T M 1981 *J. Nucl. Mater.* **103/104** 1397.

(29) Bullough R and Quigley T M 1981 *J. Nucl. Mater.* **113** 179.

(30) Bullough R and Willis J R 1984 *Proc. of the IUTAM Symposium "Fundamentals of Deformation and Fracture (1984)"* to be published.

(31) Windsor M E Bullough R and Wood M H 1982 *Harwell Res. Rep. AERE-R10683.*

(32) Nabarro F R N 1967 *Phil. Mag.* **16** 231.

(33) R Bullough M W Finnis J R Matthews and M H Wood 1983 *Proc. conf. on Dimensional Stability and mechanical behaviour of irradiated metals and alloys, Brighton 11–13 April 1983* **1** BNES, London (1983) 125.

(34) Challenger K D and Moteff J 1973 *J. Met. Trans.* **4** 749.

(35) Michel D J Moteff J and Lovell A J 1973 *Acta Metall.* **21** 1269.

(36) Cuddy L J 1970 *Met. Trans.* **1** 395.

(37) Hirsch P B Howie A Nicholson R B Pashley D W and Whelan M J 1965 *Electron Microscopy of Thin Crystals* , Butterworths.

(38) Standring J Wilson A M and Clayden M C 1984 *Proc. conf. on Dimensional Stability and mechanical behaviour of irradiated metals and alloys, Brighton 11–13 April 1983* **2** BNES, London (1984) 79.

(39) Coble R L 1963 *J. Appl. Physics* **34** 1679.

(40) Jones R B 1984 *Proc. conf. on Dimensional Stability and mechanical behaviour of irradiated metals and alloys, Brighton 11–13 April 1983* **2** BNES, London (1984) 57.

(41) McVay G L Walters L C and Hudman G D 1979 *J. Nucl. Mater.* **79** 395.

(42) Nabarro F R N Bullough R and Matthews J R 1982 *Acta Metall.* **30** 1761.

(43) Woo C H 1984 *J. Nucl. Mater.* **120** 55.

(44) Nicolis G and Prigogine I 1977 *Self Organisation in Non–Equilibrium Systems* , Wiley, New York.

(45) Ananthakrishna G and Valsakumar M C 1982 *Journal of Physics D: Appl. Phys.* **15** L171.

(46) Ananthakrishna G and Valsakumar M C 1983 *Physics Letters* **95A** 69.

(47) Krishan K 1982 *Radiation Effects* **66** 121.

(48) Valentin P and Martin G 1982 *Phil. Mag.* **A46** 971.

(49) Martin G Cauvin R and Barbu A 1982 *Solid–Solid Phase Transformations,* eds. H.I.Aaronson, D.E.Laughlin, R.F.Sekerka and C.M.Wayman, AIME, New York 1982, p.256.

COMPUTER SIMULATIONS OF FIRST ORDER PHASE TRANSFORMATIONS PART I: METASTABILITY

Dieter W. Heermann
Institut für Festkörperforschung
Kernforschungsanlage Jülich
Postfach 1913
5170 Jülich 1, West Germany

ABSTRACT. In the first notes I present Monte Carlo simulation results on nucleation theory. The model studied was one with medium-range interactions. From the simulations it is concluded that the classical nucleation theory breaks down in the vicinity of the mean field spinodal. However, near the coexistence curve classical nucleation theory fits the data.

1. INTRODUCTION

The central question which I address in these lectures is how does a system brought into a nonequilibrium state relaxe into equilibrium? Particular well known examples of phenomena associated with the relaxation process are the supercooling/heating of water or the segregation in binary alloys. In a typical situation the sample is in a homogeneous one phase state (c.f. Figure 1). Then a parameter like the temperature is suddenly changed such that the system is in the two phase region. The sample, being now in a nonequilibrium state evolves through spontaneous fluctuations toward the stable equilibrium state in which the two phases coexist.

Classically one distinguishes two kind of fluctuations; those which are localized and finite in extent and those which are infinitesimal but stretch over the entire volume. Accordingly two theories were developed, nucleation and spinodal decomposition theory. However, it has become clear that there is no sharp distingtion as suggested by the classical theories.

Why does one study first order phase transformations by computer simulations? There is yet no acceptable theory which correctly describes the phenomena occuring in quenched systems. As long as such a theory is not in existence computer simulations can

G. Jacucci (ed.), Computer Simulation in Physical Metallurgy, 245–256.

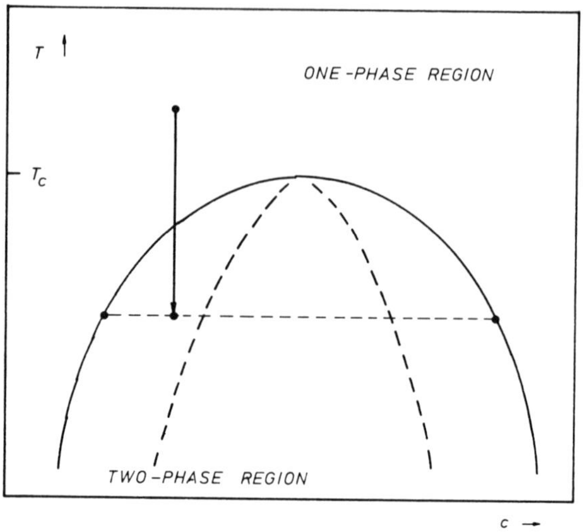

Figure 1
Typical situation of a relaxation process associated with a first order phase transformation. The system is brought from the one phase region by a sudden temperature change into a two phase nonequilibrium state. The system then evolves toward two phase coexistence.

furnish insights into the problem and thus provide a stimulation for the development of theories. But it is also a stimulus for experimentalists. Experimentalists can perform experiments to check effects found (or predicted?) by computer simulations.

One of the phenomena associated with relaxation processes is that of metastability which we discuss in Part I of these lectures. Before I present results of computer simulations of this phenomenon I shall describe the model system and the Monte Carlo simulation technique. Then I briefly review some existing theories.

2. THE MONTE CARLO SIMULATION TECHNIQUE

In this and the following lecture we concentrate on computer simulations of first order phase transformations using the Monte Carlo technique[1] (MC) as introduced by METROPOLIS[2]. What will be of interest to us is that this simulation technique is suitable for the study of relaxation phenomena. In fact, the technique is by construction a kinetic process.

In general one is interested in the evolution of a time dependent quantity $A(t)$, e.g., the magnetization or the concentration as the system evolves from nonequilibrium to equilibrium. Recall that a Monte Carlo simulation generates a Markov chain of phase points $\underline{s}_\nu$, where $\underline{s}_\nu$ describes for example the ν th configuration of spins. If we write $P(\underline{s}_\nu)=P(\underline{s},t)$, the probability of the state $\underline{s}$ at time t, then the dynamic evolution of $P(\underline{s},t)$ is governed by the master equation

$$dP(\underline{s},t)/dt = -\Sigma_{\underline{s}'}W(\underline{s}\to\underline{s}')P(\underline{s},t) + \Sigma_{\underline{s}'}W(\underline{s}'\to\underline{s})P(\underline{s},t) \tag{1}$$

where $W(\underline{s}\to\underline{s}')$ is the transition probability from the state $\underline{s}$ to $\underline{s}'$ per unit time. W is chosen such that $P(\underline{s},t)$ tends to the equilibrium distribution $P_{eq}(\underline{s})$. Hence the observable A develops according to a Markov process.

Rather than proceeding with a general description of the method we consider its application to the Ising model[3]. The Hamiltonian for this model is given by

$$-H/k_BT = (J/k_BT)\sum_{<i,j>} s_i s_j + h\Sigma_i s_i \quad . \tag{2}$$

J is the exchange interaction ($J>0$, ferromagnetic interaction), k_B the Boltzmann constant, T the temperature, h=2* magnetic dipole moment * magnetic field / k_BT will be refered to as an "applied field" and the symbol $<i,j>$ means that the sum is to be taken over nearest neighbours. s_i denotes the spin at site i which takes on the values +1 and -1 (spin up and down).

In this part of the lecture we study the Ising model with the single-spin-flip introduced by Glauber[4]. A site i is selected during the course of the Monte Carlo procedure and the spin is flipped from s_i to s_i'. Then the change $\Delta H=H(s')-H(s)$ is computed and the move is accepted if a random number $R(0<R<1)$ is less than W. The most common choice for the transition probability is

$$W = \begin{cases}(1/\tau_s)\exp(-\Delta H/k_BT) & \text{if} \quad \Delta H>0 \\ (1/\tau_s) & \text{otherwise}\end{cases}$$

τ_s is an arbitrary time constant which is generally set to 1. The above choice of W satisfies the detailed balance condition which is sufficient but not necessary[5] to ensure convergence to equilibrium.

Starting from an initial configuration of spins the above outlined algorithm is repeated many times and thus generates a sequence of configurations such that the system relaxes to thermal equilibrium. Time is measured in Monte Carlo steps divided by the number of spins (MCS).

Note that in the above model the order parameter (magnetization) is not conserved. This facilitates the simulation of metastable states since the evolution toward equilibrium is not limited by diffusion. In Part II we study the model with conserved order parameter. The dynamics of the model is then governed by the spin exchange (Kawasaki[6]).

3. BRIEF REVIEW OF THE MEAN FIELD THEORY OF PHASE TRANSFORMATIONS

Before we plunge into the results of machine simulations we briefly review the mean field picture for the phase transformation. Equilibrium thermodynamics requires that the free energy F is a convex function. Figure 2 shows a schematic phase diagram and the free energy as a function of the order parameter. The states outside the coexistence curve are thermodynamically stable. The states below the coexistence curve do not correspond to equilibrium states and a system brought from the one into the two phase region under the coexistence curve by a sudden change of, for example the temperature, will evolve towards a stable equilibrium state. Mean field theory(van der Waals type theories) suggest that the free energy can be analytically continued from the one into the two phase region (broken curve in Figure 2b). The kinetic mechanism by which the evolution toward equilibrium is based on is this continuation. The region where $\partial^2F'/\partial c^2$ is positive is called metastable (the states are also called metastable) and the evolution toward equilibrium is described by nucleation and growth. The region where $\partial^2F'/\partial c^2$ is negative is called the spinodal region (the states are called unstable) and the

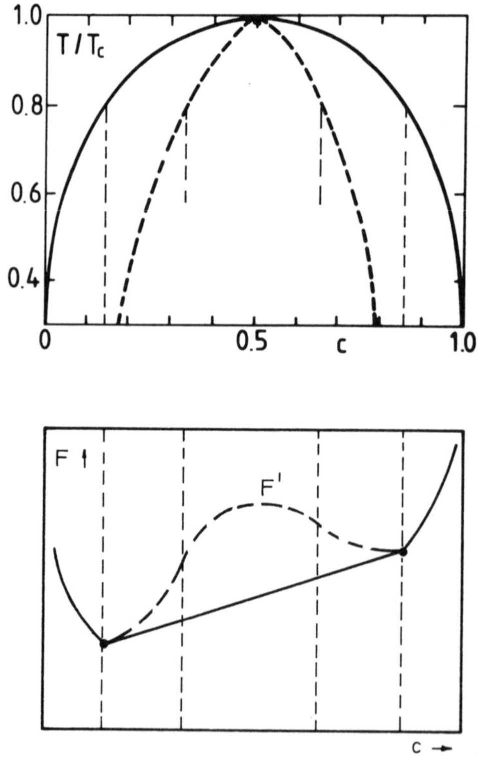

Figure 2
Schematic phase diagram and the free energy.

phase transformation takes place by spinodal decomposition. In this picture there is a sharp distinction between the metastable and the spinodal region. They are separated by the spinodal $(\partial^2 F'/\partial c^2=0)$.

Mean field theory, however, is strictly valid only for systems with infinitely long range forces[7]. In systems with finite range forces the distinction between meta- and unstable states is not as clear cut as mean field theory suggests. Monte Carlo simulations[8] showed no evidence of a spinodal and it has been pointed out[9,10,11] that there should be a smooth transition between the metastable and the spinodale region. Recent MC simulations also showed that a unique F' can not be defined[12].

4. CLASSICAL THEORY OF METASTABILITY

Suppose we suddenly cool a system which was in a state above T_c (c.f. Figure 1) to a state in the metastable region. Experimentally one observes the formation of ordered domains, the droplets (clusters) as the system tends to restore thermal equilibrium. Such nucleation, i.e., the formation of droplets is described by the classical theory of Becker and Döring[13]. They put forward a set of kinetic equations intended to describe the evolution of the droplet size distribution.

The classical nucleation theory assumes that the change $\Delta f/k_BT$ in the free energy for the formation of a droplet with radius r can be decomposed into a surface term and volume term

$$\Delta f/k_BT = 4\pi\sigma r^2-(3/4)\pi r^3 h \quad . \tag{3}$$

σ is the surface tension and taken to be that of a flat interface. The influence of the curvature of the droplet is neglected. h measures the distance to the coexistence curve. Figure 3 shows a schematic plot of $\Delta f/k_BT$ versus r. $\Delta f/k_BT$ has a maximum at

$$r^* = 2\sigma/h \tag{4}$$

and hence the relative stability of the state is governed by the height of the nucleation barrier. A droplet with a radius r smaller than r^* finds it energetically more favourable to shrink and those with $r>r^*$ predominately grow, thus transforming the state. If the nucleation barrier is high enough one can approximate the nucleation process by a steady state. This enables one to use equilibrium concepts to calculate the number of droplets of size r

$$n_r \propto \exp(-\Delta f/k_BT) \quad , \tag{5}$$

and the nucleation rate J, i.e., the number of droplets overcoming the barrier

$$J \propto n_{r^*} \propto \exp(-16\pi\sigma^3/(3h^2)) \quad . \tag{6}$$

Clearly the above can only be valid, if at all, as long as the height of the nucleation barrier is much larger than the thermal activation energy k_BT . Hence the capillarity approximation of Eq.(3) is expected

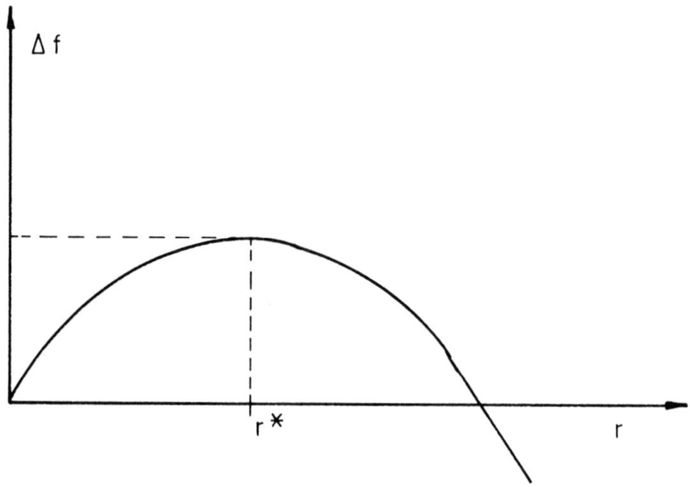

Figure 3
Free energy for the formation of a droplet.

to be valid for $r \to \infty$ and $r^* \to \infty$, i.e., at or very near the coexistence curve.

In the next section I will give some examples of computer simulations where the model is quenched into the metastable region.

5. COMPUTER SIMULATIONS OF METASTABILITY

In many cases a difficulty in experimental situations is that the coexistence curve is not precisely known. This leads to fallacious predictions for the nucleation rate. Another problem is that droplets escape detection, e.g., in binary mixtures[14] and the nucleation rate is not measured directly but infered from the volume fraction of the metastable state which has been transformed. Hence the validity of the nucleation theory is difficult to assess from the experimental data. The situation for computer simulations on the other hand is different. In machine simulations one can follow the phase transformation process under well defined conditions at all points in time. For example one can count all droplets and compare the distribution with the predictions of existing theories. There have been many computer simulations studying the phenomenon. A detailed review can be found in Gunton et al[15]. Here I describe some recent simulations of the three dimensional Ising model [16,17,18].

In these simulations one starts from a high temperature one phase state. This is usually done by initializing the system such that all spins either point up or down. Then a temperature is fixed (here $T/T_c=0.59$) and an appropriate field h is switched on opposite to the magnetization. At suitable time intervals the system is frozen and the configuration generated by the Monte Carlo procedure analyzed.

First let us look at the size distribution of the droplets produced by the simulation (c.f. Figure 4). The figure shows the number of droplets of size s (s being the number of spins in the cluster) formed after a sudden quench to various depths. The straight lines are visual fits to the data for large droplet

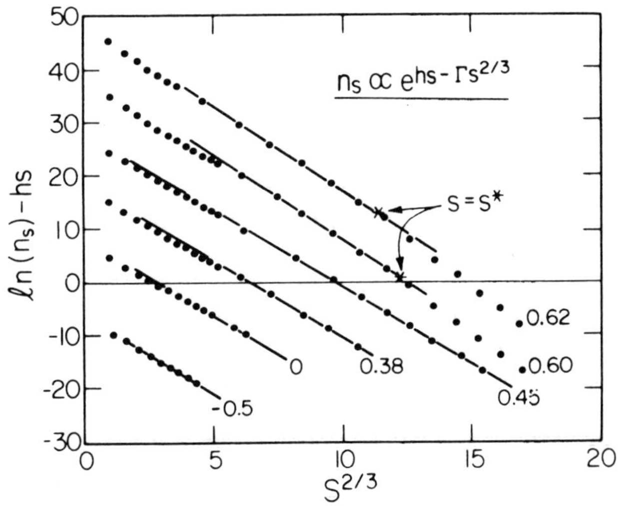

Figure 4
Observed droplet distribution generated by Monte Carlo simulations of the three dimensional Ising model at $T/T_c = 0.59$.

sizes. Overall the agreement of the observed distribution with the classical nucleation theory (Eq.(5)) is good. There are, however, deviation for the small droplets. This is not surprising since the assumption of a compact droplet (Eq.(3)) is not expected to hold for small droplets. For such small droplets, consisting of 10 or less monomers equations were developed which are in good agreement with the simulations[19,20].

The parallel lines in Figure 4 suggest that the assumption of a field independent surface tension is also quite good.

Figure 5 and Figure 6 show the critical droplet size and the nucleation rate as a function of the quench depth. Again the results show good overall agreement. The straight line in Figure 6 is the prediction of the classical nucleation theory. For small nucleation rates, determined by directly counting the number of droplets overcoming the barrier and from the number of critical droplets there are deviations. Those deviations might be due to the transition from spherical to cubical droplet shapes as the droplet size increases due to a decreasing field[21].

Simulations at higher temperatures also showed such deviations[17].

Thus far we tacitly assumed that a droplet is an <u>Ising droplet</u>. An Ising droplet is defined as a collection of up spins in a sea of down spins if they are connected by nearest neighbour bonds.

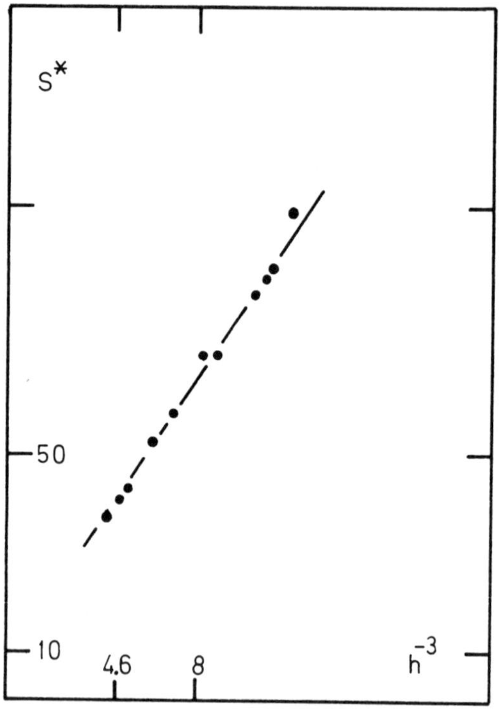

Figure 5

Critical droplet size. The straight line is a visual fit indicating good agreement with the classical nucleation theory.

This picture breaks down somewhat below the critical temperature (roughly at 0.96 T_c) due to percolation[22]. An alternatice definition has been proposed by Coniglio and Klein[23]. In their droplet model nearest neighbour up spins belong to the same droplet only if an active bond exists between them. Bonds are active with a certain probability depending on the temperature. For the low temperature simulations presented here there is almost no difference between the two models. Simulations using the Coniglio-Klein droplet model at high temperatures can be found in Reference 17.

We return now to the question of the existence of a sharp spinodal which in mean field separates the metastable from the unstable states. However, we will not discard the question of the validity of the classical nucleation theory.

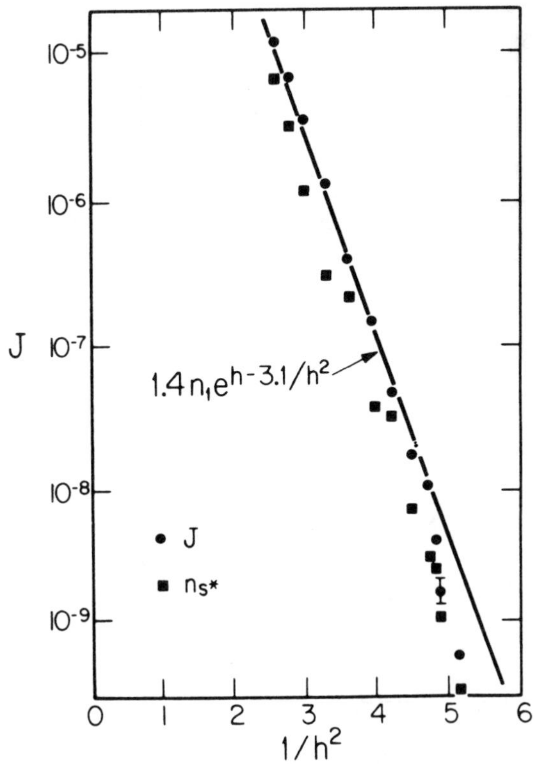

Figure 6
Nucleation rate as a function of the quench depth. The temperature was $T/T_c=0.59$. The straight line is the prediction of the classical nucleation theory.

Computer simulations of the Ising model[8] did not show the existence of a well defined spinodal. In fact, the spinodal singularity appears to be rounded and there is a smooth transition between the metastable and the unstable states[9,10,11]. On the other hand one knows that in the limit of infinite range interactions mean field theory is correct[7]. Since the usual Ising model has short range forces (nearest neighbour only) one should expect that the phenomena depend on the range of the forces. Figure 7 shows the results of computer simulations for increasing range of interaction[24]; q denotes the number of neighbours with which a given spin is interacting. Plotted is the inverse susceptibility versus the applied field, i.e., quench depth. One notes two points. First, with increasing range one can quench deeper into the metastable region before the system becomes unstable. Second, with increasing range the data follow more closely the mean field prediction which is in agreement with a recently developed Ginzburg criterion[25].

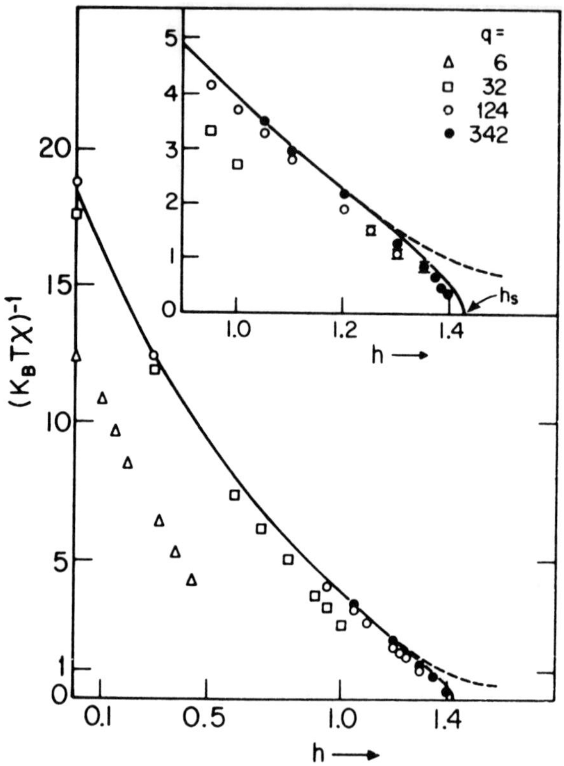

Figure 7
Inverse susceptibility as a function of the quench depth (applied field h). h_s denotes the mean field spinodal. The full curve is the mean field prediction and the dashed curve the prediction from the classical nucleation theory. The symbols denote the results of simulations with various interaction ranges. The interaction strength was $J/k_BT = 9/4\ (1/q)$.

Returning to the question of the validity of the classical nucleation theory (broken curve) we see a breakdown of this theory in the vicinity of the mean field spinodal.

6. DISCUSSION

Rather than discussing the physical results presented in this part (the discussion is deferred to Part II) I would like to discuss some general points connected to computer simulations. Although the Monte Carlo algorithm itself at first sight seems very easy non trivial difficulties present itself when actually performing simulations.

In principle one would like to study systems in the thermodynamic limit. In practice the computer storage available puts severe constraints on the size of the systems one can simulate. Also the kind of phenomenon one is interested in has to be taken into consideration. To study metastability the box should be "large enough" to avoid nucleation caused by the presence of the walls and to get droplets of large size. One therefore simulates systems of different sizes and employs periodic boundary conditions, which renders the boxes quasi infinite. The results are then used to study finite size effects, which requires intuition and experience.

Time has to be considered. The larger the system the more computer time has to be spent (and money)! Hence it is essential to have a highly optimized computer program.

To reduce the statistical error many runs have to be performed. Again intuition and experience and required to asses the quality of the data.

REFERENCES

1. K. Binder, *Monte Carlo Methods in Statistical Physics*, Springer Verlag 1984
2. N. Metropolis, A.W. Rosenbluth, M.N. Rosenbluth, A.H. Teller, E. Teller, J. Chem. Phys. 21, 1087 (1953)
3. E. Ising, Z. Physik 31, 253 (1925)
4. R.J. Glauber, J. Math. Phys. 4, 294 (1963)
5. L.D. Fosdick, Methods Comp. Phys. 1, 245 (1963)
6. K. Kawasaki, in *Phase Transitions and Critical Phenomena*, ed. C. Domb and M.S. Green, Vol. 2, Academic Press, New York 1972.
7. O. Penrose and J.L. Lebowitz, in *Studies in Statistical Mechanics*, eds. J.L. Lebowitz and E.W. Montroll, Vol. 7, North Holland, Amsterdam 1979
8. K. Binder and H. Müller-Krumbhaar, Phys. Rev. B9, 2328 (1973)
9. J.S. Langer, Physica 73, 61 (1974)
10. K. Binder, C. Billotet and P. Mirold, Z. Physik B30, 183 (1978)
11. K. Binder, Phys. Rev. B8, 3423 (1983)
12. K. Kaski, K. Binder and J.D. Gunton, J. Phys. A16, L623 (1983)
13. R. Becker and W. Döring, Ann. Physik 24, 719 (1935)
14. W.I. Goldburg, in *Scattering Techniques applied to Supramolecular and Nonequilibrium Systems*, eds. Chen, Chu and Nossal, Plenum 1981

15. J.D. Gunton, M. San Miguel and P.S. Sahni, in *Phase Transitions and Critical Phenomena*, eds. C. Domb and J.L. Lebowitz, Academic Press, London 1983
16. D. Stauffer, A. Coniglio and D.W. Heermann, Phys. Rev. Lett. 49, 1299 (1982)
17. D.W. Heermann, A. Coniglio, W. Klein and D. Stauffer, J. Stat. Phys. (1984)
18. D.W. Heermann, Acta Metall (1984)
19. O. Penrose and A. Buhagior, J. Stat. Phys. (1984)
20. O. Penrose, J.L. Lebowitz, J. Marro, M. Kalos and J. Tobochnik, preprint 1983
21. K. Kehr and K. Binder, in *Monte Carlo Methods in Stat. Phys.*, ed. K. Binder, Springer Verlag 1984
22. H. Müller-Krumbhaar, Phys. Letters A50, 27 (1975)
23. A. Coniglio and W. Klein, J. Phys. A13, 2775 (1980)
24. D.W. Heermann, W. Klein and D. Stauffer, Phys. Rev. Lett. 47, 1262 (1982)
25. K. Binder, Phys. Rev. A (1983)

COMPUTER SIMULATIONS OF FIRST ORDER PHASE TRANSFORMATIONS
PART II: SPINODAL DECOMPOSITION

Dieter W. Heermann
Institut für Festkörperforschung
Kernforschungsanlage Jülich
Postfach 1913
5170 Jülich 1, West Germany

ABSTRACT. In the second lecture notes Monte Carlo simulation results on spinodal decomposition and the transition region between nucleation and spinodal decomposition are presented. For medium-range models the classical theory of spinodal decomposition works for a certain region for short times following a quench. In the transition region a dynamical spinodal is found.

1. INTRODUCTION

In the first lecture I pointed out that there is no clear distinction between nucleation and spinodal decomposition. As in the nucleation phenomenon one observes for quenches into the spinodal region the appearance of precipitates. These domains coarsen, grow to macroscopic size and eventually like binary mixtures and binary alloys decompose. This segregation process is limited by diffusion since the order parameter is conserved. However, nucleation and spinodal decomposition have been treated differently. While nucleation was associated with an energy barrier which the localized fluctuations have to overcome, spinodal decomposition was associated with phenomena where no or a negligible small energy barrier exists. In this case the barrier for the transformation is diffusional and even infinitesimal, nonlocalized fluctua-

G. Jacucci (ed.), Computer Simulation in Physical Metallurgy, 257–268.

tions are amplified. In mean field theory the energy barrier vanishes as one approaches the spinodal from the metastable side[1]. The different treatment seems to be reasonable on or very close to the coexistence curve and for critical quenches (50-50 mixtures) for early times. In the intermediate region, however, new concepts are required.

In this lecture we study the validity of the linearized Cahn-Hilliard theory of spinodal decomposition. After briefly reviewing this theory I present computer simulations of spinodal decomposition using a system with medium range forces. Then we examine the phenomena occuring in the transition regime between nucleation and spinodal decomposition.

2. REVIEW OF THE CAHN-HILLIARD THEORY OF SPINODAL DECOMPOSITION

The starting point of the Cahn-Hilliard theory of spinodal decomposition [2,3] is the assumption that a coarse-grained free energy (Ginzburg Landau Hamiltonian)

$$F(c) = \int_V \left(\frac{1}{2}\kappa(\nabla c)^2 + f(c)\right)dr \tag{1}$$

can be defined in the two phase region. $c(r)$ is the local concentration and $f(c)$ a free energy density (doubly well potential). Note that $F(c)$ depends on the coarse graining length.

To derive a prediction for the evolution of the structure factor, which is proportional to the small angle diffuse scattering intensity[4], after a fast quench into the miscibility gap one starts from the diffusion equation

$$\begin{aligned} \frac{\partial}{\partial t}c(r) &= -M\nabla^2\frac{\delta F}{\delta c} + \xi(r,t) \\ &= M\nabla^2(-\kappa\nabla^2 c + \partial f/\partial c) + \xi(r,t) \end{aligned} \tag{2}$$

with a random noise term $\xi(r,t)$ included by Cook[5] to ensure the correct statistical description of the system dynamics. M is a mobility assumed to be a constant. Linearization of Eq.(2) leads to the prediction[6] for the structure factor $S(k,t)$

$$S(k,t) = S(k,0)\exp(\omega(k)t) + S_T(k)(1-\exp(\omega(k)t) \quad , \tag{3}$$

with $S_T(k)$ being the Ornstein-Zernike[7] equilibrium structure factor. The amplification factor $\omega(k)$ in this theory is given by

$$\omega(k) = -2Mk^2(k^2 + \partial^2 f/\partial c^2|_{c_0}) \quad . \tag{4}$$

For $\partial^2 f/\partial c^2 < 0$ (inside the spinodal region) there exists a critical wave vector

$$k_c = \sqrt{1/\kappa|\partial^2 f/\partial c^2|} \quad , \tag{5}$$

such that the amplification factor is positive for $k<k_c$. Hence the Cahn-Hilliard theory predicts that long-wave-length fluctuations will be amplified exponentially and the maximum growth occurs at $k_m = k_c/\sqrt{2}$.

Fluctuations with $k>k_c$ will be exponentially damped.

3. COMPUTER SIMULATIONS OF SPINODAL DECOMPOSITION

To gain understanding of the decomposition process occuring for example in binary alloys (Fe-Al, Zn-Al) we simulate the Ising model with conserved order parameter. The system is then governed by the Kawasaki dynamics[8] (spin exchange). Here I present a recent computer study of the Ising model with medium range forces[9]. A review of earlier simulations of the nearest neighbour Ising model can be found in Gunton et al[10] and Marro[11].

From the Ginzburg criterion[12]

$$1 \ll R^3(1-T/T_c)^{1/2}(1-c/c_{sp})^{3/2} \tag{6}$$

where R is the interaction range we expect the linearized Cahn-Hilliard theory to hold for short times after the quench for a certain concentration range. This is indicated in Figure 1 by the shaded area. Indeed, the resutls of the computer simulations show that for short times after the quench from the one phase region the long wavelength fluctuations are exponentially amplified (c.f. Figure 2). Then for later times the behaviour changes over to non-linear phenomena, i.e., the non-linear terms neglected in Eq.(3) become important. The agreement gets worse as one moves closer to the mean field spinodal concentration into the non-classical transition regime.

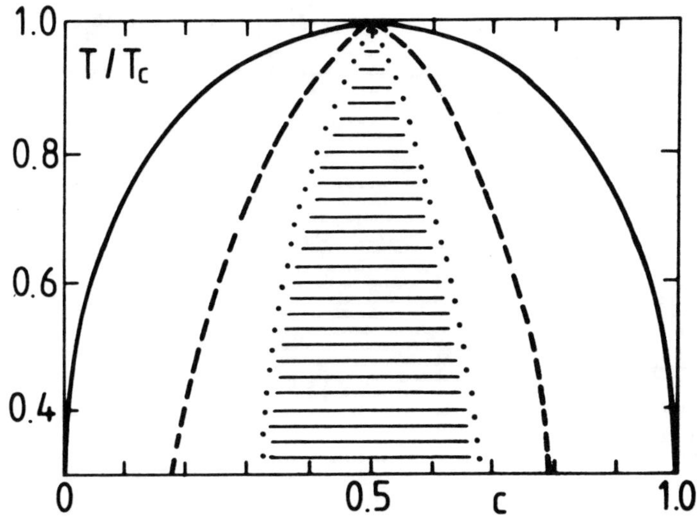

Figure 1
Schematic phase diagram. The dashed line gives the mean field spinodal. The shaded area indicates the region where according to a Ginzburg criterion the linearized Cahn-Hilliard theory is expected to hold for short times after a quench.

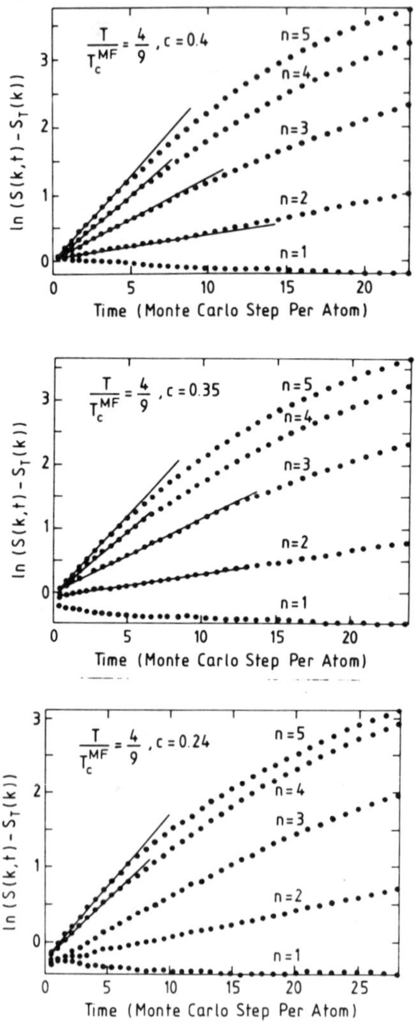

Figure 2

Evolution of the structure function for an Ising model with medium range forces. The Cahn-Hilliard theory predicts exponential growth.

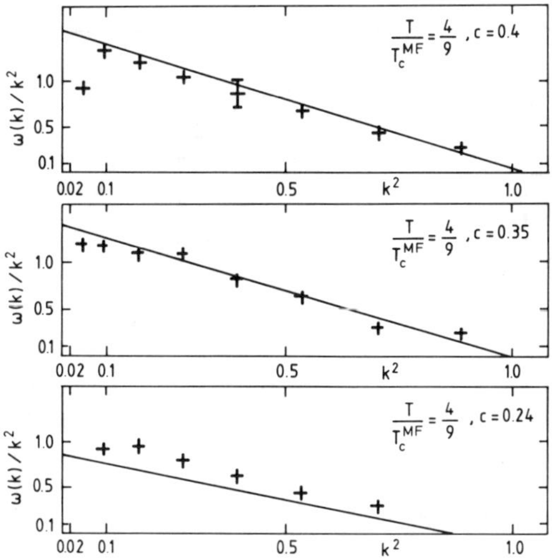

Figure 3
Amplification factor $\omega(k)$ divided by the square of the wave vector versus k^2 . The straight lines are the predictions from the linearized Cahn-Hilliard theory.

4. THE TRANSITION REGION BETWEEN THE METASTABLE AND THE SPINODAL REGION

So far we have probed by computer simulation the relaxation phenomena occuring near the coexistence curve and in the spinodal region where the linearized theory holds at least for short times. Now we turn our attention to the transition region between the metastable and the spinodal region. The shaded area of the schematic phase diagram in Figure 4 indicates this transition region and the concentrations we study. Again we use an Ising model with medium range forces in the simulations.

Recall from the first lecture that the classical nucleation theory was seen to break down for quenches deep in the metastable region. There the height of the nucleation barrier is of the order of k_BT . Field theoretic approaches[13,14] for this regime and simulations[15] have shown that the droplets in this region are ramified. Let us follow the growth of such a ramified droplet at the point P1. Classically the droplets are assumed to be compact spheres and the growth is predicted to be (Lifshitz-Slyozov-Wagner[16,17])

$$R \propto t^{1/3} \quad . \tag{7}$$

Hence the growth should follow a trajectory as depicted in Figure 5 by the straight line. What one observes, however, is a two stage growth process[18]. First the droplet compactifies, and then grows according to the Lifshitz-Slyozov-Wagner prediction. During the compactification

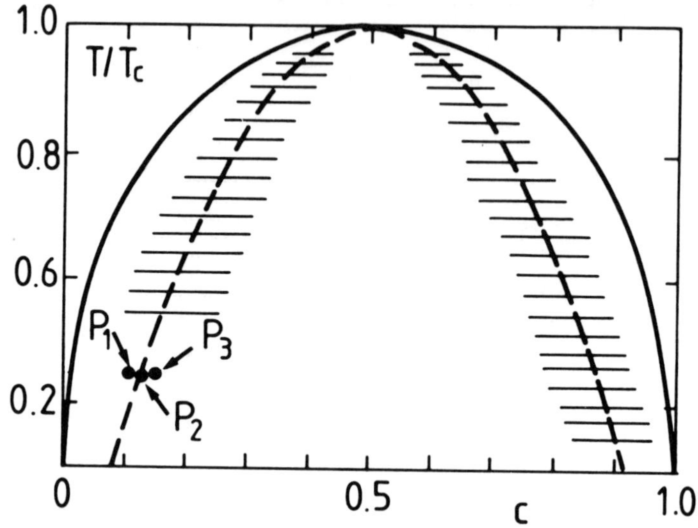

Figure 4
Schematic phase diagram. The dashed curve gives the mean field spinodal and the shaded area indicates the non-classical transition regime between nucleation and spinodal decomposition.

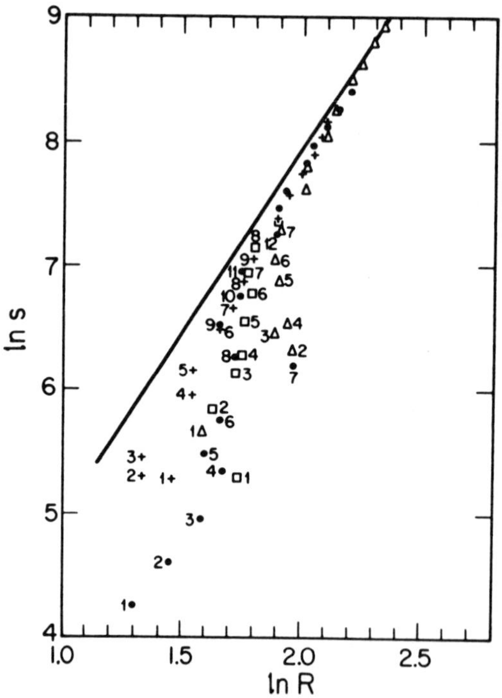

Figure 5
Growth of droplets in the transition region (P1). Plotted in in this double logarithmic plot is the number of particles in the droplet versus its radius of gyration. The different observed droplets and the numbers denote the time sequence.

period the radius of gyration remains roughly constant. Nevertheless the droplet mass increases which indicates that the droplet profile changes from very diffuse to that of nearly spherical shape[19].

The above described simulations were carried out with non conserved order parameter. In the conserved order parameter case the situation is similar. Figure 6 shows the evolution of the radius of gyration of the largest droplet in the system. Initially the radius increase steeply, reaches a maximum and decreases while the droplet mass monotonically increases (Figure 7). Again we interpret this as a compactification.

The situation for the concentration (P3) inside the mean field spinodal region is dramatically different! There the radius and the mass immediately after the sudden quench increase without bound and after a time Δt has elapsed decreases. Hence a percolating structure has formed which after a while dissolves again.

Recall that in mean field theory the susceptibility diverges at the spinodal. In other words, the mean droplet size diverges there. The simulation, however, shows that although a percolating structure forms the mean droplet size remains finite! (See Figure 8,9c).

The above leads one to postulate a dynamical spinodal[20], not to be confused with the static mean field spinodal. The definition for the dynamical spinodal reads as follows. There exists a concentration $c^*(T)$, $T<T_c$, where for the first time a droplet after the quench percolates through the system with the mean droplet size remaining finite and Δt , the time span where it percolates, becomes essentially zero.

5. SUMMARY

In these lectures I have presented computer simulation studies of the nonequilibrium part of two component systems. These simulations showed that the classical nucleation theory is a good approximation to the phenomena occuring in the metastable region for low temperatures and near the coexistence curve. The theory, however, breaks down in the vicinity of the mean field spinodal. For short range interaction systems no indication of a spinodal was found. Rather a smooth transition region between the metastable and the unstable region was found which narrows with increasing interaction range. In the transition region a transient percolation effect was seen and a dynamical spinodal proposed. In the spinodal region the computer simulations showed that the linearized Cahn-Hilliard theory holds for short times for a certain concentration range.

Clearly further computer simulation studies are needed to shed light on the phenomena occuring in the two phase region. Especially the transition region, the late stages of the droplet growth and decomposition require further investigation.

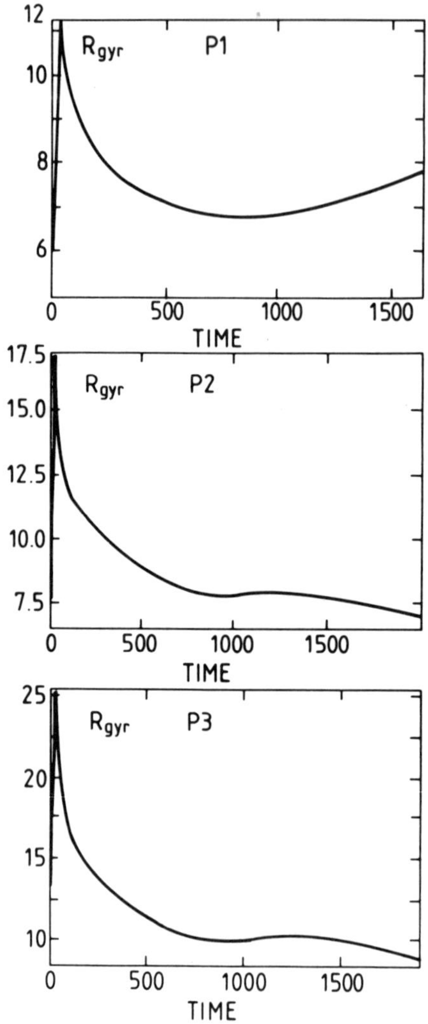

Figure 6
Evolution of the radius of gyration of the largest droplet in the system following a quench for the concentrations indicated in Figure 4.

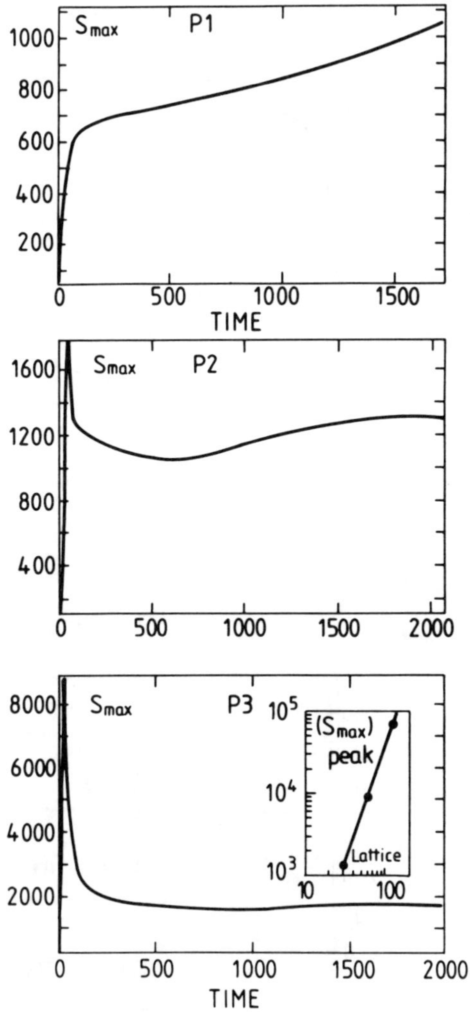

Figure 7
Droplet mass as a function of time.

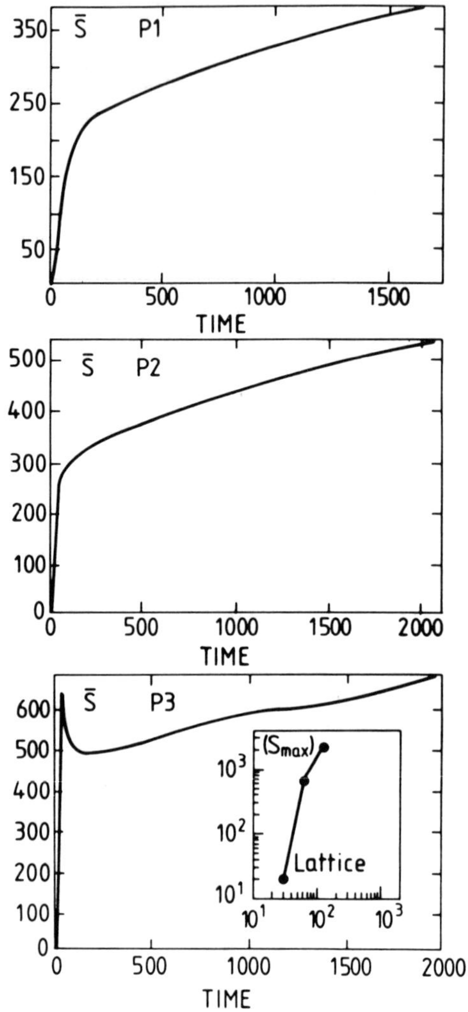

Figure 8
Mean droplet size as a function of time.

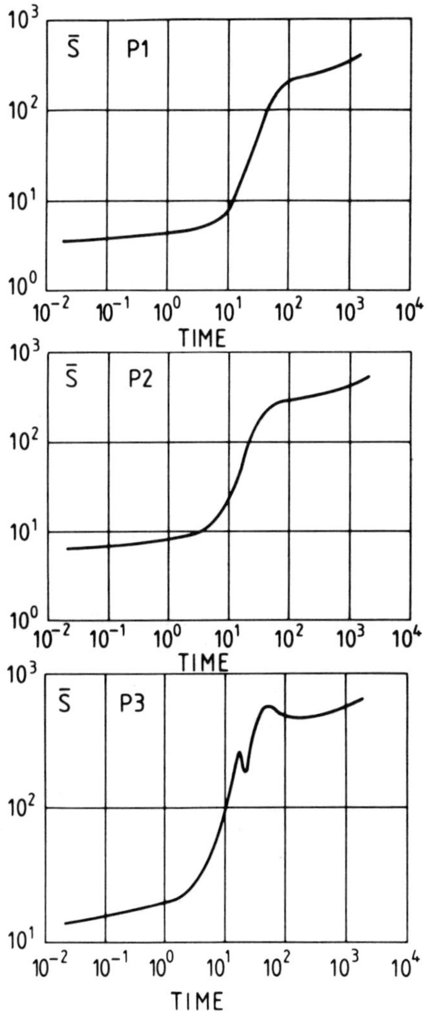

Figure 9
Log-log plot of the evolution of the mean droplet size following a quench.

REFERENCES

1. J.W. Cahn and J.E. Hilliard, J. Chem. Phys. 31, 688 (1959)
2. J.W. Cahn, Acta Metall 9, 795 (1961)
3. J.E. Hilliard, in *Phase Transitions*, ed. H.I. Aaronson, American Society for Metals, Metals Park, Ohio, 1970
4. K.B. Rundman and J.E. Hilliard, Acta Metall 15, 1025 (1967)
5. H.E. Cook, Acta Metall 18, 297 (1970)
6. J.S. Langer, M. Barron and H.D. Miller, Phys. Rev. A11, 1417 (1975)
7. L.S. Ornstein and F. Zernike, Proc. Sect. Sci. med. Akad. Wet. 17, 793 (1914)
8. K. Kawasaki, in *Phase Transitions and Critical Phenomena*, eds. C. Domb and M.S. Green Vol. 2, Academic Press, New York 1972
9. D.W. Heermann, Phys. Rev. Lett. 52, 1126 (1984)
10. J.D. Gunton, M. San Miguel and P.S. Sahni, in *Phase Transitions and Critical Phenomena*, eds. C. Domb and J.L. Lebowitz, Academic Press, London 1983
11. J. Marro and J.L. Valles, Phys. Letters A95, 443 (1983)
12. K. Binder, Phys. Rev. A,(1983)
13. W. Klein and C. Unger, Phys. Rev. B28, (1983)
14. C. Unger and W. Klein, Phys. Rev. B,(1984)
15. D.W. Heermann and W. Klein, Phys. Rev. B27, 1732 (1983)
16. I.M. Lifshitz and V.V. Slyozov, J. Phys. Chem. Solids 19, 35 (1961)
17. C. Wagner, Z. Elektrochem. 65, 581 (1961)
18. D.W. Heermann and W. Klein, Phys. Rev. Lett. 50, 1062 (1983)
19. D.W. Heermann, Acta Metall (1984)
20. D.W. Heermann, Z. Physik B, (1984)

HEAT AND PARTICLE FLUXES DURING LASER AND ELECTRON BEAMS IRRADIATION OF METALS

A. MIOTELLO
Centro Studi del C.N.R. and Istituto per la Ricerca
Scientifica e Tecnologica
38050 - Povo (TN)
Italy

ABSTRACT. Some properties of a resolidified metal (in particular the structure and the composition if atoms of impurity are embedded in it) are strongly dependent on the regrowth conditions.
High resolidification velocities may induce liquid-amorphous phase transition or trapping of impurity atoms. New regimes in the resolidification processes have been recently introduced in physical metallurgy by the advent of laser and electron beam pulses, ranging between 5 ÷ 100 nsec in duration and within 1 ÷ 15 J/cm^2 in energy density. Here we discuss some mathematical and physical aspects of the heat and particle fluxes during irradiation.

1. HEAT FLOW IN AN ALUMINIUM SAMPLE INDUCED BY A NANOSECOND LASER PULSE.

The characteristic times and lengths involved in the problem of heat flow through a semiinfinite Al sample irradiated with nanosecond laser pulses are first discussed with reference both to the underlying physics and to numerical solution of heat diffusion equation (1,2).
Any calculation of the temperature T of a laser irradiated sample based on heat equation, when thermodynamical parameters such as the thermal conductivity K or the specific heat c are involved, assumes that the lengths over which we study a given heat phenomenon are longer than the typical length over which it is meaningful to talk in terms of macroscopic parameters such as K or c. As Harrington's paper shows (3) such a length is about five times the mean free-path l_p of the heat carriers.
As a consequence the heat transport equation assumes the usual form (in a one-dimensional case, x being the coordinate of the axis normal to the sample surface with origin at the surface, t the time, ρ the Al density, A(x,t) the heat source term):

G. Jacucci (ed.), Computer Simulation in Physical Metallurgy, 269–287.

$$c\rho \frac{\partial T}{\partial t} = \frac{\partial}{\partial x} (K(T)\frac{\partial T}{\partial x}) + A(x,t) \tag{1}$$

only if we are interested in variation of T over lengths $\gtrsim 5 l_p$.
A second remark concerns the characteristic times involved in the problem. The collision time of the hot carriers (electrons in the case of a metal) with the lattice is very short in metals (4). However the thermalization time of the produced hot carriers is irrelevant for phenomena occurring over a duration greater than 1 nsec.
This point is quite relevant also for whork in semiconductors and it has been questioned, proposing a completely new approach to the understanding of electron-lattice characteristic coupling time(5).
In the metal case we assume that light, when absorbed, is instantaneously converted into lattice heat, so that the use of a single thermodynamical temperature, is a meaningful thing.
Finally, in order to carry out calculations on eq.1 we must further specificy the source term. The absorption of a laser pulse of given temporal shape F(t) (F(t) being the incident power density) is controlled by two parameters: the absorption coefficient α and the reflection coefficient R. We have:

$$A(x,t) = \alpha(1-R)F(t)\exp(-\alpha x) \tag{2}$$

Equation 1 must be solved together with boundary conditions:

$$T(x,0) = T_o \quad ; \quad T(\infty, t) = T_o \tag{3}$$

$$\left.\frac{\partial T}{\partial x}\right|_{x=0} = 0 \tag{4}$$

$$K_{sol} \left.\frac{\partial T}{\partial x}\right|_{X^+} - K_{liq} \left.\frac{\partial T}{\partial x}\right|_{X^-} = \rho H_{fus} \dot{X} \tag{5}$$

plus during resolidification:

$$\dot{X} = -V_{res} = f\lambda k_u \frac{H_{fus} M}{k_B T_M} \left(\frac{T_M - T_i}{T_i} \right) \tag{6}$$

T_o is the initial temperature of the sample.
Eq.5 refer to liquid-solid interface velocity ($\dot{X}$) which varies in such a way as to conserve the balance of ingoing and outgoing heat flux (H_{fus} is the latent heat of melting / resolidification). $\dot{X}$ must satisfy the undercooling constraints (eq.6) as required by crystal regrowth thermodynamics (T_i is the liquid-solid interface temperature and T_M is the melting temperature, k_B is the Boltzmann constant, M is the mass of an Al atom, λ is a typical length of the order of atomic size, f a pure number of order unity and k_u is the frequency of the configurational rearrangement taking place at the interface ($k_u \simeq 10^{13}$ sec^{-1} (6)). Eq.6 may also written, according to a simple argument (see ref.(7) and ref. therein)

$$\frac{(T_M - T_i)}{T_i} \simeq \frac{v_{res}^{Stefan}}{V_s}$$

where V_s is the metal sound velocity and v_{res}^{Stefan} is the resolidification velocity as determined by the pure heat flux conservation (i.e.eq.5 which determines a Stefan boundary problem if acting alone (7)). Since $v_{res}^{Stefan} \sim 20$ m/sec $\Delta T_i = T_M - T_i \sim 0{,}5\%$ of T_M i.e. negligible.
The resolidification velocity in metals is then controlled only by heat flux.
In the case of high absorbed energy, involving surface vaporization, the heat flow equation must satisfy some additional boundary conditions (8).
Denoting by S the istantaneous evaporating front position with respect to a reference frame fixed in the bulk of the sample, then it is:

$$\rho H_{vap} \dot{S} = K_{liq} \left.\frac{\partial T}{\partial x}\right|_{x=S} \tag{7}$$

H_{vap} being the (constant) heat of vaporization.
The thermodynamical condition for the surface front velocity $\dot{S}$ is given by one of the following two relationships, stated in terms of the surface temperature T_S:

$$\dot{S} = \begin{cases} U \exp\left(\dfrac{-MH_{vap}}{k_B T_s}\right) & T_S < T_{eb} \\ \text{as given by eq.7 plus eq.1 discretized} & T_S = T_{eb} \end{cases} \tag{8}$$

U is a velocity of the same order as sound velocity in metals.

2. THE NUMERICAL CALCULATION.

2.1. The Finite Differences Method

If we know the values T_j of a function $T(x)$ at regular intervals $x=j\Delta x$ $j=\ldots, -2, -1, 0, 1, 2\ldots$ of its argument, than the first, second ... forward differences, $\Delta T_j, \Delta^2 T_j, \ldots$ of T_j are defined to be:

$$\Delta T_j = T_{j+1} - T_j \tag{9}$$

$$\Delta^2 T_j = \Delta T_{j+1} - \Delta T_j = T_{j+2} - 2T_{j+1} + T_j \tag{10}$$

These are called "forward" differences because they involve successive values T_j, T_{j+1}, ... working forward, or in the direction of j increasing from T_j. In the same way, backward differences (working backward from T_j) may be defined.

It may be convenient to have a notation which involves points symmetrical about j, and so the central difference notation is introduced. In this notation, the first, second, ... central differences are defined as:

$$\delta^1 T_{j+\frac{1}{2}} = \Delta T_j = T_{j+1} - T_j \tag{11}$$

$$\delta^2 T_j = \delta^1 T_{j+\frac{1}{2}} - \delta^1 T_{j-\frac{1}{2}} = \Delta T_j - \Delta T_{j-1} = T_{j+1} - 2T_j - T_{j-1} \tag{12}$$

and so on . It will be seen that, as in eq.11 differences of odd order involve points distributed symmetrically about $j+\frac{1}{2}$ and not about j; in order to have symmetry with respect to j, the so called mean differences are introduced:

$$a\,\delta^1 T_j = \frac{1}{2}\left[\delta^1 T_{j+\frac{1}{2}} + \delta^1 T_{j-\frac{1}{2}}\right] = \frac{1}{2}\left(T_{j+1} - T_{j-1}\right) \tag{13}$$

and so on. Here"a" denotes the operation of taking the mean of values for $j\pm\frac{1}{2}$. The whole of theory of the finite differences method is based on the assumption that the values ..., T_{-2}, T_{-1}, T_0, T_1, T_2, ... are known and that a "difference table" of them and their successive differences 9, 10, ... is formed. To get a completely numerical method it is necessary to replace in eq.1 the time derivative by a difference. To do this , we choose an interval of time Δt, and write $T_{j,i}$ for $T(j\ \Delta x, i\ \Delta t)$.

There are various formulae for $\frac{\partial T}{\partial t}$ which may be used and, of these the simplest is:

$$\left.\frac{\partial T}{\partial t}\right|_{t = i\,\Delta t} = \frac{T_{j,i+1} - T_{j,i}}{\Delta t} \tag{14}$$

As a conclusion eq.1 may be written according to the finite differences scheme as:

$$c\,\rho\,\frac{T_{j,i+1}-T_{j,i}}{\Delta t} = \left[-K_{j-1,i}\,\frac{T_{j,i}-T_{j-1,i}}{\Delta x} + K_{j,i}\,\frac{T_{j+1,i}-T_{j,i}}{\Delta x}\right]\cdot \frac{1}{\Delta x} + F_i\,(\exp\,(-\alpha x_{j-1}) - \exp\,(-\alpha x_j))\,\frac{1}{\Delta x} \qquad (15)$$

where $K_{j,i}$ is the thermal conductivity of the layer j, at the time step i, $x_j = j\,\Delta x$ and F_i is the light beam intensity at time $i\,\Delta t$.

When the index j in $K_{j,i}$ refers to a layer partially melted at T_M, then $K_{j,i}$ is calculated as an average value obtained by the values of the thermal conductivity referring to the liquid and solid phase,weighted with the fraction of melted and solid layer respectively. It is now necessary to discuss the size of Δx and Δt.

In order to get a solution in a reasonable number of steps, it is desiderable to choose Δt as large as is safe but some restriction on its value is necessary according to the following argument.

Suppose that when $t = i\,\Delta t$ the maximun error in any of the $T_{j,i}$ is η, then if $dT_{j,i+1}$ is the small change in $T_{j,i+1}$ caused by small changes in the $T_{j,i}$, and if $\frac{K}{c\rho}\,\frac{\Delta t}{\Delta x^2} = \Lambda = \text{const}$

$$\left|dT_{j,i+1}\right| \leq \Lambda \left|T_{j+1,i}\right| + \Lambda \left|T_{j-1,i}\right| + \left|2\Lambda - 1\right| \left|dT_{j,i}\right| \leq \left[2\Lambda + \left|2\Lambda - 1\right|\right]\eta \qquad (16)$$

Now if the method is useful, errors must not grow, that is, we must have $\left|dT_{j,i+1}\right| \leq \eta$ and by eq.16 this requires:

$$\Lambda \leq \frac{1}{2} \qquad (17)$$

This is the required restriction on Δt, usually called the stability condition. It is a sufficient condition, but not a necessary one.

2.2. Some Results Obtained by Integration of eq.1 with Boundary Conditions 3-5.

Fig.1 shows the general trend of the results (1) when the energy density of the pulse is such that melting (but not vaporization) of the surface layer occurs. We used the following Al thermodynamical parameters: c= 0.94 J/gr K, ρ = 2.7 gr/cm^3 (for both solid and liquid phase); for the thermal conductivity K_{sol} = 2.25 W/cm K in solid phase and K_{liq} = = 1 W/cm K in liquid phase, H_{fus} = 395.6 J/gr.

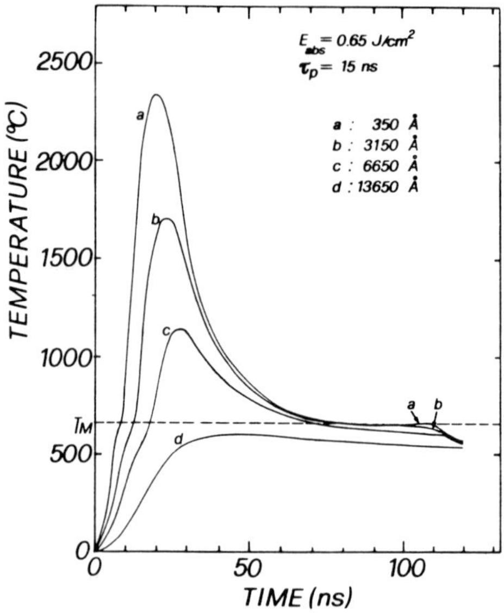

Figure 1. Time dependence of the temperature, at different depths, of a laser irradiated Al sample (1).

The temperature curve of the first layer may be used to characterize the general dynamical behavior of the sample under laser irradiation. We may roughly distinguish five time interval, corresponding to different physical processes: heating in solid phase, heating in liquid phase, cooling in liquid phase, resolidification, cooling in solid phase. As one may easily note, the temperature rise in the liquid phase is steeper than the one in the solid phase, due to lower thermal conductivity; the maximum temperature is reached at a time t_{max}, which depends on τ_p (the half duration of the laser pulse) and not on the E_p (the energy pulse), according to the law:

$$t_{max} \simeq \sqrt{2}\,\tau_p \qquad (18)$$

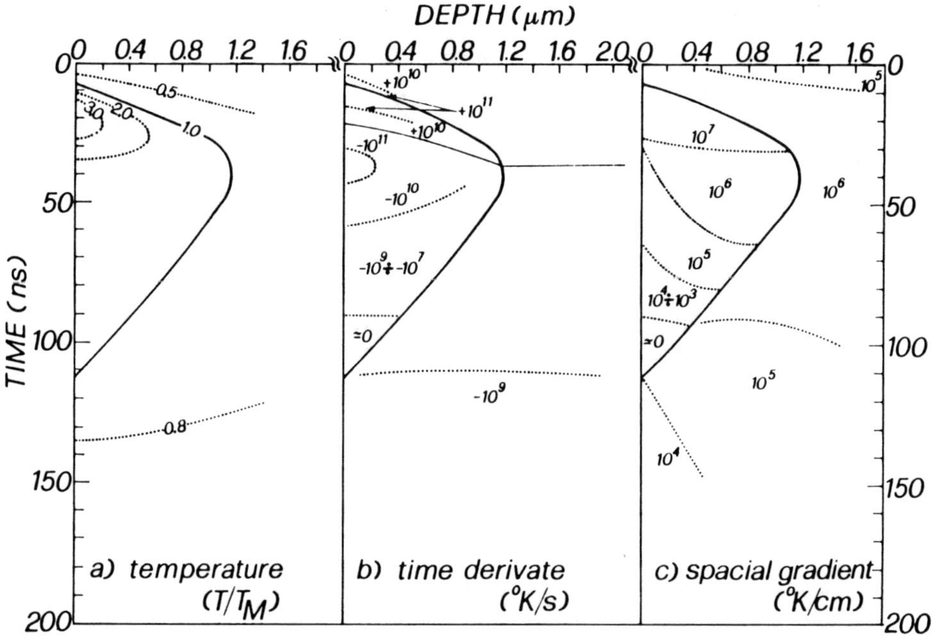

Figure 2a, b, c. General behavior in the plane x, t for a) the temperature T, b) its time derivative, and c) its spacial gradient in a laser heated Al sample (E_p = 3.1 J/cm^2, τ_p = 15 nsec). The solid line, corresponding to the position of the liquid-solid interface as a function of time, is shown in all three plots for the sake of comparison. Each isocontour of fig. 2a is labeled by the corresponding value of T/T_M, while in the case of fig. 2b, 2c the orders of magnitude in the different regions are shown (1).

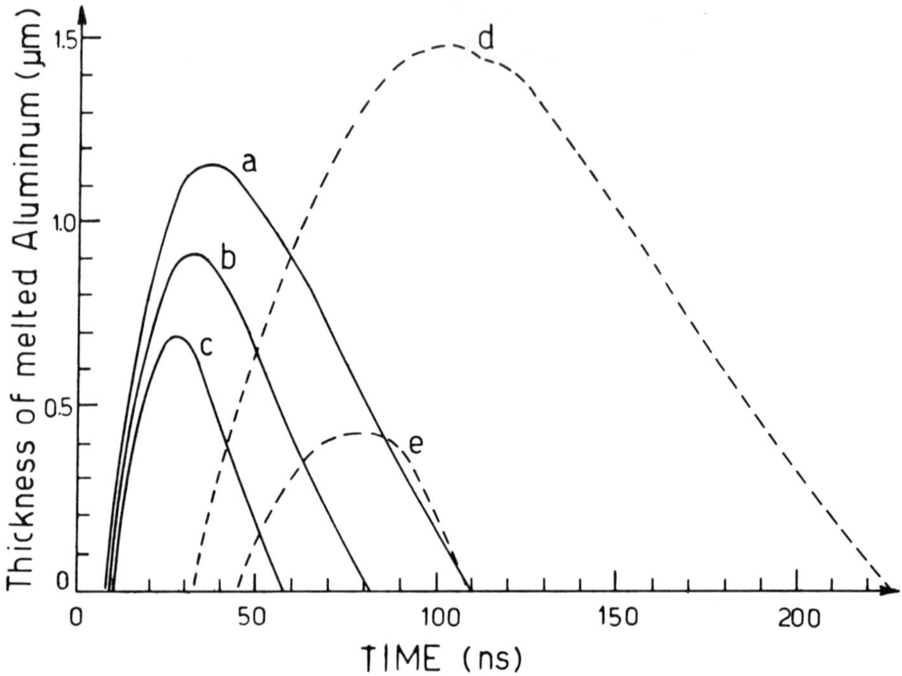

Figure 3. Time dependence of the thickness of melted Al for different pulse energy density and durations: (curve a, b, c: $\tau_p = 15$ nsec, $E_p = 3.1, 2.6, 2.1$ J/cm^2 respectively; curve d, e: $\tau_p = 50$ nsec, $E_p = 4.3, 2.5$ J/cm^2) (1).

3. ELECTRON BEAM INDUCED HEAT FLOW TRANSIENT IN ALUMINIUM

The main difference between laser and electron-beam lies in the energy deposition mechanism (9) which in the case of electrons involves both "free and bound" atomic electrons. In the typical range of electrons energy (5 ÷ 30 keV) all electrons of irradiated system are seen as free. The energy loss depends only on the atomic number Z and on the sample density ρ . Energy loss trough plasmon excitation are irrelevant. Moreover the bremsstrahlung losses are negligible for incident electron energies up to 60 MeV in a target with Z = 13. At the energies of interest the deposition range and the depth dose function are the result of a statistical summation of the energy loss along the electrons paths, taking into account all the deviations from travelling direction due to elastic collisions with nuclei. For $10 < Z < 15$ Everhart and Hoff (10) propose a depth dose function which for a monochromatic electron beam with electron energy equal to E_o is:

$$-\frac{dE}{dx} = f\,\frac{E_o}{R_G}\,\lambda(y)$$

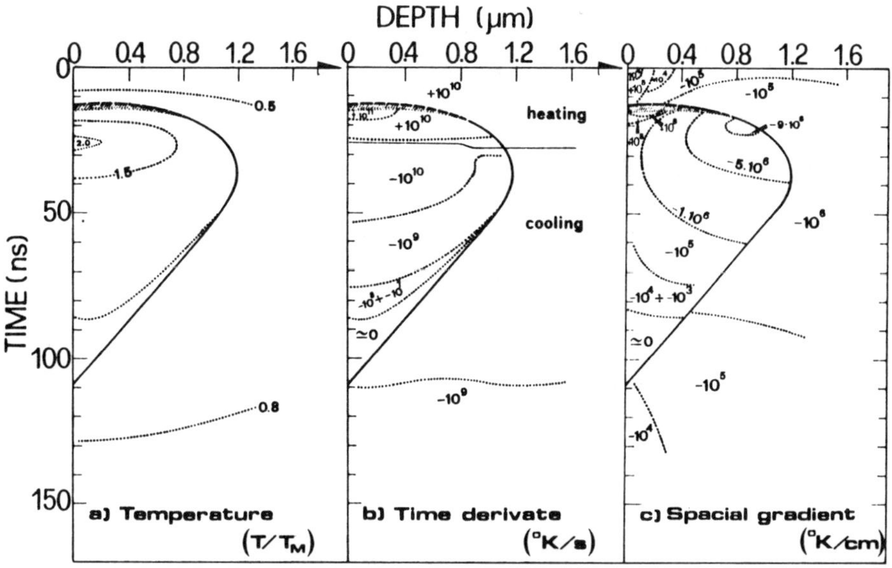

Figure 4. Depth-time plots for a) the temperature T, b) its time derivative, and c) its spacial gradient in an electron beam heated Al sample (E_o = 12keV, E_{abs} = 0.65 J/cm^2).

In fig.4 (9) the depth - time plots for the sample temperature T (x,t), $\dot{T}$, $\frac{\partial T}{\partial x}$ are shown for electrons having E_o = 12 keV (a good average value in many experiments) and for E_{abs} = 0.65 J/cm^2: in the case of a ruby laser pulse, this latter value corresponds to a maximun surface temperature just below the (normal) boiling one, the appropriate depth - time plots being shown in fig. 2. By comparing the two sets of plots, we notice that the cooling stage of the transient is roughly the same with both energy sources. The two heating stages, on the contrary, differ in many details: in particular, while the laser induced heating and melting starts from the metal surface with high negative thermal gradients, in the case of an electron source a layer of about one half the electron penetration range R_G, 13095 Å in this case, is simultaneously melted (dashed region); a proper melt front builds up only when this process is over. The heating rate may on the contrary be quite comparable with both energy sources, at least in the near surface region and during the pulse, since the more dilute energy deposition mechanism appropriate to electrons is compensated by the constraint that the heat, flowing from the deposition peak region to the surface, cannot escape from the surface itself.

$$\lambda(y) = \begin{cases} 0.6 + 6.21y - 12.4y^2 + 5.96y^3 & 0 \leq y = \dfrac{x}{R_G} \leq 1.1 \\ 0 & \text{elsewhere} \end{cases} \tag{19}$$

where x is the depth from the surface into the sample, $R_G = 4.57\ 10^{-6}\ E_o^{1.75}/\rho$ cm is the Grün projected range (11) and $f = 1 - R_{el}$ plays the role of an e - beam absorption coefficient in order to take into account the backscattered electrons which leave the sample. The heat source term (assuming that all the electrons energy losses are immediately converted into lattice heat at place where they actually occur, neglecting any diffusion process of the produced hot carriers) may be written in terms of the beam flux $\phi(t,E)$ of electrons impinging on the sample with energy between E and E + dE according to the relationship:

$$A(x,t) = \int dE\, \phi(t,E)\, \frac{dEabs(x,E)}{dx} \tag{20}$$

where on the basis of eq.19 $\dfrac{dEabs}{dx} = \dfrac{f(E)E\,\lambda(y)}{R_G(E)}$

In simple cases, the electron pulse generator is characterized by the current and voltage waveforms J(t) and V(t).

Then $\phi(t,E) = \dfrac{J(t)}{e}\, \delta[E - E^*(t)]$

with e = electron charge and $E^*(t) = eV(t)$ and

$$A(x,t) = \frac{J(t)}{e}\, \frac{dEabs(x,E^*(t))}{dx} \tag{21}$$

the total absorbed energy E_{abs} is given by:

$$E_{abs} = \int_0^{t_f} dt \int_0^{\infty} dx\ A(x,t) \tag{22}$$

t_f being the pulse duration. When eq.21 applies E_{abs} becomes

$$E_{abs} = \int_0^{t_f} dt\ J(t)\cdot V(t)\ f[\,eV(t)] \tag{23}$$

In Al the projected range R_G varies between 0.28 µm at E = 5keV and 4.73 µm at E = 25 keV. Thus the e - beam generators used at present never yield deposited energy profiles in metals comparable to the laser pulse case, where deposition occurs within 200 Å from the surface.
For a given material, the reflectivity parameter R_{el} may in principle depend both on the energy of the beam electrons and on their angle of incidence to the sample surface. In the following, calculations are limited to the case of normal incidence ($Rel = 0.15$).

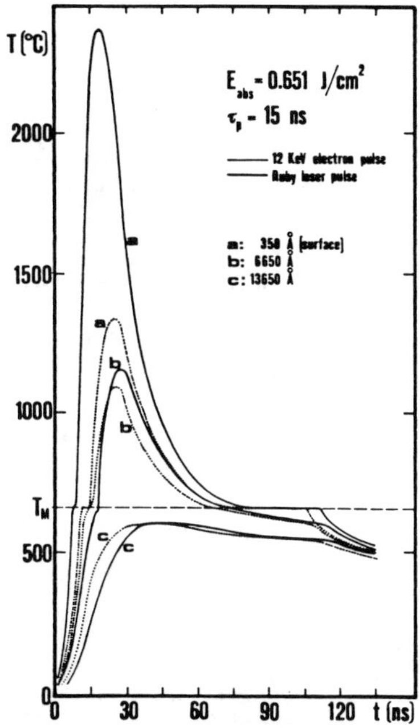

Figure 5. Temperature evolution of an Al sample, at different depths, as obtained with laser and electron beam sources (9)

4. SORET EFFECT

As it is known, in a binary solution under a thermal gradient one component diffuses preferentially to the hotter end and the other to the colder one (this is called the Soret effect). The magnitude and sign of the Soret effect can be expressed in terms of heat of transport Q_1^*.

In a binary alloy one measures the difference in the heats of transport of solute and solvent Q*. The heat of transport Q*(in a dilute solution limit) is connected to the thermal migration coefficient D' through the relationship (12):

$$Q^* = S_T\, R\, T^2, \quad S_T = \frac{D'}{D} \tag{24}$$

where T is the temperature, R the gas constant, D the ordinary diffusion coefficient and S_T the Soret coefficient of the process. Usually S_T is very small and therefore thermal migration may be neglected as compared to ordinary diffusion.

In a laser induced heat flow transient, however, temperature gradients of the order of 10^7 K/cm occur and a contribution due to thermal migration

cannot be excluded, even though we may quite safely assume that the symmetric cross coefficient does not affect the heat flow.
The heat of transport depends on the precise nature of the mass transport process. None of the existing models in generally accepted. In the case of liquid metals Bhat and Swalin (13) proposed a model according to which Q* is made of two parts, one of which arises from the interaction of the conduction electrons with solute ions and the other from the interaction of the diffusing ion with solute ones. The latter can be calculated by assuming that the liquid is a dense gas and applying the thermotransport theory in binary gas mixtures.
The sign of this part is determined primarily by the mass difference, the heavier impurity migrating to the colder end. In the limit of a dilute impurity in a host metal (this is a typical situation) the fluxes with respect to the center of mass velocity reduce to fluxes in the laboratory frame.
The impurity flux is then (12):

$$J = - D\frac{\partial n}{\partial x} - DS_T\, n\, \frac{\partial T}{\partial x} \tag{25}$$

where $n = n(x,t)$ is the impurity concentration (in atoms/cm^3).
The continuity equation for the impurity concentration n together with equation 25 and in the assumption of constant D and S_T, yields (14, 15):

$$\frac{\partial n}{\partial t} = - \mathrm{div} J = D\frac{\partial^2 n}{\partial x^2} + DS_T \frac{\partial n}{\partial x}\frac{\partial T}{\partial x} + DS_T\, n\, \frac{\partial^2 T}{\partial x^2} \tag{26}$$

with boundary condition:

$$D \left.\frac{\partial n}{\partial x}\right|_{x=0} + DS_T\, n \left.\frac{\partial T}{\partial x}\right|_{x=0} = 0 \tag{27}$$

which prevents the impurity to escape through the surface and

$$D \left.\frac{\partial n}{\partial x}\right|_{x=X(t)} + DS_T\, n \left.\frac{\partial T}{\partial x}\right|_{x=X(t)} = 0 \tag{28}$$

which prevents the escape of impurities at $x = X(t)$.
$X(t)$ is the position of the liquid - solid interface.

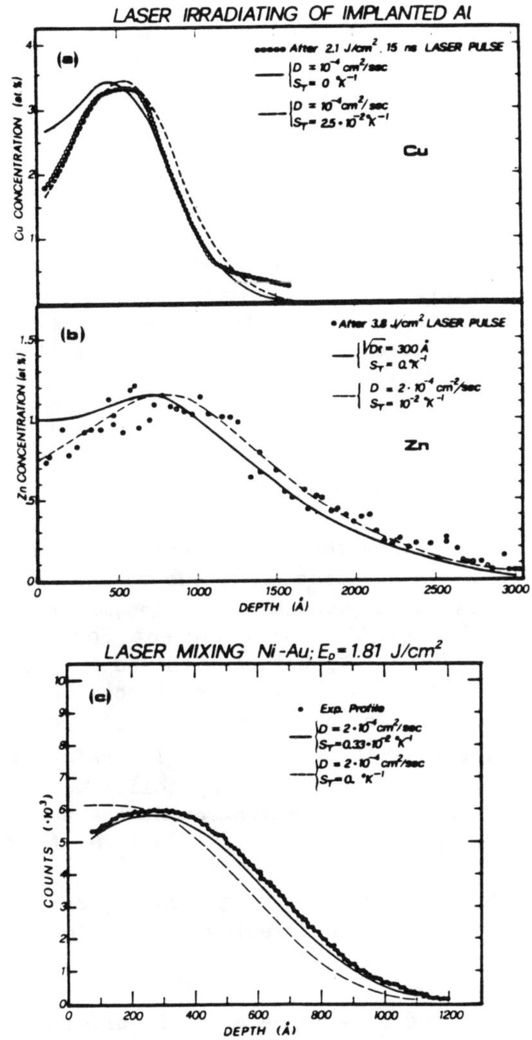

Figure 6. Experimental evidence for Soret diffusion a) Al (Cu) system of ref. 16, b) Al (Zn) system of ref. 17, c) Ni (Au) system of ref. 18.

The Soret coefficient reported in fig.6 and obtained by a best fit procedure allow a determination of the heats of transport, relating to these processes, through eq.24.
One obtains respectively (14, 15):

$$Q^*_{Cu} = 8 \times 10^4 \text{ cal/gr-mole}$$

$$Q^*_{Zn} = 2.54 \times 10^4 \text{ cal/gr-mole}$$

$$Q^*_{Au} = 2.85 \times 10^4 \text{ cal/gr-mole}$$

These values are higher than the few ones usually quoted in the literature for other liquid metals (13, 19) but might be related to non - linear effects which occur in these transients because of the very high thermal gradients (about five orders of magnitude larger than the ones used for standard measurements).

5. NON - EQUILIBRIUM SEGREGATION EFFECTS IN A FREEZING LIQUID ALUMINIUM LAYER AFTER A NANOSECOND LASER PULSE IRRADIATION.

Much of recent literature explains the redistribution of impurities in implanted crystals under laser irradiation as due to a combined melting and resolidification process during which impurities may redistribute in the liquid because of the high diffusion coefficient. For impurities with equilibrium segregation coefficent k_o between 0.3 and 1 (e.g. B, P, As in Si (20)) the tendency is towards a uniform distribution across the molten layer (effective segregation coefficient $k_e = 1$). For impurities which are very little soluble in the solid ($k_o << 1$) preferential segregation of impurities in the liquid phase during solidification with a consequent enhancement of impurity concentration at the surface may occur (e.g. other III - V impurities in Si (21) Cu in Si (22,23), Fe in Si (25)).
Such an enhancement always corresponds to a k_e much higher than k_o and it may be explained in terms of the high freezing rate following the laser pulse .

5.1. Formulation of the Diffusion - Segregation Problem: the Case of Pb in Al

The equilibrium value k_o of Pb in Al is vanishing small ($\lesssim 10^{-4}$). In the following we show that experimental Pb impurity profiles detected in an Al sample after laser irradiation (16) may be analyzed on the basis of a diffusion - segregation model in which $k_e = 1$ when the resolidification front velocity is of the order of 40 m/sec (i.e. a typical resolidification front velocity in laser induced heat flow transient in metals (26)).
We assume an ordinary diffusion process of initial Pb profile (i.e. the

Pb profile obtained by ion implantation), in liquid phase (induced by a ruby laser pulse of $E_p = 1.3\ J/cm^2$, $\tau_p = 15$ nsec).

Continuity equation for the impurity concentration n(x,t)) at depth x and time t (x=0 is the Al surface position) is:

$$\frac{\partial n}{\partial t} = D \frac{\partial^2 n}{\partial x^2} \quad , \qquad 0 < x < X(t) \tag{29}$$

X(t) is the position of the liquid - solid interface, D is the Pb diffusion coefficient assumed constant.
We have in addition the boundary conditions:

$$\left.\frac{\partial n}{\partial x}\right|_{x=0} = 0 \quad , \qquad \left.\frac{\partial n}{\partial x}\right|_{x=X^+(t)} = 0 \tag{30}$$

which prevent the escape of impurity at x = 0 and x = X(t) before that resolidification occurs.
During resolidification the segregation process is decribed by requiring that:

$$\lim_{x \to X^+(t)} n(x,t) = k_e \lim_{x \to X^-(t)} n(x,t) \tag{31}$$

In the formulation of the diffusion - segregation problem the function X(t) is known in general from previous calculations of heat flow induced in Al by a laser pulse.
We assume that no Pb diffusion occurs at x>X(t).
Eq.29 with boundary conditions 30, 31 has been numerically integrated with the condition for the spacial and time steps:

$$\frac{D\ \Delta t}{\Delta x^2} < \frac{1}{2}$$

for the convergence of the numerical solution (see above).
Moreover Δ x, during the resolidification process must satisfy:

$$\Delta x < \frac{2D}{v_{res}} \tag{32}$$

Condition 32 assures that segregation occurs if the residence time of impurity in the Δ x layer is less than the regrowth time. In metals $2D/v_{res}$ may be of the order of a monolayer thick or less (see the typical high resolidification front velocity).

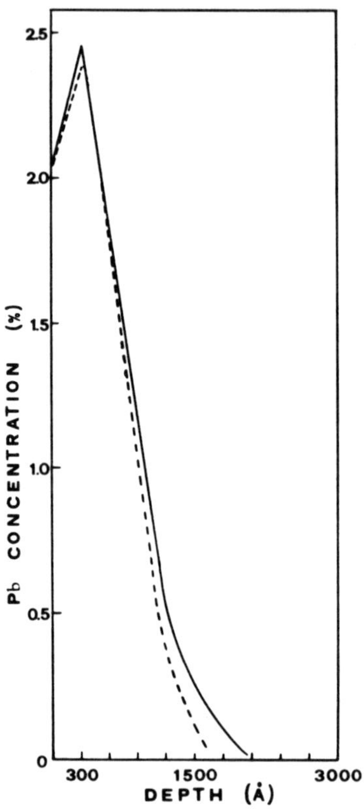

Figure 7. Experimental Pb profile (full line) and the theoretical one (dashed line) as obtained with $k_e = 1$, $D = 2 \cdot 10^{-4}$ cm^2/sec, $v_{res} = 40$ m/sec.

For $E_p = 1.3$ J/cm^2 the heat flow calculation show a molten layer in which the temperature is at all times near the melting one and so the use of a constant D value is justified.

The unity value for the K_e segregation coefficient is typical for impurities in metallic system under laser irradiation experiments. This is due to the fact that the impurity residence time τ_{res} at the regrowing interface is much too long, as compared to the regrowth time τ_{reg}, in order to avoid trapping. As a consequence, if D_i is the impurity diffusion coefficient just at the regrowing interface, the condition for segregation to occur:

$$\frac{\tau_{res}}{\tau_{reg}} = \frac{\lambda\, v_{res}}{D_i} \ll 1 \qquad (33)$$

(λ being the monolayer characteristic length) is not satisfied under laser irradiation, since v_{res} is quite high as compared to semiconductor (Si in particular) in similar irradiation conditions. Indeed metals have higher thermal conductivity and lower latent heat of recrystallization). We may finally quote the fact that Sb, Cr, Cu, Sn, all show an effective segregation coefficient in Al equal to 1, even though the equilibrium segregation coefficient is either very large (Sb, Cr) or very small (Cu, Sn), (14, 27, 28, 29).

Acknowledgements

I would like to thank L.F. Donà dalle Rose and P. Mazzoldi for useful discussion during the preparation of the manuscript.

REFERENCES

1. L.F. Donà dalle Rose and A.Miotello, Rad. Eff. 53, 7(1980).

2. L.F. Donà dalle Rose and A. Miotello, Rad.Eff. 53, 19(1980).

3. R.E. Harrington, J. Appl. Phys. 38, 3266(1967).

4. N. Bloembergen, "Laser-Solid Interactions and Laser Processing" 1978, Materials Research Society, Boston, AIP Conference Proceedings (AIP, New York, 1979) p.1.

5. J.A. Van Vechten, R. Tsu, F.W. Saris and D. Hoonhout, Phys. Lett. 74A, 417(1979); and J.A. Van Vechten, R. Tsu and F.W. Saris, Phys. Lett. 74A, 422(1979).

6. F. Spaepen and D. Turnbull, "Laser-Solid Interactions and Laser Processing", 1978, Materials Research Society, Boston, AIP Conference Proceedings (AIP, New York, 1979) p.73.

7. L.F. Donà dalle Rose, Journal de Physique 10 (C5), 469 (1983).

8. F.V. Bunkin; M.J. Tribel'skii, Sov. Phys. Uspekhi 23, 105(1980).

9. L.F. Donà dalle Rose, A. Miotello and R. Brotto, Rad. Eff. 69, 1(1983).

10. T.E. Everhart, P.H. Hoff, J.Appl. Phys. 42, 5837(1971).

11. A.E. Grun, Z. Naturforsch. A12, 89(1956).

12. R.S. De Groot and P. Mazur, "Non-equilibrium Thermodynamics" (North-Holland, Amsterdam, 1962).

13. B.N. Bhat and R.A. Swalin, Acta Met. 20, 1387(1972).

14. A. Miotello, L.F. Donà dalle Rose and A. Desalvo, Appl. Phys. Lett. 40, 135(1982).

15. A. Miotello, L.F. Donà dalle Rose,Phys. Lett. 87A, 317(1982).

16. G. Della Mea, L.F. Donà dalle Rose, P. Mazzoldi and A. Miotello, Rad. Eff. 46, 133(1980).

17. W.R. Wampler, D.M. Follstaedt and P.S. Peercy in "Laser and Electron Beam Solid Interaction and Material Processing", eds. J.F. Gibbons, L.D. Hess, T.W. Sigmon (North-Holland, New York, Oxford 1981)p.567.

18. P.P. Pronko, H. Wiedersich, K. Seshan, A.L. Helling, T.A. Lograsso and P.M. Baldo in: "Laser and Electron Beam Solid Interaction

and Materials Processing", eds. J.F. Gibbons , L.D. Hess, T.W. Sigmon (North-Holland, New York, Oxford 1981) p. 599.

19. S.P. Murarka,T.Y. Kim, M.Y. Hsieh and R.A. Swalin, *Acta Met*. **22**, 185(1974).

20. C.W. White, W.H. Christie, B.R. Appleton, S.R.Wilson,P.P. Pronko and C.W. Magee, *Appl. Phys. Lett*. **33**, 662(1978).

21. C.W. White, S.R. Wilson, B.R. Appleton and F.W. Young Jr., *J. Appl. Phys*. **51**, 738(1980).

22. P. Baeri, S.U. Campisano, G. Foti and E. Rimini, *Appl.Phys.Lett.* **41** 1246(1978).

23. C.W. White, J. Narayan, B.R. Appleton and S.R. Wilson, *J. Appl. Phys*. **50**, 2967(1979).

24. C.W. White, S.R. Wilson, B.R. Appleton and J. Narayan in: "Laser and Electron Beam Processing of Materials", edited by C.W. White and P.S. Peercy (Academic, New York, 1980) p.124.

25. P. Baeri, S.U. Campisano, M.G. Grimaldi and E. Rimini in: "Laser and Electron Beam Processing of Materials", edited by C.W. White and P.S. Peercy (Academic, New York, 1980) p. 130.

26. A. Miotello and L.F. Donà dalle Rose, *Rad. Eff*. **55**, 235(1981).

27. W.R. Wampler, D.M. Follstaedt and S.T. Picraux, *Appl. Phys. Lett*. 36, 366(1980).

28. G. Battaglin, A. Carnera, G. Della Mea, P. Mazzoldi, E. Jannitti, A.K. Jain and D.K.Sood, *J. Appl. Phys.* **53**, 3224(1982).

29. S.T. Picraux, D.M. Follstaedt, J.A. Knapp, W.R.Wampler and E.Rimini in: "Laser and Electron Beam Solid Interaction and Materials Processing", eds. J.F. Gibbons, L.D. Hess, T.W. Sigmon (North-Holland, New York, Oxford, 1981) p.575.

APPLICATIONS OF MOLECULAR DYNAMICS TO THE SIMULATION OF SPUTTERING

M. W. Finnis
Theoretical Physics Division
A.E.R.E. Harwell
Oxfordshire OX11 ORA
United Kingdom

ABSTRACT. Molecular dynamics is the numerical solution of the classical equations of motion of a system of interacting particles. Two applications are described in this lecture. The first is to the sputtering of gold atoms by electrons in the high voltage microscope, the second is to their sputtering by argon ions. By studying the atomic trajectories and constructing angular and energy distributions respectively we have been able to explain features of the experimental results in terms of focussing and steering by surface neighbours of the sputtered atoms. In the case of electron beam sputtering we were able to estimate the contribution of collision sequences compared to direct sputtering from the surface layer.

1. INTRODUCTION

Sputtering is the emission of atoms from a surface due to bombardment by neutrons, ions or electrons. Although it has been studied for over thirty years, its metallurgical importance has increased more recently with the assessment of potential first wall materials for fusion reactors, in which sputtering would cause erosion and contamination of the plasma. The process of vacancy injection which accompanies sputtering may also modify void swelling behaviour, as for example in Ni, in which there is evidence that vacancy injection from voids stops their growth beyond a certain radius[(1)].

The earlier theories or models of sputtering were of the type known as 'statistical theories'. Their aim was to predict the yield of sputtered atoms as a function of the energy and intensity of the incident particles, and as the name suggests they worked with angular and energy distribution functions and the mathematical machinery of transport theory. At their best, statistical theories were able to predict the energy and angular distribution of sputtered atoms[(2–3)]. However, by their nature these statistical theories alone could not describe any effects of the lattice structure of the target which became evident in later experiments, such as channelling, focussed collision sequences and lattice steering. The latter typically give rise to spot patterns of sputtered intensity[(4–5)]. The effects of lattice geometry at lower energy are also important in the process of vacancy injection. These special effects were described explicitly, outside the statistical approach, with reference to the early molecular dynamics studies.

It was not long before the field was attacked directly by methods of atomistic simulation, with considerable success. This remains an active area, partly because its success is not complete – there are still problems to be solved – but also because the continuing advances in computer technology allow the problems to be tackled in more sophisticated and realistic ways.

G. Jacucci (ed.), Computer Simulation in Physical Metallurgy, 289–311.

The student of sputtering by computer simulation has to contend with some special difficulties which are not usually faced by the simulator of processes in bulk materials. A fundamental difficulty is the description of the interatomic potential. The short-ranged repulsive part of the atom-atom potential must be characterised to much closer interatomic separations than are required for studying thermal defect properties, because we are dealing with collisions involving hundreds or thousands of eV. At the same time the attractive part of the potential affects details like the sputtering threshold energy and the sputtered angles. The full range of the potential is therefore sampled, and – what is worse – at a surface. The nature of interatomic potentials is known to be different at a surface, because of the local redistribution of electrons, but no quantitative theory is available to tell us how different. Other difficulties in the simulation of sputtering are technological and common to the atomistic simulation of radiation damage in general : the large range of the energy scales, for example from a 20keV incident ion to a 6eV sputtered atom, the size of the damaged region on an atomic scale and the statistical nature of the results (distribution functions) all make for CPU–hungry programs.

Partly for the latter reason, there have been broadly two levels of realism in computer simulation models of sputtering. They have been reviewed recently by Harrison[(6)]. The fastest simulations in terms of CPU are performed with so-called Binary Collision Models, of which the most famous is perhaps MARLOWE[(7)], which approximate the dynamics of the system as a sequence of binary collisions, scattering cross-sections for which can be computed and stored once and for all. They suffer the drawback that the many-body dynamics at lower energies is poorly described by this approximation; there are furthermore certain characteristic threshold and surface barrier energies to be defined, the physical meaning of which is unclear. Nevertheless the fact that such models accurately describe the target lattice structure, besides their computational speed, has made them a valuable tool[(8)].

The second, in principle more realistic, type of models are of the molecular dynamics type, in which the classical equations of motion of an interacting system of particles are solved. The scope of these is limited only by a knowledge of the interatomic potentials and by the available computing power, not by any other uncontrolled approximations.

In this lecture, I will describe two applications of molecular dynamics to the interpretation of sputtering experiments. I make no apology for describing only the work with which I have been directly associated, which happens to be on gold, sputtered by two very different irradiations : 500–1100 keV electrons in one set of experiments with the high-voltage electron microscope (HVEM) and 20 keV Ar^+ ions in the other. I hope to show by means of these examples how molecular dynamics simulation can be combined with experiment to elucidate sputtering mechanisms.

2. SIMULATION OF SPUTTERING BY 500 – 1100 keV ELECTRONS

Experiments in the HVEM were perfomed to investigate sub-threshold sputtering, in which the energy of the incident particles is insufficient to cause damage to the bulk crystal. The experimental arrangement is shown schematically in figure 1. The results were used to derive the angular distribution of sputtered atoms $I(\Omega)$ as a function of electron energy, shown in figures 2 and 3.
A full description of the experiment and the simulation is given in ref.(9).

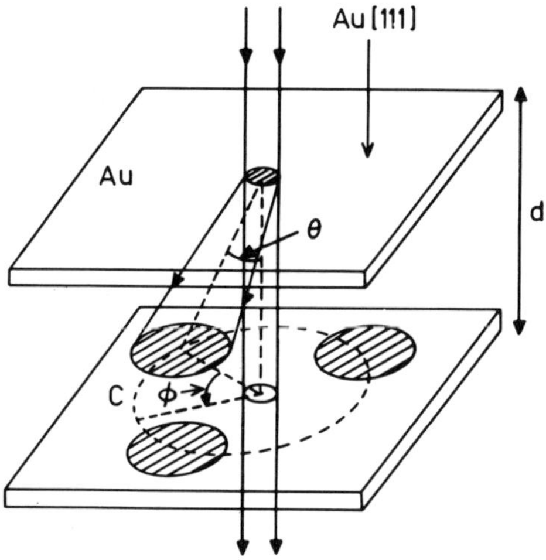

Figure 1 : Diagram of experimental arrangement for investigating sputtering from gold foils in the HVEM. Gold atoms transmission-sputtered by electrons in the direction (θ,ϕ) were collected on a carbon foil situated at a distance d below the gold. At energies > 600keV, gold deposit thicknesses showed maxima at angles $\phi = 120^0$ apart as illustrated schematically.

2.1 The molecular dynamics model

The computer program we used for all the simulations described in this lecture is the Harwell molecular dynamics program documented in ref.(10). In the present case our dynamic model comprised 9 planes of atoms. There were 36 atoms in each plane arranged on a hexagonal lattice bounded by a rectangle with sides along <110> and <112> directions, and the planes were stacked to form a rectangular block of f.c.c crystal with (111) surfaces top and bottom. These surfaces remained free while periodic boundary conditions were imposed across the other pairs of faces to simulate an infinite slab. Such boundary conditions are selected by specifying the values of three parameters in the input data[(10)].

The atoms interacted by a simple nearest-neighbour potential :

$$\Phi(r) = D[exp(-2\alpha(r-r_0)) - 2exp(-\alpha(r-r_0))] \quad for\ r \leqslant r_0, \qquad (1)$$

where :

$r_0 = 2.88\mathring{A}$ = the nearest-neighbour distance,
$D = 0.638eV$
$\alpha = 1.581\mathring{A}^{-1}$.

For $r > r_0$, $\Phi(r)$ was interpolated smoothly to zero within the second neighbour distance using a cubic spline. The parameters D and α were fitted to the cohesive energy and bulk modulus of Au. The slope of the potential is compared to previous potentials which have been used for radiation damage studies in figure 4.

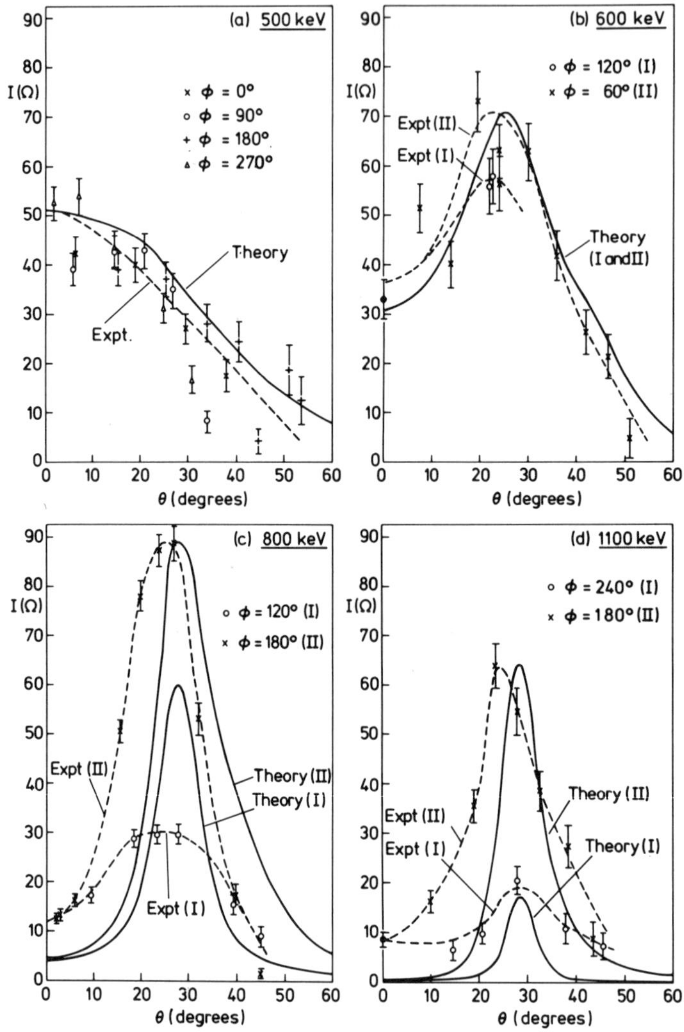

Figure 2 : The variation of intensity $I(\Omega)$ with θ. The points joined by broken curves show experimental results. The solid curves are from the theory described here.

The length of the time step for iterations was chosen such that the fastest atom would move less than 0.1Å per time step. Typical runs for calculating a sputtering trajectory would take a minute or two on the IBM 370 which we were using at Harwell in 1976. The present IBM 3081 is faster.

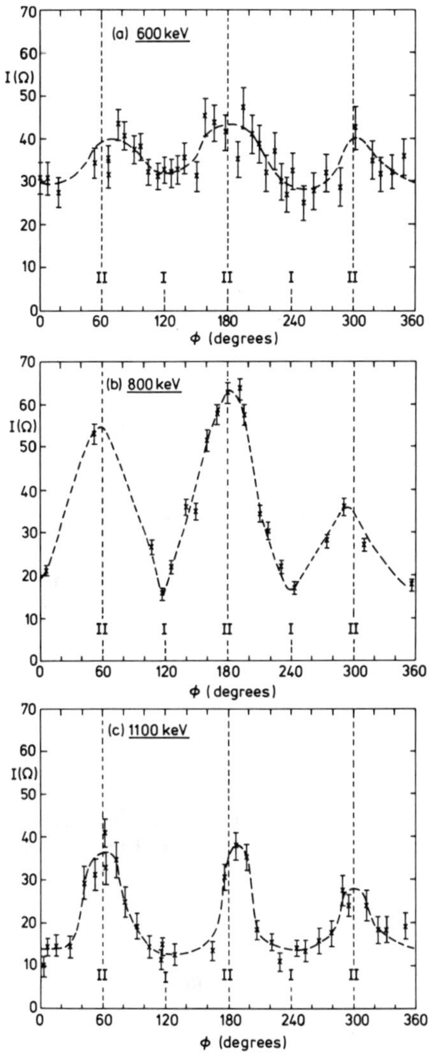

Figure 3 : Experimental variation of $I(\Omega)$ with ϕ (arbitrary units).

2.2 Simulation of sputtering yields and angular distributions

The probability of an atom recoiling in a given direction from a collision with an incident electron is given by the differential scattering cross section $I'(\Omega')$, which is a function of electron beam energy. $I'(\Omega')d\Omega'$ is the probability per unit electron flux that an atom will recoil in the range of solid angle $d\Omega'$ about the direction Ω'. We obtained $I'(\Omega')$ from the relativistic scattering theory of Mott[(13)] by the method of Oen[(14)]. The task for simulation was then to calculate the final distribution of sputtered atoms $I(\Omega)$, where :

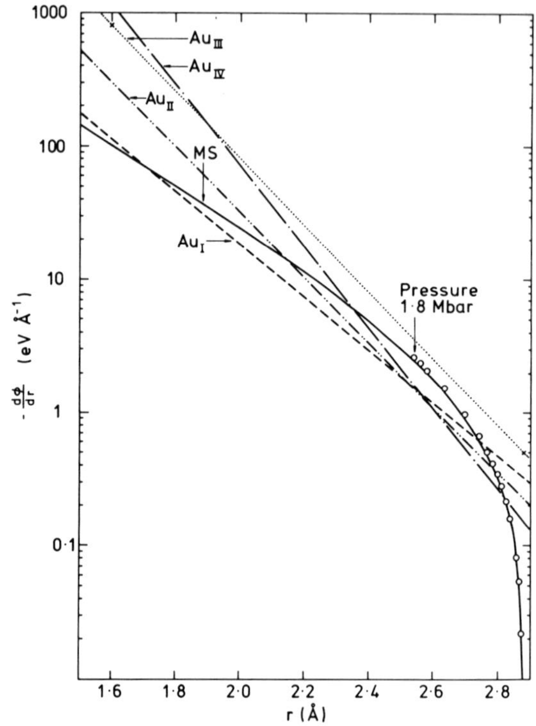

Figure 4 : A comparison of the slopes of various Au–Au potentials. Au_I – Au_{IV} are from Bauer, Anderman and Sosin[11]. The present Morse–Spline potential is labelled MS. The circles are effective values deduced from the experimental pressure-volume relation[12].

$$I(\Omega)d\Omega = I'(\Omega')d\Omega', \tag{2}$$

or more explicitly :

$$I(\Omega) = I'(\Omega')\left[\frac{d(cos\ \theta')}{d(cos\ \theta\ )}\ \frac{d\phi'}{d\phi}\right]. \tag{3}$$

In principle there are two ways of proceeding. The most general approach would be to generate a large number of trajectories, with initial angles selected from the distribution $I'(\Omega')$ and initial energies given by :

$$E = E_{max} \cos^2\theta', \tag{4}$$

$$E_{max} = 2E_e(E_e + 2m_0c^2)/Mc^2, \tag{5}$$

where :

E_e = energy of incident electron,
m_0 = rest mass of electron
M = mass of a gold atom, and
c = velocity of light.

From a histogram of the final directions of the sputtered atoms, the distribution $I(\Omega)$ would be obtained directly.

Although this procedure is feasible, and it is indeed the usual approach to sputtering simulations, it would require several hundred trajectories and is certainly a costly option. We adopted a compromise approach, by only calculating selected trajectories in order to obtain the derivatives in equation (3). This requires far less computation; its only drawback is that $I(\Omega)$ is only obtained for selected values of ϕ.

To define the origin of ϕ we note that the three <110> directions emerge from the (111) surface at $\phi = 0^0$, 180^0 and 300^0. We shall refer to these as directions II in the plane of the collecting surface (see figures 2 and 3). They contain the spot maxima. The intermediate positions, $\phi = 0^0$ etc. we refer to as directions I, and these correspond to the minima between spots.

2.3 Results

Figure 5 shows two typical trajectories.

Plots of θ versus θ' made from 10–20 trajectories per layer are shown in figure 6.

Estimates of the total yield from experiment and simulation are shown in figure 7. The calculations of yield were based on the cut-off angles determined by the trajectory studies, in the following way. For the surface contribution, it is simply necessary to integrate $I'(\Omega')$ out to θ_{max}, which at a given energy is a threshold angle for sputtering. The ϕ variation for the surface layer was at most 4%, as determined by numerical differencing with trajectories at $\phi + 5^0$ for directions I and II. For the sublayers there was a lower and upper threshold θ_1^n and θ_2^n (see figure 6). We assumed that sputtering occurs if the primary velocity is within any of the three cones centred on $\frac{1}{2}(\theta_1^n + \theta_2^n)$, and integrated $I'(\Omega')$ over the solid angle of the cones. This approximation avoided a detailed scan of trajectories as a function of ϕ.

The final angular distributions calculated are shown in figure 8. They have been convoluted with a Gaussian to include the significant effect of thermal broadening, as discussed in ref.(9).

2.4 Discussion

The qualitiative features of the experimental results are explained by the simulations, namely :

(1) The steep rise of yield with energy indicates the contribution of recoils in the sub-surface layers, as suggested by Cherns *et al.* [(15)]. At 1100keV the contribution of focussed collision sequences to sputtering is about 40% of the total yield with components from up to 3 sublayers.

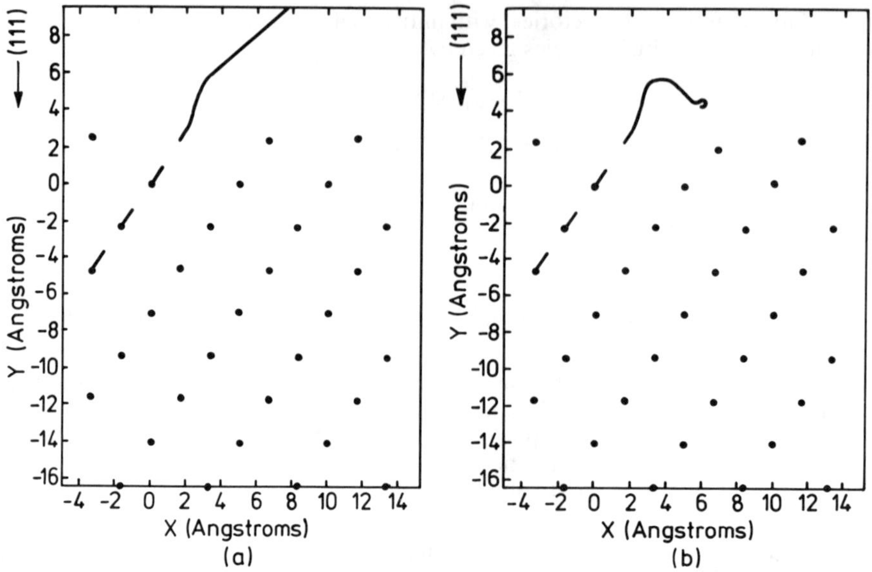

Figure 5 : Theoretical trajectories of atoms in a 110 plane intersecting the gold surface, during an interval of 435 time steps following a primary recoil in the third sublayer directed along a 110 direction.
(a) Primary energy = 16.5 eV.
(b) Primary energy = 15.5 eV.

(2) At 500keV only surface atoms contribute. They are steered away from the normal by the surface potential, which broadens the sputtered distribution.

(3) The radial peak in the sputtered distribution at 600keV is a reflection of the scattering cross-section and has nothing to do with focussing.

(4) With increasing energy the spot pattern becomes sharper. The simulations demonstrate that at least four factors contribute to this, namely :

(i) The sharpening of the Mott scattering cross-section $I'(\Omega')$.
(ii) Less thermal broadening as the atoms emerge faster.
(iii) A stronger rainbow effect, that is the singularity in $d(cos\theta')/d(cos\theta)$ becomes stronger at higher energy.
(iv) Focussing increases $d\phi'/d\phi$ as the energy increases.

Discrepancies between the simulations and the experimental results are attributed to :

(1) The contribution of surface atoms on ledges.

(2) Thinning by surface redistribution : a process which affects the experimental estimate of the total sputtering cross section (the yield). This estimate (fig. 7) is obtained from the time to sputter through the sample, and because of the formation of pits it is an overestimate of the true sputtering rate[16].

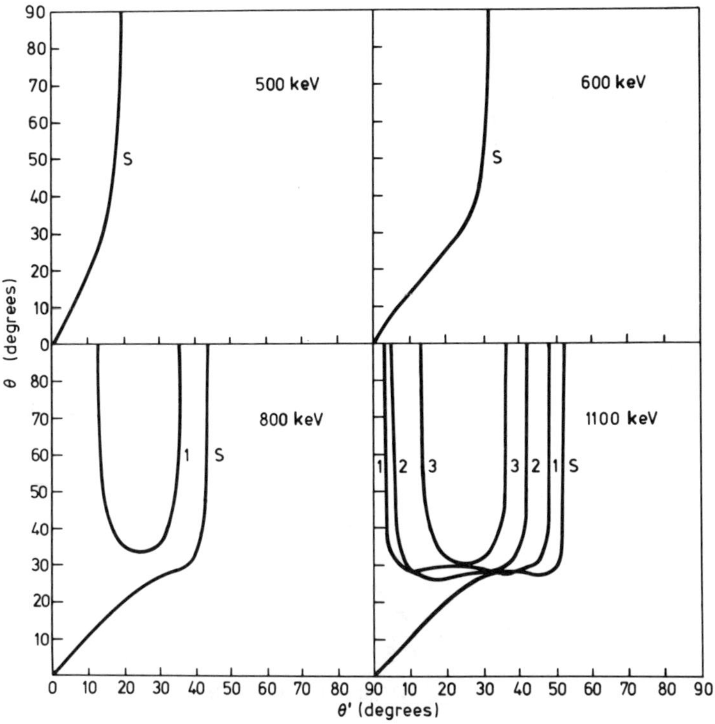

Figure 6 : Theoretical curves relating primary angle θ′ to the angle of emergence of the sputtered atom θ. Results are for primary recoils in the surface layer *S*, and first (1) second (2) and third (3) sublayers.

(3) Inadequate modelling of the surface binding forces. Our ignorance of the nature of the interatomic forces and how to model them at surfaces means that comparisons with experiment, especially near the threshold for sputtering, remain qualitative.

3 SIMULATION OF SPUTTERING BY 20 keV ARGON IONS

3.1 Introduction

The experiments we consider in this second application are measurements of the time-of-flight (TOF) spectrum[(17)], and hence the energy spectrum. We consider the case of a (100) gold surface, for which the TOF spectrum is shown in figure 9. Measurements of this type on targets bombarded by heavy ions show a variety of structure, depending on the crystallinity and orientation of the target and the energy and orientation of the incident beam. The molecular dynamics simulations described here are discussed in more detail in reference (18), which contains further references to experimental and theoretical studies of TOF spectra.

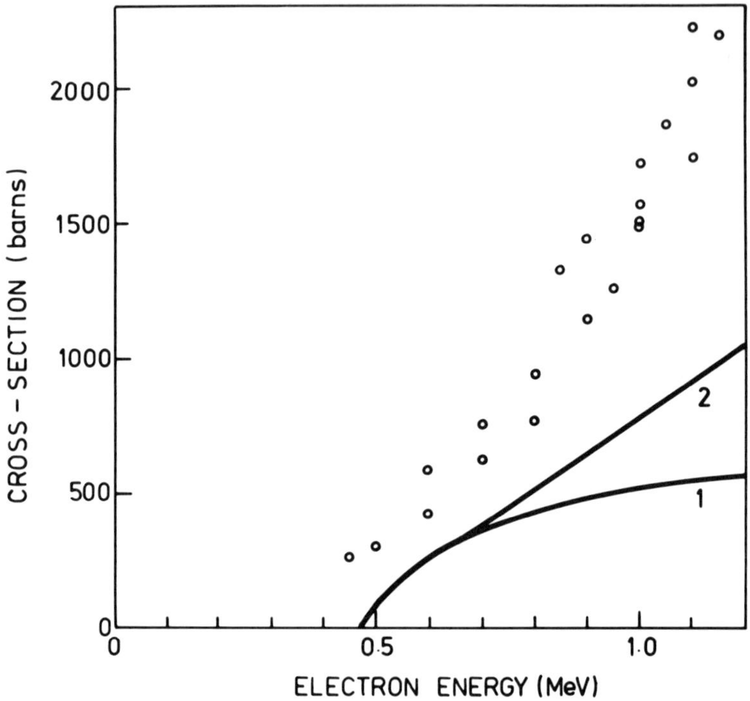

Figure 7 : The total cross-section for sputtering of (111) gold foils as a function of electron energy. Curves (1) and (2) show surface and total cross-sections from the theory. Experimental results are from Cherns et al.

In this case the incident ion causes a cascade of damage, so that the atoms involved in the last few collisions before a sputtering event have a distribution of initial energies and angles. Compared to the previous case of electron beam sputtering, the energy represents an extra dimension in the distribution to be considered prior to the sputtering event. It would therefore be orders of magnitude more time consuming to attempt to simulate the angular and energy distribution of sputtered atoms by a 'brute-force' histogram method. Each direction cell of the histogram would have to be further subdivided into energy cells. Instead, a compromise of the type employed previously is unavoidable, in which only selected critical trajectories are used to map out the distribution. The following method was used, which also gives physical insight into the important processes.

sample, where $i = 1$ is the surface layer, $i = 2$ is the first sublayer etc. At the start of its trajectory, before it left the metal, each atom in the distribution $\Phi_i(E,\Omega)$ was in layer i where it was a member of a distribution $\Phi_i'(E',\Omega')$. The measured final sputtered atom distribution is the sum of contributions from all layers :

$$\Phi(E,\Omega) = \sum_i \Phi_i(E,\Omega). \tag{6}$$

In general computer simulations have shown that significant contributions come only from atoms within one layer of the surface $i \leqslant 2$. The controversy was about whether the observed structure in the sputtered spectrum was due to structure in the distribution Φ' within the bulk[19], arising from, for example, focussing, or to structure introduced by refraction or steering effects right at the surface layer[20–21], which map the distribution Φ' onto the distribution Φ. A major conclusion of this study is that the structure and intensity of the sputtered energy spectrum in the <100> normal direction is dominated by the ejection of sublayer atoms which are focussed strongly by their neighbours in the surface layer, so that the latter point of view now seems most correct.

If we assume a one-to-one correspondence between the velocity and direction of an atom within the bulk metal and its final velocity and direction after it has emerged, the mapping can be written formally in terms of a sputtering Jacobian :

$$\Phi_i(E,\Omega)dEd\Omega = \Phi_i'(E',\Omega')dE'd\Omega', \tag{7}$$

$$\Phi_i(E,\Omega) = [J_i(E,\Omega;E',\Omega')]^{-1}\ \Phi_i'(E',\Omega'). \tag{8}$$

The Jacobian can be written explicitly in spherical polar coordinates referred to the surface normal :

$$J_i(E,\Omega;E',\Omega') = det\begin{vmatrix} \frac{\partial E}{\partial E'} & \frac{\partial \theta}{\partial E'} & \frac{\partial \psi}{\partial E'} \\ \frac{\partial E}{\partial \theta'} & \frac{\partial \theta}{\partial \theta'} & \frac{\partial \psi}{\partial \theta'} \\ \frac{\partial E}{\partial \psi'} & \frac{\partial \theta}{\partial \psi'} & \frac{\partial \psi}{\partial \psi'} \end{vmatrix} \times \left|\frac{\sin\theta}{\sin\theta'}\right|. \tag{9}$$

In reality there is no unique mapping from (E',Ω') to (E,Ω). Thermal vibrations are an obvious cause of convolution of the mapping. Even without thermal vibrations, the final sputtered energy will depend to some extent on the history of collisions leading to E'.

The model crystallite which was used for the simulations is illustrated in figure 10. Justification that the Jacobian mapping is reasonable at low temperature was provided by the calculations of figure 11.
Comparison of 11(b) and 11(c) shows that for the chosen geometry at least, the binding energy is insensitive to within about 2% to the precise sequence of collision events which gave the initial atom its initial kinetic energy of E'. Further justification for using this mapping was provided by the calculations of figure 12, which show that the sputtered angle θ is also rather insensitive to the history of collisions leading to the initial angle θ'.
Also, as figure 13 shows, it was verified that during a collision sequence, all memory of where the collision started is lost by the time of the second collision.
The energy loss at the first collision is several eV higher than at any subsequent collision;

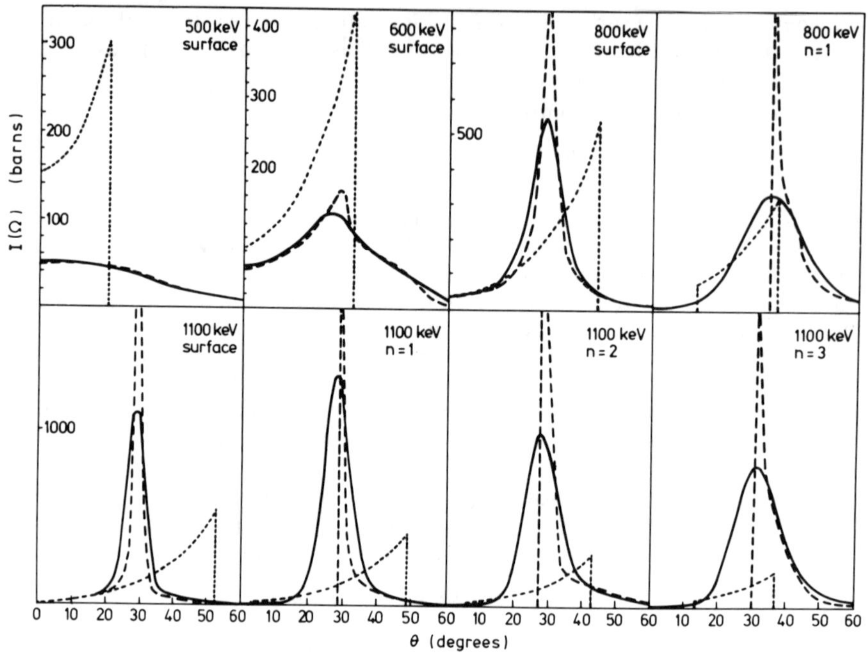

Figure 8 : Theoretical curves for the Mott scattering cross-sections (dotted), the unbroadened sputtered intensities (dashed) and the thermally broadened intensities (solid curve) as functions of θ at $\phi = 60^0$. The Mott scattering cross-section is plotted for the range of recoil energies which lead to sputtering, viz. $\theta_1'' < \theta' < \theta_2''$.

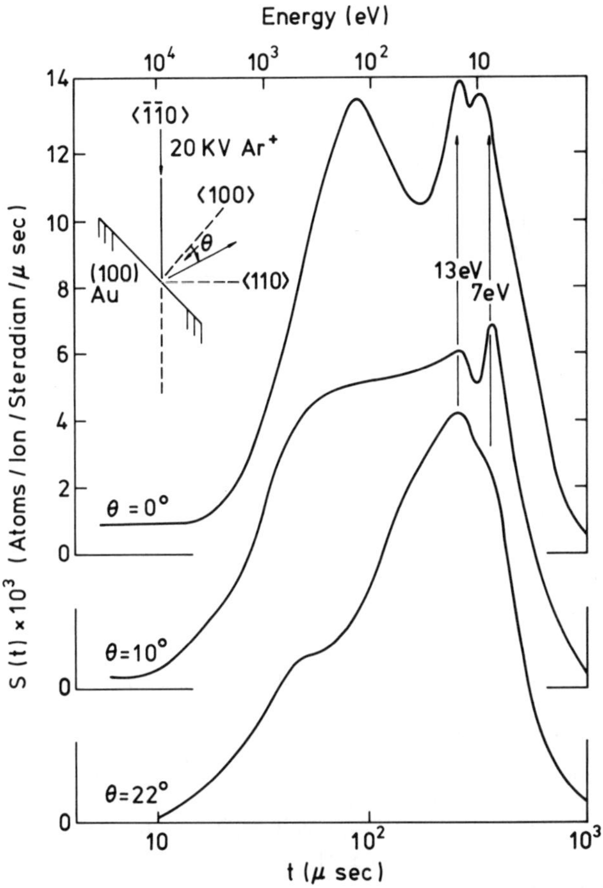

Figure 9 : The time-of-flight spectrum measured by Thompson, Reid and Farmery [17] for sputtering from a gold (100) surface. The angle θ has been redefined from the original figure of Thompson *et al.* to conform to the present notation.

3.2 The Sputtering Jacobian

The parallel improvements in statistical theories and computer simulation led to an interesting controversy over the origin of peaks in the spacial and energy distribution of sputtered atoms. To explain the point at issue, let us define the energy and angular distribution $\Phi_i(E,\Omega)$ as the distribution of atoms sputtered from layer i of the

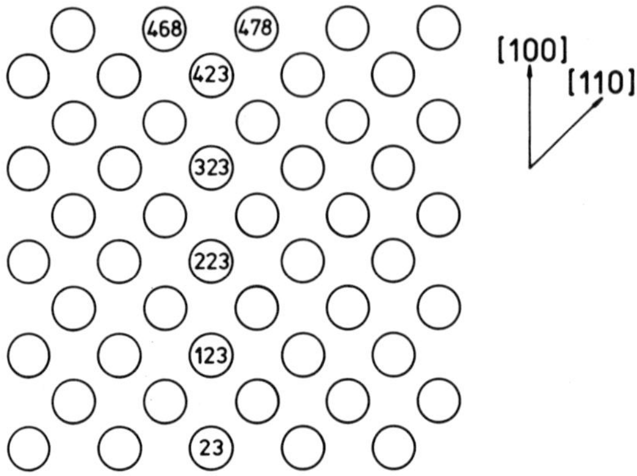

Figure 10 : A (001) cross-section of the crystallite used in the molecular dynamics simulations of sputtering from a (100) surface.

this is due to the injection of potential energy into the collision sequence. The energy loss per collision increases thereafter slowly by about 1eV per collision, from 14eV to 17eV; in the same interval the kinetic energy decays from 77 to 34eV.

After these preliminary calculations to test the methodology we made relatively short calculations of individual trajectories in order to estimate the Jacobian by numerical differencing. Then we made a simple hypothesis about the distribution $\Phi_i'(E',\Omega')$, assumed independent of the surface, with reference to the statistical theory[(2)], in arbitrary units :

$$\Phi_i(E',\Omega') = E'^{-2}. \tag{10}$$

The resulting sputtered distribution thereby includes the role of the surface layer and first sublayer in determining the energy and angular distribution, but the effect of previous collisions is only approximately taken into account.

For the calculations by numerical differencing it is convenient to simplify the Jacobian, which for sputtering in the <100> normal direction becomes :

$$J_i(E,0;E',0) = \left| \left(\frac{\partial E}{\partial E'}\right)_{\theta',\psi'} \left[\left(\frac{\partial \theta}{\partial \theta'}\right)_{E',\psi'}\right]^2 \right|. \tag{11}$$

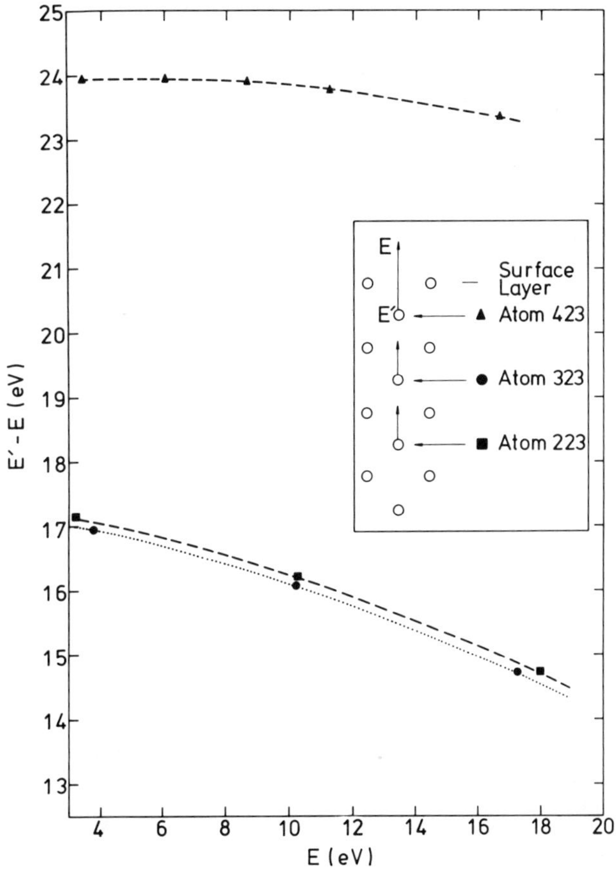

Figure 11 : The effective binding energy $E' - E$ of a sublayer atom, labelled atom 423, ejected in the <100> direction when an initial velocity is given to
(a) atom 423 itself,
(b) the atom below, labelled atom 323,
(c) the second atom below, labelled atom 223.

3.3 Results

The results of running many tens of trajectories are summarised in figures 14–18.

The curve of effective binding energy, figure 14, and the analogous results at different angles, figure 15, told us that the structure in the sputtered energy spectrum was not to be found in the factor $\partial E/\partial E'$ but must be sought in the angular part of the Jacobian. The interesting structure turned out to be caused by the steering of atoms originating in the first sublayer.

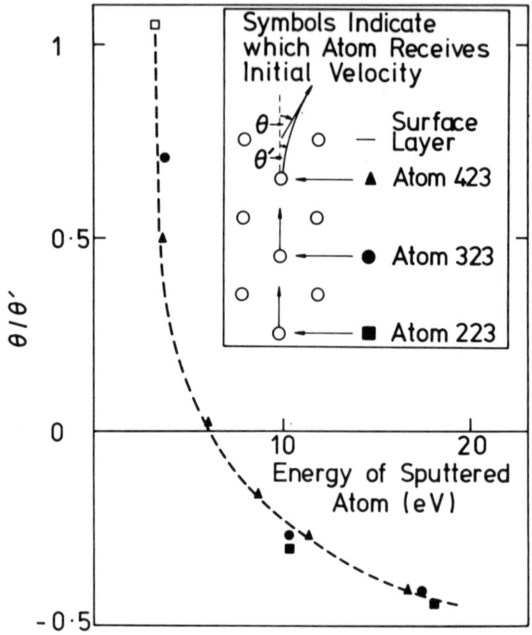

Figure 12 : The ratio θ/θ′ for small angle sputtering $\theta' = tan^{-1}(0.02)$ in the <100> direction for the same three cases as described in figure 9.

A critical energy of the sublayer atom was about 30eV, at which it emerges at about 6eV. At this energy, for small deviations in its initial direction it still emerges normal to the surface. This is a rainbow effect, $\partial\theta/\partial\theta' = 0$, which must cause a low energy peak in the time-of-flight spectrum. Snapshots of the turning points in its oscillating trajectory ar shown in figure 16.

At energies above this critical value we find $\partial\theta/\partial\theta' < 0$. However, we can be sure that in the limit of high energy, the atom must have a straight trajectory, i.e. $\partial\theta/\partial\theta' = 1$. So we sought and found another zero in $\partial\theta/\partial\theta'$ which turned out to be at a sputtered energy of 156eV. This is a second rainbow effect, and may be pictured as the sublayer atom starting at the focal point of a 'lens' formed by the square of surface atoms which it has to pass through.

For angles other than the surface normal, there are other complications at low energy, which appear when the mapping of E',θ' onto E,θ is explored in more detail. The contour map of figure 17 can be examined to look for the peculiarities in the Jacobian. In fact the inverse mapping is not unique; for emerging energies $E < 6eV$ two E',θ' values sometimes correspond to one E,θ value.

The calculated TOF spectrum in figure 18 was based on only the lowest branch of the mapping, close to $\theta' = 0$.

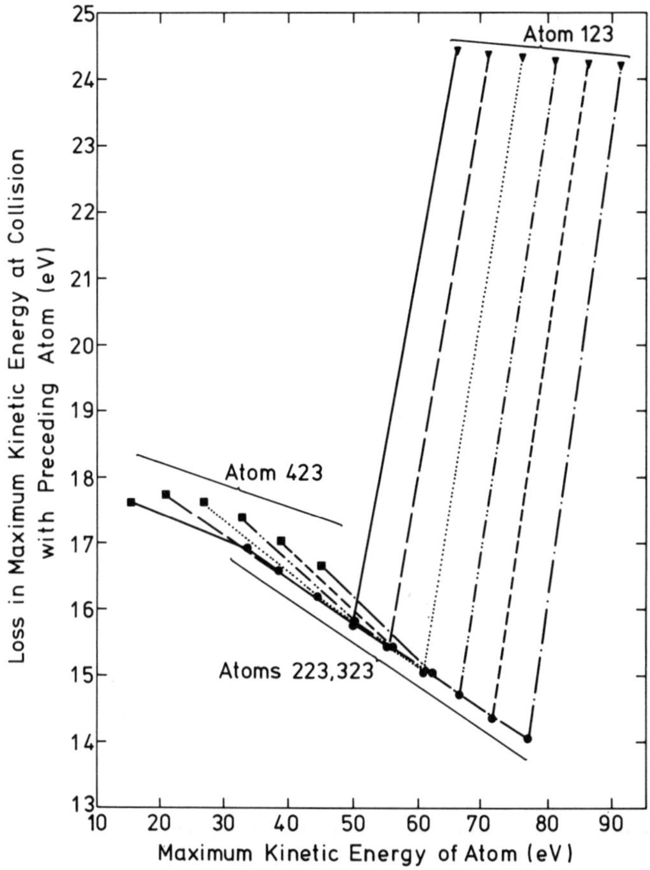

Figure 13 : The loss in maximum kinetic energy after each of the four collisions for a sequence in the <100> direction from atom 23 to 423. The six values of the initial kinetic energy range from 90 to 115 eV at 5 eV intervals.

3.4 Discussion

By evaluating the sputtering Jacobian for a model of a gold (100) surface using trajectories generated with the molecular dynamics program MOLDY a new theory for the structure in the experimental TOF spectrum has been developed.

It was concluded that :

(1) Most of the sputtering in the <100> direction is from sublayer atoms which are focussed towards <100> by their surface neighbours. The observations which follow refer to these sublayer atoms and the Jacobian for their sputtering.

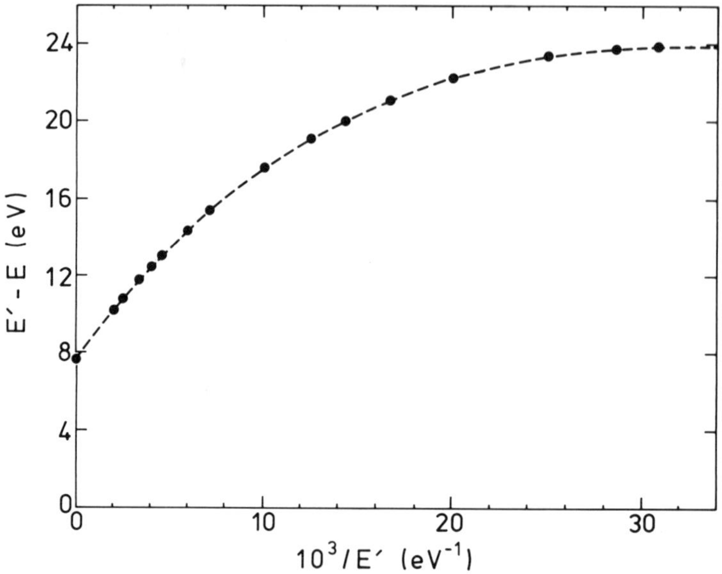

Figure 14 : The effective binding energy $E'-E$ of a sublayer atom, labelled atom 423, ejected in the <100> direction when an initial velocity is given to atom 423 itself, as in figure 11, replotted as a function of $1/E'$.

(2) The experimentally observed peak at about 150eV is due to a zero in the sputtering Jacobian and occurs in the simulation independently of the nature of the bulk spectrum, even if it is isotropic.

(3) At lower energy the topology of the sputtering Jacobian becomes more complex, and it has another zero. Of the two low energy peaks observed experimentally it seems most plausible to assign the 13eV peak to the 6eV peak in the simulation which corresponds to $\partial\theta/\partial\theta' = 0$, because it disappears as the angle is varied away from the normal[21]. Such a discrepancy is well within the uncertainty of the potential.

(4) It was conjectured[18] that the experimental 7eV peak corresponds to the fold in the contours of θ' at positive θ as θ' is increased from about 5^0, combined with anisotropy in the distribution of initial energies of the sputtered atom. This has not yet been tested.

5. SUMMARY

Two examples have been presented which show how the technique of molecular dynamics can assist in the interpretation of sputtering experiments. Neither was of the completely brute-force variety, which would follow the dynamics of the system from the instant of impact of the bombarding particle for a distribution of impact parameters. The philosophy was rather to follow selected trajectories of the atoms

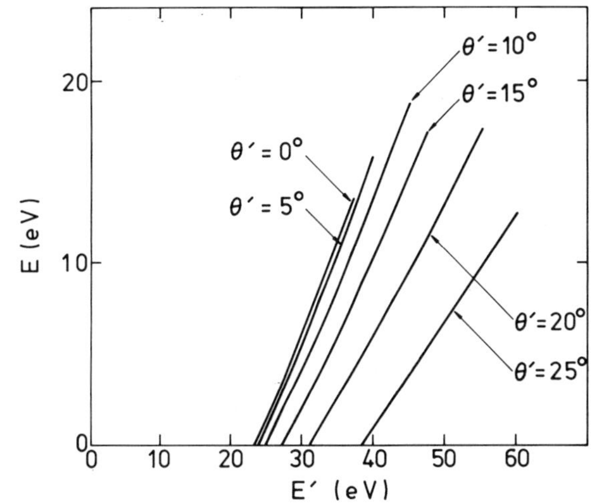

Figure 15 : *E* versus *E*′ for sputtering of atom 423 in various initial directions θ′.

immediately prior to the sputtering event, in order to study the mapping of initial to final energies and directions.

As a result a semi-quantitative picture of the origin of the angular distributions (for electron bombardment) and energy distributions (for time-of-flight spectra) could be derived. Particularly in the latter case the mechanisms of the focussing of emerging sublayer atoms by their surface neighbours were quite unexpected, and provided the first satisfactory explanation of the main features of the experimental results.

In spite of recent improvements in computing power since these calculations, which have enabled more ambitious molecular dynamics simulations to be undertaken, the uncertainties in the interatomic potentials remain a bar to obtaining more quantitative results by techniques of this kind, particularly near the threshold for damage. However, theoretical development of interatomic potentials continues and sputtering simulations may be a valuable testing ground for new models in the future.

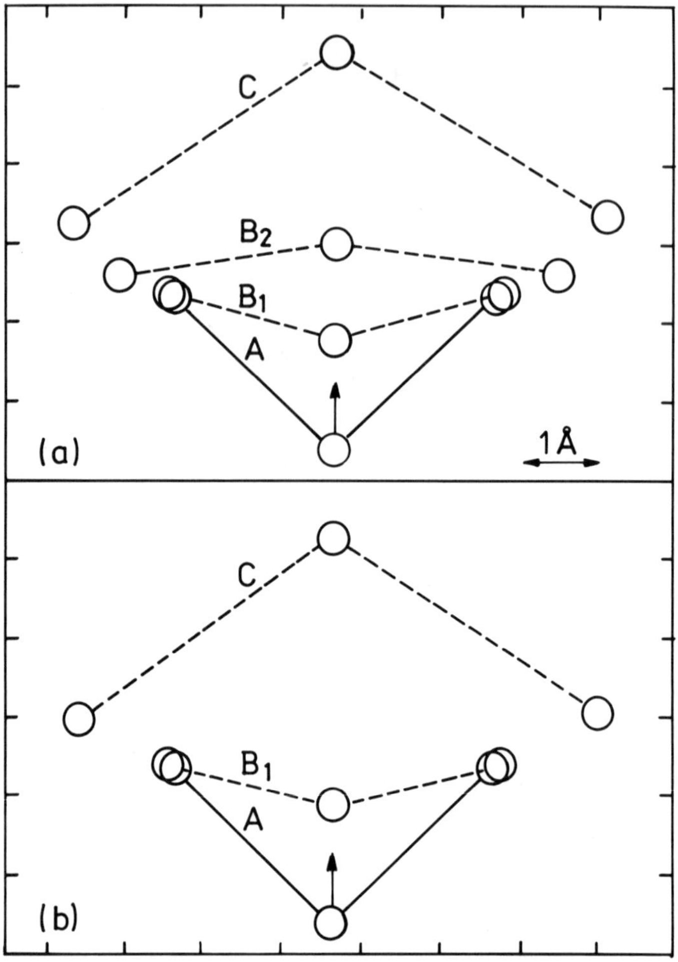

Figure 16 : Snapshots of the sputtered atom 423 and its two surface neighbours 468 and 478 in the $\phi = 0$ plane for $\theta' = tan^{-1}(0.02)$ and
(a) $E' = 28eV$ – below the energy at which $\partial\theta/\partial\theta' = 0$.
(b) $E' = 32.5eV$ – above the energy at which $\partial\theta/\partial\theta' = 0$.
The labelled points have the following meaning :
A – initial configuration.
B_1 – transverse velocity of atom 423 vanishes.
B_2 – transverse velocity of atom 423 vanishes again.
C – atom 423 becomes fully detached from the surface.

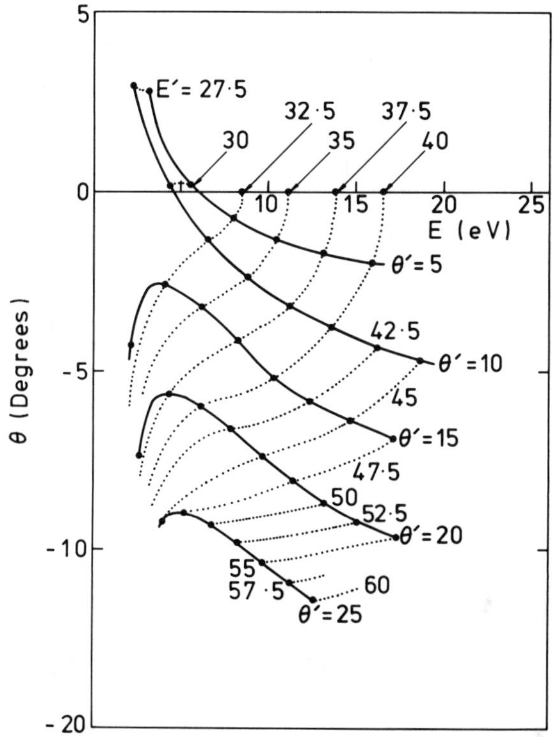

Figure 17 : Contour map of E' and θ' for a range of values of E' (dotted curves) and $\theta' > 0$ (full curves). Because of the symmetry about $\theta = 0$, the signs of θ and θ' are arbitrary provided they are both reversed together.

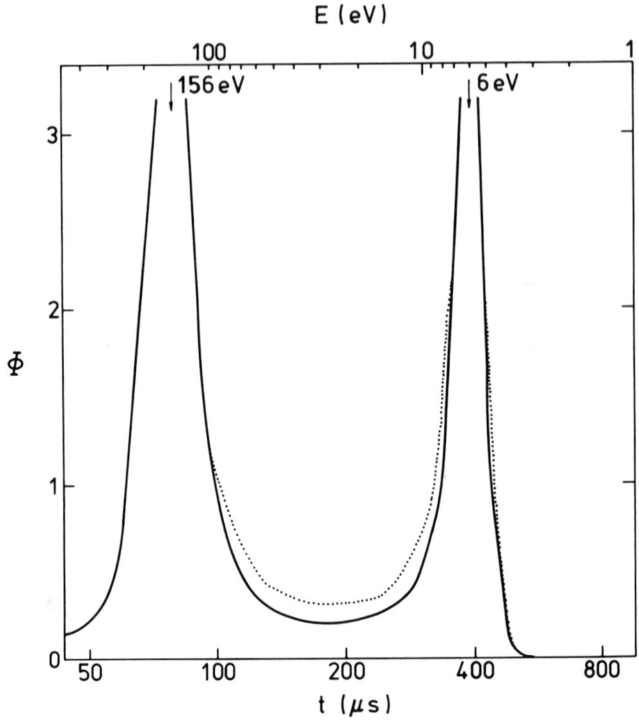

Figure 18 : Calculated time-of-flight spectrum Φ for sputtering in the normal <100> direction :

$$\Phi = E^{\frac{3}{2}}/(E' d\theta/d\theta')^2,$$

where the normalisation is arbitrary. The solid curve is the result of sublayer sputtering alone. The dotted curve includes the surface contribution.

ACKNOWLEDGEMENTS

I am grateful for discussions of their experiments with Dr. D.Cherns and Prof. M.W.Thompson.

REFERENCES

(1) Nelson R S 1975 *J. Nucl. Mater.* **57** 77.

(2) Thompson M W 1968 *Phil. Mag.* **18** 377.

(3) Sigmund P 1969 *Phys. Rev.* **184** 383.

(4) Silsbee R 1957 *J. Appl. Phys.* **28** 1246.

(5) Gibson J B Goland A N Milgram M and Vineyard G H 1960 *Phys. Rev.* **120** 1229.

(6) Harrison D E 1983 *Radiation Effects* **70** 1.

(7) Robinson M T and Torrens I M 1974 *Phys. Rev.* **B9** 5008.

(8) Shulga V I 1983 *Radiation Effects* **70** 65.

(9) Cherns D Finnis M W and Matthews M D 1977 *Phil. Mag.* **35** 693.

(10) Finnis M W and Harker A H 1983 *A User's Guide to MOLDY, a Molecular Dynamics Program, Report AERE-R 8824.*

(11) Bauer W Anderman A I and Sosin A 1969 *Phys. Rev.* **185** 870.

(12) Keeler R N and Kennedy G C 1972 *American Institute of Physics Handbook, 3rd edition* McGraw–Hill, New York, Table 4.101

(13) Mott N F and Massey H S W 1965 *The theory of atomic collisions in solids, 3rd edition* O.U.P. Chapter IX.

(14) Oen O S 1973 *ORNL–4879*

(15) Cherns D Minter F J and Nelson R S 1976 *Nuclear Instruments and Methods* **132** 369.

(16) Cherns D 1977 *Phil. Mag.* **36** 1429.

(17) Thompson M W Reid I and Farmery B W 1978 *Phil. Mag.* **38** 727.

(18) Finnis M W 1980 *Phil. Mag.* **43** 1321.

(19) Thompson M W 1959 *Phil. Mag.* **4** 139.

(20) Harrison D E Johnson J P and Levy S 1966 *App. Phys. Lett.* **8** 33.

(21) Lehmann C and Sigmund P 1966 *Phys. Stat. Sol.* **16** 507.